SOIL CLASSIFICATION

A GLOBAL DESK REFERENCE

SOIL CLASSIFICATION

A GLOBAL DESK REFERENCE

EDITED BY
HARI ESWARAN
THOMAS RICE
ROBERT AHRENS
BOBBY A. STEWART

CRC Press is an imprint of the
Taylor & Francis Group, an informa business

Cover art courtesy of USDA–Natural Resources Conservation Service

CRC Press
Taylor & Francis Group
6000 Broken Sound Parkway NW, Suite 300
Boca Raton, FL 33487-2742

First issued in paperback 2019

ISBN 13: 978-0-367-45462-3 (pbk)
ISBN 13: 978-0-8493-1339-4 (hbk)

Visit the Taylor & Francis Web site at
http://www.taylorandfrancis.com

and the CRC Press Web site at
http://www.crcpress.com

Library of Congress Card Number 2002035039

Library of Congress Cataloging-in-Publication Data

Soil classification : a global desk reference / edited by Hari Eswaran ... [et al.].
p. cm.
Includes bibliographical references and index.
ISBN 0-8493-1339-2 (alk. paper)
1. Soils—Classification. I. Eswaran, Hari.

S592.16 .S643 2002
631.4′4—dc21

2002035039
CIP

Preface

Activities in soil classification reached their zenith in the 1990s. In the 1950s, the increasing demands on land use and developments in soil management required a new assessment of soil resources. Many national classification systems were developed, and the initiation of the World Soil Map project by the Food and Agricultural Organization provided a forum for scientists to develop the needed new approaches. Perhaps more than anybody else, the United States Soil Conservation Service decided to develop a totally new approach, and this decision propelled new research and documentation of soil information. By the 1970s, many of the systems had matured and during the next decade, the systems were validated and enhanced. In the 1980s and 1990s, other countries continued the trend of modifying their national systems, relying on the new concepts and approaches, modifying them for local applications.

At the dawn of the new century, there were still many systems but unlike before, there was a tremendous congruence of the systems. There are differences in nomenclature, but in general, many of the national systems were built on the framework designed by Soil Taxonomy. Notable exceptions are the French and Russian systems. The purpose of this monograph is to bring together these systems so that students of soils have a common source in one language to consult. Many of the papers in this monograph were presented at a special symposium held in Charlottesville, NC, in 2001 under the auspices of the Soil Science Society of America.

During the last two decades, the International Union of Soil Sciences (IUSS) also sponsored a Working Group to develop a World Reference Base (WRB) for Soil Resources. WRB has evolved into a classification system, but is yet to be widely accepted as the international soil classification system. The divergence of views, as shown in this monograph, illustrates some of the difficulties in obtaining a global consensus. However, the editors hope that this monograph will help to rally soil scientists around the world to develop an acceptable classification system for soils, for it is only when the global soil science community agrees to such a system that we can truly say that we have a science.

Hari Eswaran
Washington, D.C.

Thomas Rice
San Luis Obispo, California

Robert Ahrens
Lincoln, Nebraska

Bobby Stewart
Amarillo, Texas

About the Editors

Dr. Hari Eswaran is the National Leader for World Soil Resources at USDA Natural Resources Conservation Service. He provides the Agency's leadership in information and documentation of global soils resources and their use and management. He is also the Chairman of the Working Group on Land Degradation and Desertification of the International Union of Soil Sciences and a former Chairman of the Commission on Soil Genesis and Classification of the Society. As a member of the Soil Science Society of America, Dr. Eswaran has assisted in organizing several symposia and editing their proceedings. He is a Fellow of the Soil Science Society of America and has received the International Award from the Society. He has also received the USDA Superior Service Award and other awards from many countries. Email: *hari.eswaran@usda.gov.*

Dr. Robert Ahrens is Director of the National Soil Survey Center in Lincoln, Nebraska. The Center provides technical leadership for soil interpretations, soils data, laboratory analyses, soil survey investigations, and standards and soil classification. He led the effort to revise soil taxonomy and was responsible for several editions of the "Keys to Soil Taxonomy." Dr. Ahrens has been a regional correlator and state correlator, and was a field soil scientist for several years. He provided quality assurance for the soil survey of Kuwait. He is a member of the Soil Science Society of America and the International Union of Soil Sciences. Email: *bob.ahrens@usda.gov.*

Dr. Thomas J. Rice is a Professor of Soil Science and Chairman of the Soil Science Department, California Polytechnic State University, San Luis Obispo, where he is responsible for teaching soil science, land use planning, soil geomorphology, soil resource inventory, and advanced land management. He has been a Certified Professional Soil Scientist (C.P.S.S.) since 1982. Dr. Rice has published numerous journal articles, research reports, and popular press articles. He has been the project director for funded studies involving soil taxonomy updates in California, Nevada, and Utah; a comprehensive soils database for California; non-point source pollution study in western range-lands; mercury pollution study in a California watershed; soil mapping of a national wildlife refuge; and soil map unit interpretation record updates for California. Email: *trice@calpoly.edu.*

Dr. B.A. Stewart is Distinguished Professor of Agriculture and Director of the Dryland Agriculture Institute at West Texas A&M University. Before joining West Texas A&M University in 1993, he was the Director of the USDA Conservation and Production Research Laboratory, Bushland, Texas. Dr. Stewart is a past president of the Soil Science Society of America, and was a member of the 1900–1993 Committee on Long Range Soil and Water Policy, National Research Council, National Academy of Sciences. He is a Fellow of the Soil Science Society of America, American Society of Agronomy, and Soil and Water Conservation Society, a recipient of the USDA Superior Service Award, and a recipient of the Hugh Hammond Bennet Award of the Soil and Water Conservation Society. Email: *bstewart@mail.wtamu.edu.*

Contributors

Mario Luiz D. Aglio
Ciencia do Solo
EMBRAPA Solos
Rua Jardim Botanico
Rio de Janeiro, Brazil
mario@cnps.embrapa.br

Robert J. Ahrens
National Soil Survey Center
USDA-NRCS
Lincoln, NE
bob.ahrens@usda.gov

Richard W. Arnold
Retired Soil Scientist
USDA Natural Resources Conservation Service
Fairfax, VA
ct9311@aol.com

Fred H. Beinroth
Department of Agronomy and Soils
University of Puerto Rico
Mayaguez, PR
F_beinroth@rumac.uprm.edu

Frank Berding
Land and Water Development Division
FAO
Rome, Italy

Winfried E.H. Blum
Institute for Soil Research
Universität für Bodenkultur
Wien, Austria
IUSS@edv1.boku.ac.at

Johan Bouma
Member, Scientific Council for Government Policy
The Hague Professor of Soil Science
Wageningen University
Wageningen, The Netherlands
J.Bouma@alterra.dlo.nl

Ray B. Bryant
USDA-ARS
Pasture Systems and Watershed Management Research Unit
University Park, PA
rbb13@psu.edu

Stanley W. Buol
Department of Soil Science
North Carolina State University
Raleigh, NC
stanley_buol@ncsu.edu

Zhi-Cheng Chen
Institute of Soil Science
Academy of Sciences
Nanjing, China
ztgong@issas.ac.cn

Jozef Deckers
Institute for Land and Water Management
Catholic University Leuven
Leuven, Belgium
scppe.deckers@agr.kuleuven.ac.be

Craig A. Ditzler
National Soil Survey Center
USDA-NRCS
Lincoln, NE
craig.ditzler@usda.gov

Humberto G. dos Santos
Ciencia do Solo
EMBRAPA Solos
Rua Jardim Botanico
Rio de Janeiro, Brazil
humberto@cnps.embrapa.br

Paul Driessen
International Institute for Aerospace Survey and Earth Sciences
Enschede, The Netherlands

Rudi Dudal
Institute for Land and Water Management
Katholic University Leuven
Leuven, Belgium
Rudi.dudal@agr.kuleuven.ac.be

Robert J. Engel
National Soil Survey Center
USDA-NRCS
Lincoln, NE
bob.engel@usda.gov

Hari Eswaran
USDA
Natural Resources Conservation Service
Washington, DC
hari.eswaran@usda.gov

R.W. Fitzpatrick
CSIRO Land & Water
CRC for Landscape Environments and Mineral Exploration
Glen Osmond, South Australia
rob.fitzpatrick@csiro.au

John M. Galbraith
Crop and Soil Environmental Sciences Department
Virginia Polytechnic Institute and State University
Blacksburg, VA
ttcf@vt.edu

Maria I. Gerasimova
Geographical Department
Moscow State University
Moscow, Russia

Idarê A. Gomes
Ciencia do Solo
EMBRAPA Solos
Rua Jardim Botanico
Rio de Janeiro, Brazil

Zi-Tong Gong
Institute of Soil Science
Academy of Sciences
Nanjing, China
ztgong@issas.ac.cn

Sergey Goryachkin
Institute of Geography
Russian Academy of Sciences
Moscow, Russia
pedology@igras.geonet.ru

Allen E. Hewitt
Landcare Research
Lincoln, New Zealand
hewitta@landcareresearch.co.nz

D.W. Jacquier
CSIRO Land and Water
Canberra, Australia
david.jacquier@csiro.au

Marcel Jamagne
Institut National de la Recherche Agronomique (INRA)
Olivet, France
Marcel.Jamagne@orleans.inra.fr

Dominique King
Institut National de la Recherche Agronomique (INRA)
Olivet, France
dominique.king@orleans.inra.fr

Michiel C. Laker
Department of Plant Production and Soil Science
University of Pretoria
Pretoria, South Africa
mclaker@mweb.co.za

Irina I. Lebedeva
Dokucharv Soil Science Institute
Moscow, Russia

José F. Lumbreras
Ciencia do Solo
EMBRAPA Solos
Rua Jardim Botanico
Rio de Janeiro, Brazil
jflum@cnps.embrapa.br

D.J. Maschmedt
Department of Water, Land and Biodiversity Conservation
Adelaide, South Australia
maschmedt.david@saugov.sa.gov.au

N.J. McKenzie
CSIRO Land & Water
Canberra, Australia
neil.mckenzie@csiro.au

Freddy O.F. Nachtergaele
Land and Water Development Division
FAO
Rome, Italy
Freddy.nachtergaele@fao.org

Francesco Palmieri
Ciencia do Solo
EMBRAPA Solos
Rua Jardim Botanico
Rio de Janeiro, Brazil
palmieri@cnps.embrapa.br

B. Powell
Department of Natural Resources and Mines
Indooroopilly, Queensland, Australia
powellb@dnr.qld.gov.au

Thomas J. Rice, Jr.
Soil Science Department
California Polytechnic State University
San Luis Obispo, CA
trice@calpoly.edu

N. Schoknecht
Agriculture Western Australia
South Perth, Western Australia
nschoknecht@agric.wa.gov.au

Lev L. Shishov
Dokuchaev Soil Science Institute
Moscow, Russia

Otto Spaargaren
International Soil Reference and Information Center
Wageningen, The Netherlands

Victor O. Targulian
Institute of Geography
Russian Academy of Sciences
Moscow, Russia
targul@centro.ru

Valentin D. Tonkonogov
Dokuchaev Soil Science Institute
Moscow, Russia

Goro Uehara
Tropical Plant and Soil Sciences Department
University of Hawaii
Honolulu, Hawaii
goro@hawaii.edu

Gan-Lin Zhang
Institute of Soil Science
Academy of Sciences
Nanjing, China
glzhang@issas.ac.cn

Contents

Part 1

Concepts and Innovations in Soil Classification

CHAPTER 1

Philosophies of Soil Classifications: From Is to Does

S.W. Buol

CONTENTS

ABSTRACT

Philosophies of soil classification are guided by existing knowledge and pragmatic circumstances. Soil is a natural entity that connects the inorganic minerals of earth to the organic organisms of life, and is therefore germane to several academic disciplines. Each of these disciplines seeks to identify and classify soil in relation to its entity of study. Pedologists have established, through definition and classification, that soil *is* an entity worthy of independent academic recognition. People who seek nourishment and utilitarian support from soil identify and classify it by criteria that relate to what it *does* as it interacts with their attempts to utilize land resources. Political entities seek guidance in evaluating the impact of policies and regulations on soil-related natural resources. Soil classification provides a link between soil samples and the natural entities on the land surfaces of earth, and is a communicator of soil properties as spatially represented on maps of diverse scales and in soil property aggregation. No single classification can equally serve all who seek to study and obtain sustenance from soil. Several classifications, each guided by a philosophy of service to an identified audience, are required.

0-8493-1339-2/03/$0.00+$1.50

INTRODUCTION

Show me your [classification] system and I will tell you how far you have come in the perceptions of your research (Kubiëna, 1948).

Soil is so universal that all people know of its existence, and therefore have individual concepts of what soil is and what soil does. The universality of soil, the individuality of people, and the limited geographic exposure to soil each person has experienced ensure an infinite array of philosophies about how to classify soils. To some people, soil is a singular entity, but many have some appreciation for more than one kind of soil, abstractly referring to simplistic classifications of fertile, wet, black, red, sandy, clayey, etc. Others classify soil by association with geology, geography, climate, and vegetation. Philosophies of soil classification are fascinating.

INDIGENOUS CLASSIFICATIONS

Everyone who uses soil classifies it. Most indigenous classifications are in reference to a specific use. Indigenous people are directly interested in what a soil does with respect to their intended use of the land. Indigenous classifications are often quite colorful and informative. At an early age, I understood that "push-dirt" in southern Wisconsin referred to soil that adhered to the plow, and, if slightly too wet, would not be inverted by a moldboard plow. I later learned "push-dirt" was an apt classification for some Newglarus (fine-silty over clayey, mixed, superactive, mesic Typic Hapludalfs) soils, wherein a moldboard plow sometimes reached into the upper part of the clay-textured 2Bt2 horizon that, early in the spring, was often wetter than was predicted by looking at the surface soil. The wet, sticky 2Bt2 material adhered to the moldboard and pushed forward, rather than inverting as intended. "Black Jack," "pipe clay," and "Beeswax" land (Hearn and Brinkley, 1912) referred to distinct soils on the piedmont of North Carolina. Local farmers were said to also call them "dinner-bell" soils because they were too wet to plow in the morning and so dry and hard in the afternoon that a mule could not pull the plow. Plowing was best done at noon, i.e., dinner-time on the farm. We now know "dinner-bell" or "Black Jack" soils as an Iredell (fine, mixed, active, thermic Oxyaquic Vertic Hapludalfs).

Indigenous people who till the soil are concerned with what a soil *does* when it is interacting with their activities. Soil scientists attempt to identify what a soil *is* and to classify it using measurable chemical and physical criteria.

ANTHOLOGY OF SOIL CLASSIFICATIONS

Scientists in the latter part of the 19th century were actively seeking to define soil in a context that separated soil science from geology. Dokuchaev (1883) related soil properties to elements of vegetation and climate, in addition to the influence of geologic parent material. Vegetation and climate were not considered to influence geology, and soil could be defined as a scientific entity apart from geology. In 1927, Glinka (quoted in Jenny, 1941) reported that Russian soil scientists regarded soil type as a summary of the external and internal properties of a soil, so as to include climate, vegetation, and animal life as a unit for soil classification, rather than as volumes of soil, i.e., modern day pedons.

Although he clearly recognized soil as an independent entity, Hilgard (1914) defined soil as "the more or less loose and friable material in which plants, by means of their roots, may or do find a foothold and nourishment, as well as other conditions of growth." Thus he considered soil primarily to be a means of plant production. Perhaps the inscription on Hilgard Hall on the University of California-Berkeley campus best relates his concern for the concepts of soil held by indigenous people: ***"TO RESCUE FOR HUMAN SOCIETY THE NATIVE VALUES OF RURAL LIFE."***

The Bureau of Soils in the United States recognized the need to communicate soil science to indigenous people with regard to how it interacted with various crops and management practices, and produced practical reports that "deal with everyday problems (of the indigenous farmer) and are written with as little use of technical terms as possible" (Whitney, 1905, p. 11).

While concerns for what a soil *does* for human endeavors to obtain sustenance dominated soil mapping efforts in the United States, soil classification efforts sought to establish what *is* a natural entity worthy of independent recognition. Perhaps the infant science of soil reached a status of scientific adulthood, independent of its geologic parent, with the widely read treatise of Jenny (1941). The expression $S = f(cl, pm, r, o, t)$ served to identify soil (S) as a distinct natural entity, but identified it in terms of climate (cl), parent material (pm), relief (r), organisms (o), and time (t). Jenny (1941) concluded that the requirements necessary for establishing correlation of soil formation factors to soil properties were possible only under controlled experimental conditions, and in the field we must be satisfied with approximations and general trends. When the soil-forming factors are considered over time, each becomes a criterion that cannot be measured. Climate consists of weather averages, but one extreme weather event may leave an indelible imprint in a soil. Parent material may be the material below the soil, but incremental aerosol additions made to the soil are often not represented. Tectonic movements and erosion alter relief. Organisms in and on the soil change with natural succession, fire, wind, and human intervention. Jenny (1941) reasoned that the use of soil-forming factors in classification was handicapped by an inability to determine the exact composition of the initial (parent material) and the final (mature soil) stages of soil development. However, in view of the many correlations between soil properties and soil-forming factors, many soil classifications identified soils by association with soil-forming factors, i.e., forest soils, prairie soils, tropical soils, tundra soils, etc.

PHILOSOPHY OF SELECTING A UNIT OF SOIL TO CLASSIFY

Cursory observation of a soil reveals a vertical arrangement of soil components that change, often gradually, as one traverses the landscape. Our understanding of soil is limited without the use of chemical, mineralogical, biological, and physical quantification of soil samples. Soil can be dismembered, sampled, and autopsied, but only if we know where each soil sample is located within a body of soil do the analyses help us understand both what a soil is and what a soil does. Identification of the depth and vertical arrangement of a soil's component parts through several degrees of sophistication, i.e., topsoil, subsoil, generic horizon nomenclature, and diagnostic horizons, is necessary in all attempts to classify soil.

Perhaps no single problem has plagued soil classification more than the identification of the spatial boundaries of a soil individual on the landscape that is to be sampled for study and used as a unit of classification. As soil science struggled to claim adulthood as an independent science, a basis of identifying a soil individual suitable for classification was at the root of many philosophies. Soil classification is closely allied with soil mapping, and most practitioners intertwined pragmatic realities of mapping into their philosophies of identifying an appropriate volume of soil for classification.

From the inception of soil surveys in the United States, three categories were employed in classifying soils in the field: series, type, and phase. Series names were place names of cities, towns, etc. in the area where the soil was first defined. Type identified the texture of the surface horizon, and phase referred to slope, rockiness, and other features of the area being identified. The grouping of soils into categories above series was not discussed in the 1937 *Soil Survey Manual* (Kellogg, 1937). At that time, a series was defined as "a group of soils having genetic horizons similar as to differentiating characteristics and arrangement in the soil profile, and developed from a particular type of parent material" (Kellogg, 1937, p. 88). It is clear from the discussion that differentiae from higher categories were not imposed on series. Instead, "All mappable differences in the profile significant to the growth of plants should be recognized in the classification" (Kellogg, 1937, p. 89). Clearly, a spatial area that could be identified on a soil map was considered germane to the concept of series. Most soil mapping in the United States at that time was done on 1:63,360 scale maps.

The concept of "mappable differences" as classification criteria caused severe confusion. Mappable differences in one area (in which soils belonged to a named series because of "genetic horizon" similarities) often were not "mappable" in other areas, or at different mapping scales. By 1951, problems associated with aligning "genetic horizon" criteria with "mappability" were recognized, but no clear protocol was established to overcome the problems (Soil Survey Staff, 1951). It was simply stated that the time and expense involved in separating two soil series could not always be justified. Moreover, the criteria for using only similarity of "genetic horizons" to define series were extended to other, often geologic, features below the soil profile. This was pragmatically justified in response to increasing use of soil survey information for engineering purposes that relied on soil properties not always present in the genetic horizons but violating a basic tenet of classification, i.e., that criteria used to classify an object should be observable within the object classified.

Identification of an individual volume of soil that could be used for classification had to be addressed before progress could be made in the development of quantitative soil classification. The conceptual philosophy for identifying an individual soil paralleled the concept of the unit cell used to identify minerals. The unit cell of a mineral identifies all the elements present in that mineral, and is the basic unit for identification and classification. That exact unit cell composition, however, may not be present in each mineral particle belonging to that mineral class. In the process of developing Soil Taxonomy, the concept of pedon, "the smallest volume that can be called 'a soil,' " was developed and used as a basic unit for soil identification (Soil Survey Staff, 1960; Johnson, 1963). Individual pedons can be identified, sampled, and studied much like the study of an individual mineral particle. But more than one pedon has to be included to represent the range of properties defined by a soil individual, just as many mineral particles belonging to the same mineral class must be studied to observe the entire composition of that mineral class. Soil series as a unit of classification was defined in the 7th Approximation as "…a collection of soil individuals that, within defined depth limits, are uniform in all soil properties diagnostic for series. …Soil individuals are real things, but series are conceptual." (Soil Survey Staff, 1960, p.15).

By 1993, soil series were defined "as a class, a series is a group of soils or polypedons that have horizons similar in arrangement and in differentiating characteristics" (Soil Survey Staff, 1993, p. 20). Mappability as a criterion was dropped from earlier concepts of series and it was emphatically stated that, "When the limits of soil taxa are superimposed on the pattern of soil in nature, areas of taxonomic classes rarely, if ever, coincide with mappable areas" (Soil Survey Staff Division, 1993, p. 21). Unfortunately, the task of conveying the fundamental differences between soil classification and soil mapping has proven difficult (Cline, 1977) and remains so today (Soil Survey Staff, 1999).

PHILOSOPHY OF CLASSIFICATION TO ACCOMMODATE SOIL MAPPING

Every attempt to convey spatial distribution of soils using a soil map results in the inclusion of more than one pedon of the same soil class, and invariably pedons of other soil classes within each delineation. The fact that many different pedons are included within every map unit delineation on a soil map should not be any more disturbing than finding individual deciduous trees of different ages, sizes, and morphologies, with perhaps a few coniferous trees and some grass species growing within a delineation of deciduous forests on a vegetative map.

Maps are representations of the spatial distribution of classified objects identified at a reduced scale. Soil maps are of many scales and are therefore best served by a classification system that provides aggregation of the objects mapped in several levels of generalization—this is one obvious function of a hierarchical system of classification. The philosophies that guide the structure of hierarchical systems for classifying objects subject to spatial mapping are in large part determined by the mapping experience of the people constructing the classification system.

Comparisons among all soil classification systems are not possible, but one example illustrates philosophies guided by a desire to reflect soil properties vs. a desire to accommodate pragmatic soil mapping concerns. All soil classification systems recognize poorly drained soils and well-drained soils as distinct entities. Most soil scientists will agree that there are major differences in composition and function associated with drainage characteristics. Most humid areas of the world have a mosaic of well-drained and poorly drained soils spatially related to landscape features. Hierarchical systems driven by a philosophy that identifies major contrasting soil properties have poorly drained soils identified at a high categorical level. Systems driven by experience with detailed soil mapping at rather large scales tend to recognize poor drainage in subordinate categories, and use higher hierarchical categories to identify soil properties spatially related to parent material or broad physiographic regions, rather than local topography. This latter arrangement is compatible for both small-scale maps that use taxonomic units of the higher categories for soil identification and large-scale detailed maps that are able to delineate spatially intricate landscape positions and use the taxonomic units of the lower, more detailed, categories to identify mapping units.

UTILITARIAN EXPECTATIONS OF SOIL CLASSIFICATION

Human need for soil information takes many forms. Each person can prioritize the following needs, and perhaps others, that are expected of soil classification. Individual needs for information shape individual philosophies of what a classification should entail.

- Analytical needs: Analytical need-based philosophy of classification seeks to accommodate the needs of scientists as they seek fuller understanding of soil. A sample of soil material is a discrete entity, but not a soil. Information obtained from the analysis of a soil sample contributes to our knowledge of soil only if the location of that sample can be identified both within an individual soil and among all soils. Soil is a basic reagent for all land use. Quantitative transfer of technology and research results to other sites requires that basic soil components be identified with the same rigor as the purity and normality of reagents are identified in chemical experiments.
- Political needs: The philosophy of politically-based classification identifies projected response to policy initiatives. Soil is a natural resource claimed by all political entities. Expenditures from governmental coffers for soil-related technologies need to be directed to soils in which policy-directed technology is applicable. The ability of soil scientists to correctly guide political decisions has a pragmatic impact on financial support, and thus the sustainability of classification. Governmental priorities arise, then wane. For example, assistance for installation of drainage systems in "wet lands" was a goal of U.S. government policy in the 1950s, while preservation of "wet lands" became the policy in the 1990s. Concerns for erosion control led to the land capability classification system that guided the expenditure of funds for terraces, strip crops, and other soil conservation practices to soils where erosion could be reduced, and avoided soils where such practices were of little or no value. Political needs often require that classifications identify soil and land subject to specific laws and regulations. Regulatory requirements are of all spatial scales. Some are site-specific and require identification of soil properties within pedons, i.e., elemental concentrations in specific horizons or depths. Land use regulations often relate to transient soil conditions, such as saturation and flooding. In the absence of statistically significant long-term observations of sporadic events, pedogenic features are frequently used as evidence.

- Single property economic evaluations: When certain soil properties are of concern to the economic evaluation of a soil for a specific use, that property dominates the philosophy of the persons seeking to classify soils. For example, construction engineers digging a pipeline or constructing a highway are concerned with the depth to hard rock, where excavation with a drag-line or bulldozer is not feasible and blasting may be required. Economic concerns for specific soil properties foster philosophies of simple, single-property classification systems.
- Management response classifications: Vegetative growth relies on many soil parameters. Some soil conditions can be effectively ameliorated by cost-effective management practices, whereas others are cost-prohibitive. Some management objectives require initial capital investments that must be recouped over the life of the project, whereas others require recurring expenditures to sustain the project. Classifications guided by soil response to specific management practices are often transient, as alternative technologies are implemented.
- Ecological interactions: Soil is the site of interaction between the organic and inorganic chemistries on the land areas of earth. Soil classification based solely on undisturbed or ambient vegetative-soil associations is useful, but it often fails to adequately convey alternative associations, especially in human-dominated ecosystems.

BENCHMARK PHILOSOPHIES OF SOIL CLASSIFICATION

With the many and diverse expectations humans have of soil classification systems, there can be little doubt that a single classification will not satisfy all who desire systematic identification of soil properties. Experience in dealing with expectations, both internal and external, of soil science has influenced philosophies of soil classification.

- Indigenous classification: Tillers of the land classify the soil they till in terms of response to the tillage practices they employ. Within the geographic confines of their individual experience, most indigenous civilizations are adept at harmonizing soil properties and locally available management technologies. Experience with sporadic weather conditions has honed their ability to time their management operations with temporal soil conditions. Indigenous classifications are locally useful, but they lack insight into technologies not practiced in that locale. Experienced soil scientists have learned to communicate with and carefully evaluate indigenous classifications, but must expand on that base of knowledge to evaluate applicability of technology not experienced by indigenous populations.
- Political pragmatism: A classification system that fails to deliver practical responses to a significant portion of people who use soil is doomed to obsolescence, regardless of scientific innovation. Numerous scientifically sound soil classification systems have been developed by individuals and groups of individuals, but have failed to provide the practical information necessary to capture the sustained financial support necessary to expand those systems beyond the initial expertise of the individuals involved. Most political bodies are concerned with inventories of soil resources. Broad groupings and small-scale maps often satisfy their needs, and they then withdraw support for further soil study. Sustained support for soil science is best obtained when masses of indigenous people are served by spatially detailed information to which they can readily relate within their local community, and then communicate their satisfaction to political entities.
- Expediency of change: All classification systems are subject to demise if they do not respond to new information and user needs. No single individual or group of individuals has equal experience with all soils. Intensively studied soils tend to be identified by minute differences, while unfamiliar soils tend to be broadly grouped. Classification systems mature, and as more scientists participate, more uniformity results. User needs are often closely related to changes in technology related to soil use, or to politically driven land use regulations, and involve only a geographically limited number of soils. Rapid response to user needs must be locally addressed, but each should be carefully analyzed within the context of all that is known about soil before incorporation into more geographically extensive classification systems.

MULTIPLE SYSTEMS

Several indigenous proverbs can be paraphrased: "Serve all masters and serve none well." Philosophies of classification may themselves be classified into three categories:

- Natural or universal systems: Such systems provide stability through time so that scholars can relate past experience with new experience, as the science of soil continues to develop. A natural or universal classification system is analogous to a major library: A major library must accommodate ever-expanding needs, as new books are cataloged and arranged in an orderly fashion without destroying old books. From time to time old books may be moved to different locations within the library as new books are assimilated, but they are never destroyed. Continued professional effort and sound financial backing are required to maintain universal systems of classification.
- Geographic subsets: Geographic subsets of a natural or universal soil classification system are often required by the pragmatic requirements of acceptance and funding by political bodies. Soil is a natural resource, and is viewed as such by political entities. Soil scientists dependent on government and ultimately indigenous support are not likely to be viewed favorably if they must engage in time-consuming dialogue before responding to requests for new categories of soil properties within the political domain that funds their operations. While geographic subset classifications of soil are worrisome to the larger body of soil scientists who seek a universal format upon which to communicate their research, they are a necessary reality. With time and understanding, they can contribute to more universal classifications.
- Technical subsets: Like geographic subsets, technical subsets are vital to the well-being of soil science and soil scientists. Technical subsets are defined by those who seek specific information about the interaction of soils and specific technologies. Soil scientists are seldom in a position to determine categories within technical classification systems. Technical systems must address the needs of the technicians and scientists most knowledgeable about the needs of specific technologies. Soil scientists apply their expertise by identifying soil-based criteria for the categories outlined by the technical experts. Technical systems are numerous and transient, but soil scientists must devote considerable time and energy to technical classifications to insure indigenous and political support for soil science.

CONCLUSIONS

Smith (1986) credited John Stuart Mill (8th edition, 1891) for pointing out that objects had to be classified for a purpose, and that if there were different purposes, there could be several classifications of the same objects. After over 100 years of formal soil science and fathomless ages of indigenous efforts the surviving philosophy of soil classification is based on the use of several classification systems that, in concert, serve two distinct objectives: scientific and technical (Cline, 1949). Among the systems used, at least one must provide wholeness for categorization of all accumulated knowledge of soil properties, thereby providing a basis to study relationships with other biological and physical entities of earth. This scientific system must provide exact and quantitative definitions of what an individual kind of soil *is*.

A multitude of technical systems must be pragmatic in providing discrete categories that are meaningful and useful to a host of practitioners whose technology interacts with what a soil *does*. Given the diversity of technologies that interact with soil, no single technical classification is sufficient.

"One cannot say that one taxonomic classification is better than another without reference to the purposes for which both were made, and comparisons of the merits of taxonomies made for different purposes can be useless" (Soil Survey Staff, 1999, p. 15). Scientific classification systems that incorporate detailed identification of all soils may receive great acclaim from scientists, but sustained financial support is difficult to assure. Technical classifications that relate specific interactions of only a limited range of soil properties are transient, as political concerns and technologies change, but they garner public support for soil science. There is mutual dependence. Technical classifications rely on the existence of sound identification of soil properties provided by a scientific system. The scientific system relies on the technical system to garner financial support and political confidence in the value of soil science expertise. Without a philosophy that encompasses both scientific wholeness of what soil is and temporal response to technological and political needs by identifying what soil does, soil classification is incomplete.

REFERENCES

Cline, M.G. 1949. Basic principles of soil classification. *Soil Sci.* 67:81–91.

Cline, M.G. 1977. Historical highlights in soil genesis and classification. *Soil Sci. Soc. Am. J.* 41:250–254.

Dokuchaev, V.V. 1883. Russian Chernozem (Russkii Chernozem). (Transl. from Russian by N. Kaner.) Israel Prog. for Sci. Transl., Jerusalem, 1967.

Hilgard, E.W. 1914. *Soils.* Macmillan, New York.

Jenny, H. 1941. *Factors of Soil Formation: A System of Quantitative Pedology.* McGraw-Hill, New York, p. 281.

Johnson, W.M. 1963. The pedon and polypedon. *Soil Sci. Soc. Am. Proc.* 27:212–215.

Kubiena, W.L. 1948. *Entwicklungslehre des Bodens.* Springer-Verlag, Wien.

Kellogg, C.E. 1937. Soil Survey Manual. USDA Miscellaneous Publ. No. 274, p. 136.

Mill, J.S. 1891. *A System of Logic.* 8th ed. Harper Bros., New York.

Smith, G.D. 1968. The Guy Smith Interviews: Rationale for Concepts in Soil Taxonomy. SMSS Tech. Monogr. No. 11. USDA-SCS and Cornell University, p. 259.

Soil Survey Division Staff. 1993. Soil Survey Manual. USDA Handbook No. 18, U.S. Government Printing Office, Washington, DC, p. 437.

Soil Survey Staff. 1951. Soil Survey Manual. USDA, Agr. Res. Adm. Agric. Handbook No. 18, U.S. Government Printing Office, Washington, DC, p. 503.

Soil Survey Staff. 1960. Soil Classification: A Comprehensive System—7th Approximation. USDA-SCS, p. 265.

Soil Survey Staff. 1999. Soil Taxonomy: A Basic System of Soil Classification for Making and Interpreting Soil Surveys (2nd edition). Agric. Handbook No. 436. USDA-NRCS, U.S. Government Printing Office, Washington, DC, p. 869.

Whitney, M. 1905. The Work of the Bureau of Soils. USDA, Bur. of Soils. Circular No. 13, p.15.

CHAPTER 2

How Good Is Our Soil Classification?

R. Dudal

CONTENTS

Although the applications of soil science are many, few analyses have been made on the actual impact of soil classification on the optimization of land use. The question arises as to whether sufficient use has been and is being made of soil survey and soil classification. Many feasibility studies include soil survey information, but it is frequently overlooked in subsequent development plans. When soils information is sought, data supplied through a medium of soil classification are difficult to interpret for specific purposes. This situation has been ascribed to inadequate presentation of results; lack of communication between soil scientists, agriculturists, and economists; difficulties with specialized terminology; or lack of interest by planners. One may wonder, however, if the root cause is more fundamental, and if soil surveys and soil classification, as currently conceived, are entirely suitable to serve the needs of potential users.

An assessment of the adequacy of a classification system has to be made regarding the purposes for which it has been designed. Soil classification systems generally aim at establishing a taxonomy, making soil surveys, serving as a tool for interpretation, and being a means of communication. Can a single system effectively meet these different objectives? "In any system of classification, groups about which the greatest number, most precise and most important statements can be made for the objective serve the purpose best. As the things important for one objective are seldom important for another, a single system will rarely serve two objectives equally well" (Cline, 1949).

A TAXONOMY

Taxonomy is that branch of science dedicated to discovering, characterizing, naming, and classifying objects or organisms so as to understand relationships between them and with the factors of their formation. Taxonomy is about identification and recognition, as well as the establishment

0-8493-1339-2/03/$0.00+$1.50

of a hierarchy of classes which allows an orderly overview of the diversity of the objects concerned. The success of taxonomic classifications of organisms has prompted its application to soils. However, unlike plants and animals, which can easily be identified, soils constitute a continuum that needs to be broken into classes by convention. In the early days of soil science, "genesis" was considered to be the basis for classifying soils. The concept, borrowed from the biosciences, was actually not appropriate for soils that do not have phylogenetic relationships and that cannot be grouped by descent from common ancestors.

The successive systems of soil classification that were developed reflected different concepts of soil formation, as well as the state of knowledge at the time. Various hypotheses included much conjecture. Soil formation spans long time intervals during which soil-forming factors vary considerably. Hence links of soil formation to present climate were not proven to be applicable in different parts of the world. Soil classification schemes based mainly on "genesis" have retarded rather than enhanced the development of a comprehensive soil taxonomy. Recent systems of soil classification reflect considerable progress in this respect. The USDA Soil Classification (Soil Survey Staff, 1960) was a breakthrough toward classifying soil in terms of its own properties and toward defining taxa on the basis of quantitative differentiae that could be observed or measured. The selection of differentiating properties took soil formation into account, but soil-forming processes as such were no longer criteria for separating classes. Significance to plant growth also had a bearing on the selection of differentiae. As the number of soil taxa far exceeds the amount of information that the mind can comprehend and remember, a hierarchy of six categories was established, from orders down to series, somewhat analogous to the hierarchy from phylum to species in plant science. The branching sequence of organisms has evolution as an objective basis. For soils, however, the hierarchy is a matter of opinion (Swanson, 1993).

Although soil-forming processes are no longer taxonomic differentiae as such, it appears that concepts of soil formation still influence the ranking of the hierarchical categories. The genetic significance and the importance to use ascribed to the argillic horizon have led to the distinction of two orders in the USDA Soil Taxonomy (Soil Survey Staff, 1999): the Alfisols and the Ultisols. The overriding weight given to the argillic horizon has been questioned because of the constraints presented by strong weathering in soils of the tropics—ferralic attributes—that are more pronounced than the apparent advantage of clay increase with depth (Eswaran, 1990). In the South African Soil Taxonomy (Soil Classification Working Group, 1991), textural differentiation is a major differentiating characteristic, however, with much more strongly expressed textural differences than those of the argillic horizon. While Alfisols and Ultisols are separated at order level by a difference in base saturation in the subsurface, base saturation in South Africa is taken into account only at family level. The wide geographic extension of steppes in North America and Europe justified the distinction of Mollisols or Humus–Accumulative Soils at order level in Soil Taxonomy, and in the Russian Soil Classification (Shishov et al., 2001), respectively. An equivalent higher category is absent in the Australian (Isbell, 1996) and South African soil classification systems. Hydromorphic soils are distinguished at order level in Australia and Russia, in contrast to the suborder ranking to which they are assigned in the USDA Soil Taxonomy. In the Russian soil classification, Halomorphic soils are recognized at order level, while in South Africa, no distinct provision has been made for saline soils. Both the Australian and the Russian classification systems acknowledge strong human influence on soil formation at order level, as Anthroposols and Agrozems.

There are 12 orders in Soil Taxonomy, 14 in Australia, 23 in Russia, and 73 "soil forms" in South Africa, reflecting different opinions as to the weighting of soil characteristics as well as to the impact of the geographic setting in which the classifications were developed. The rationale for the distinction and number of classes at different levels of generalization is not always apparent, and neither is the justification for the very high number of classes in the higher categories, which is in the range of thousands for the subgroups. Considering that their interpretation value is limited, a clarification of the rationale for this elaborate hierarchy is called for.

Recently developed soil classification systems have greatly contributed to the taxonomy of soils, i.e., their identification, recognition, and relationships to the factors of their formation. Further consultations will, however, be required in order to ensure that the weighting and hierarchy of differentiae be harmonized, and that taxonomy concurrently becomes a means of communication.

MAKING SOIL SURVEYS

Soil surveys started well before the creation of soil taxonomies. In the United States, the phrase "soil types" was adopted as a label for mapping units set apart in the first surveys made in 1899 (Simonson, 1997). A few years later, the soil series was introduced in order to group types. Distinguishing kinds of soils was based on features that appeared to influence their relations to crops. Primary emphasis was given to texture. A secondary purpose of the surveys was to distinguish kinds of soils that differed in their formation, such as those derived from different kinds of rocks. The impact of the Dokuchaev's concepts of soil formation and distribution (Glinka, 1914) led to the creation of a more comprehensive system of soil classification, including great soil groups as a higher category (Marbut, 1928). The lowest categories, the soil series and types, created on the basis of the work of the soil survey, were considered to be the base upon which the whole structure rested. It was decided that all the criteria for classes in higher categories were to be series criteria as well. The justification for the higher categories was that grouping the very large number of series into a smaller number of classes would allow concepts and relationships to be more easily understood, and soils in different areas could be compared. Early efforts to place the series into great soil groups on the basis of existing descriptions were not successful. Improvements in the categories above the soil series were needed to complete the scheme. It was realized that higher categories, although helpful in organizing knowledge, had a decreasing value for interpretation and for transferring experience. Hence the series remained the principal units of mapping, not only on a large and intermediale scale, but also on a smaller scale. In the latter case, mapping units were expressed in terms of associations of series, rather than on the basis of higher categories. One of the reasons for this approach was the realization that the use of taxa, higher in the hierarchy, sacrifices information because of the greater heterogeneity of the classes.

The question arises as to whether the higher categories have been designed to group the lower categories, or if the higher categories have been developed separately, imposing on the lower categories differentiations with no clearly known significance in utility. The development of the taxonomy actually proceeded at both ends of the level of generalization, resulting in a difficult matching of differentiae and in a hiatus between the higher, more genetically inspired categories and the more pragmatic lower ones, geared toward responses to management and manipulation for use. Cline (1980), in his assessment of the use of Soil Taxonomy in the United States, felt that the precision and detail of field description had improved, but that standards of mapping had not markedly changed. With the emphasis on lower categories as mapping units, the role of the higher categories of a soil taxonomy for making soil surveys requires clarification.

Current soil surveys record soil classes that are characterized by the description and analysis of sampling units. The soil cover is represented in terms of one class, in homogenous areas, or of an association of classes when contrasting kinds of soils occur. However, associations as presently conceived do not necessarily reflect the relationships between the soils they contain, the sequence in which they occur, or the proportion of the area that they occupy. They seldom reflect the structure and the dynamics of the soil-landscape. Soils are characterized not only by a vertical succession of horizons, but also by spatial variations in their properties, related to lateral movements at the surface and within the soil. The soil moisture regime along a slope, the lateral flow of nutrients and weathering products, runoff and erosion, and the development of salinity and waterlogging are functions of the dynamics of soil formation in the landscape. The three-dimensional nature of soil is barely reflected by a 10-m^2 pedon. Its fourth dimension, time, is mostly overlooked. The properties of a soil profile, which we have become so accustomed to seek and appreciate, may not, in the final analysis, be as important as their spatial or temporal variance. Alternative approaches advocate a more comprehensive representation of the soil pattern. Soil sociology (Schlichting, 1970) rather than soil taxonomy has been proposed as a basis for mapping soil scapes (Buol et al., 1975), soil landscape (Hudson, 1995), soil cover (Fridland, 1980), genons (Boulaine, 1978), or "ensembles pédologiques" (Ruellan, 1984). The approach is a dynamic rather than a static characterization of soils. The leaching and accumulation processes in the solum have their equivalent in the discharge and recharge flows in the landscape (Richardson et al., 1992). The emphasis on the use of taxa in soil surveys, carrying precise criteria for separating them, may have reduced awareness of the lateral and gradual links in the soil landscape. Boundaries between taxa are sometimes difficult to establish in the field, as real soils may not abide by the rules that govern a taxonomic hierarchy. Boundaries based on the visible landscape might better reflect the three-dimensional nature of the soil cover. "There is a need for devices that will adapt quantitative limits of criteria more realistically to soil variation in the field in order to reconcile the conceptual framework of taxonomy with the reality of soils in nature" (Cline, 1980).

INTERPRETATION

Interpretation of soil surveys and predictions of the effect of management practices require that soils be grouped according to their qualities and limitations for different types of use. It is assumed that basic taxonomic classes can be grouped, or subdivided and regrouped, to permit specific predictions. The shift from a qualitative genetic approach in soil classification to a quantitative expression of properties of the soil itself has led to a more effective assessment of soil-use relationships. Most soil classification systems use a limited number of class differentiae, preferably those that are stable, and can be measured or observed. Transient properties of the surface layers, which vary with soil management practices, are generally avoided to accommodate both cultivated and undisturbed soils. In the early stage of the development of the USDA Soil Taxonomy, it was decided that soil management should not change the classification.

The emphasis on stable subsoil characteristics and the genetically inspired selection of taxa differentiae do not necessarily provide the criteria relevant to soil use. It is striking that parameters of soil hydraulics are given little consideration in soil classification, even though they are of prime importance to plant growth and to the assessment of water flows, which strongly affect environmental issues. Soil texture, as a distinctive characteristic, appears only in the lower categories, even though it is a main indicator of moisture retention, workability, and trafficability. It is significant that a number of technical classifications have been designed in which only a part of the data gathered in a soil survey is combined with added information collected for the specific purpose. The "soil fertility capability classification" (Buol et al., 1975) is an example of a classification based on characteristics that have been specifically selected for their relevance to soil fertility management. Similar groupings have been made to accommodate land capability, suitability for irrigation, erosion control, engineering, wildlife habitats, drainage design, land remediation, and others. Boundaries of these utilitarian groupings may not correspond to those of the mapping units of a soil survey, and are often more accessible to users than the taxonomic nomenclature. However, taxonomy may provide a scientific base for the recognition and characterization of the soil pattern.

In many instances, phases rather than taxonomic differentiae are the factors that determine land use choices, such as slope, landform, stoniness, flooding, and erosion. Special attention is called to the properties of the surface layer, of paramount importance to plant growth. Soil surface layers influence germination, are the seat of biological activity, store a large part of plant nutrients, contain a major portion of the roots of annual crops, and are determining factors for tillage practices. Yet these properties are not reflected in the definition of taxa because they are transient. The human factor of soil formation is underemphasized, though it is a major component of soil production potential. To overcome the gap between the interpretative value of soil classes and actual production potential, a comprehensive characterization of topsoils (Purnell et al., 1994) has been proposed for the practical purpose of assessing and monitoring fertility status and guiding management practices. Topsoil characterization should, at an appropriate scale, become an adjunct to current soil classification and provide a way to link soil classes with soil management. The proposal identifies about 60 main topsoil properties related to organic matter status, biological activity, and physical, chemical, and moisture conditions, allowing for the characterization of topsoil over and above the diagnostic horizons and features required for defining a soil class. This approach could help overcome criticisms by other disciplines and by land users regarding shortcomings in the application value of taxonomic systems. Recently, a proposal was even made (Krogh and Greve, 1999) to adopt a two-tier soil classification system, the first tier consisting of largely undisturbed soils, while a second tier would accommodate significantly changed cultivated soils with anthric properties.

Soil classification has traditionally been considered a means to optimize land use and to transfer technology between comparable soils in different areas. This approach is based on the assumption that soil classification stratifies the environment in sufficient detail that transfer by analogy can take place. This assumption may be valid when the emphasis is on broad assessments, but it should be fully realized that technology transfer should rely not only on soil attributes, but on other factors as well, such as climate, relief, hydrology, level of inputs, and socioeconomic conditions. If any of these factors vary between locations, similarity of soil classes will not result in successful transfers. Hence soil survey interpretation has progressively evolved into land evaluation (FAO, 1976), encompassing soil, atmosphere, geology, hydrology, plant and animal populations, and the results of past and present human activity.

Special attention needs to be given to the climatic conditions that determine the upper limits of crop production. Since soils with similar morphology and chemical characteristics can occur under different climates, some soil classification systems have introduced climatic data at various levels of generalization. The number of climatic subdivisions that can be made must of necessity be limited if the number of taxa in the classification system is to remain manageable. However, annual means of soil moisture and temperature do not provide sufficient information for the transfer of site-specific experience. An overlay of more detailed climatic data, independent of soil classification, may be preferable. An attempt in this direction was made in a study of the land use potential of major agro-ecological zones (FAO, 1978) in which the climatic inventory was expressed in terms of the length of the growing period, based on moisture availability, temperature, and radiation at intervals of 30 days (e.g., 90–119 days, 120–149 days, 150–179 days, etc.). Combining climatic and soil requirements made it possible to establish and map suitability ratings for specific types of land use. The adoption of climatic phases, rather than broad temperature and moisture regimes, could enhance the application of soil surveys for development purposes.

APPROACHES FOR THE FUTURE

The continuum of the soil cover at the earth's surface can hardly be apprehended in its entirety. Hence a subdivision is required in order to recognize and remember the different components of this continuum and to understand the relationships with the factors of their formation. Classes are constructions of the mind, and differentiae are selected and weighted according to their significance in soil formation and land use.

Opinions regarding the rationale of establishing and ordering classes, and regarding the importance of differentiating characteristics, vary widely. The assumption that a soil classification should provide for a taxonomy, for making soil surveys, and for interpreting them led to the development of elaborate hierarchies to capture and oversee the great number of differentiae involved. It appears, however, that the three purposes can hardly be served equally well. Concerns in this respect have been voiced previously, but no major adjustments have been made so far (Dudal, 1986).

A future approach could be inspired from the plant sciences, which clearly separate plant taxonomy, phytosociology, and phytotechnology. Plant taxonomy aims at producing a system of classification that best reflects the totality of similarities and differences among plants. Phytosociology handles the distribution of plant communities, reflecting the interdependence of the species and the environmental relationships that influence their distribution. Phytotechnology classifies plants in terms of their use (cereals, tubers, fiber, fruits, oil crops, etc.) across taxonomic boundaries and on the basis of specifically utilitarian differentiae. With regard to soil classification, a taxonomy in a global context could consist of two levels, a first one comprising reference soil groups, which would accommodate the major high-level classes of different national systems, and a second level of qualifiers, allowing for a more precise description and definition of subdivisions of the reference groups. Several qualifiers could be used to classify a soil in a sequence that best suits the purpose, or in accordance with an agreed-upon ranking. The development of such a World Reference Base for Soil Resources started in the 1980s (Dudal, 1980; 1990) and has recently materialized in a concrete proposal (FAO/ISRIC/ISSS, 1998; Deckers, 2000; Nachtergaele et al., 2000) that was endorsed by the IUSS in 1998. It has allowed for a balanced overview of the world's soil resources and serves the purpose of small-scale soil resource inventories (FAO, 1991). The reference base serves as a means of communication and as a bridge between national systems that, at lower levels of generalization, can accommodate geographically specific differentiae.

Increased attention should be given in soil surveys to the mapping of soil landscapes, and highlighting temporal and spatial dynamics within a landform, rather than merely recording the distribution of taxa as single units or as unqualified associations. The spatial structure of the soil cover does not usually correspond with the hierarchy in systems based on taxonomic differentiae. Environmental issues require information of variation within the landscape, which is not provided by the prevailing soil survey procedures. A World Soils and Terrain database (SOTER) (UNEP/ISSS/ISRIC/FAO, 1995) is being established, containing digitized map units and their attributes composed of sets of files for use in relational database management systems. In addition to soil and terrain boundaries, observation points should be recorded and georeferenced, making full use of tools and techniques of information technology.

Interpretation of soil characteristics toward land use requirements should continue to rely on soil-landscape models of variation. However, modern demands on soils require that data, as well as the differentiae of taxa, be measured and recorded. The limits of the human mind are no longer constraining; data storage and processing can now be handled digitally. Systematic surveys are not required to obtain the necessary information. Georeferenced databases lend themselves to selecting and grouping clusters of relevant data in function of demand. They allow for a flexible weighting of properties in accordance with purpose (Burrough, 1993). Hence the boundaries in technical groupings often cut across taxonomic subdivisions.

Although the three purposes of soil classification—taxonomy, soil survey, and interpretation—would benefit from being handled more independently, it is imperative that the experience gained with the relationship among soils, landform, and climate be blended with the new methods of data processing. Blind mathematical approaches could lead to stratifications of doubtful significance. A geographic dimension of the soil cover is essential for ensuring sound applications of soil science.

Our soil classification has served us well. It has enabled us to explain and characterize soil diversity in function of different sets of soil-forming factors. However, with increasing demands for targeted soil information and with the advent of modern tools for data storage and processing, soil classification now needs to be addressed more specifically to the purposes that it is meant to serve. The rationale of its hierarchy and of the selection and weighting of its differentiae has to be clearly spelled out. Additional research is required on the relationships among soil characteristics, their effects on plant growth, and their use for different purposes.

FAO (1998) initiated a database for identifying plant species appropriate for given environments and uses, and for assessing crop response to environmental and management factors. Soil classification needs to be made more accessible and relevant to a wider range of users to remedy a widely prevailing soil resource illiteracy.

REFERENCES

Buol, S.W., Sanchez, P.A., Cate, R.B., and Granger, M.A. 1975. Soil fertility capability classification, in *Soil Management in Tropical America.* E. Bornemisza and A. Alvarado, Eds. North Carolina State University, Raleigh, 126–141.

Burrough, P.A. 1993. The technologic paradox in soil survey: New methods and techniques of data capture and handling. Enschede, *ITC J.* 3:15–22.

Cline, M.G. 1949. Basic principles of soil classification. *Soil Sci.* 67:81–91.

Cline, M.G. 1980. Experience with soil taxonomy of the United States. *Adv. Agron.* 33:193–226.

Deckers, J. 2000. World Reference Base for Soil Resources (WRB), IUSS endorsement, World-Wide Testing and Validation. Letter to the Editor. *Soil Sci. Soc. Am. J.* 64:2187.

Dudal, R. 1980. Towards an international reference base for soil classification. *Bull. Int. Soc. Soil Sci.* 57:19–20.

Dudal, R., 1986. The role of pedology in meeting the increasing demands on soils. Transactions 13th International Congress of Soil Science. Plenary papers. DBG, Göttingen, Vol. I: 80–96.

Dudal, R. 1990. An international reference base for soil classification, in Recent Development in the Preparation of an International Soil Reference Base. Transactions 14th International Congress of Soil Science, Kyoto, Japan, Symposium V2: 37–42.

Eswaran, H. 1990. Soils with ferralic attributes, in Recent Development in the Preparation of an International Soil Reference Base. Transactions 14th International Congress of Soil Science, Kyoto, Japan, Symposium V2: 65–70.

Food and Agricultural Organization of the United Nations. 1976. A framework for land evaluation. Soils Bulletin 32. FAO, Rome. p. 72

Food and Agricultural Organization of the United Nations. 1978. Report on the Agro-Ecological Zones Project. World Soil Resources Reports, 48. Vol. I-IV, FAO, Rome.

Food and Agricultural Organization of the United Nations. 1998. ECOCROP, the crop environmental requirements (1) and crop environment response (2) data base. Land and Water Digital Media Series, 4. FAO, Rome.

Food and Agricultural Organization of the United Nations/ISRIC/ISSS. 1998. World Reference Base for Soil Resources. World Soil Resources Report 84. FAO, Rome. p. 88

Fridland, V.M. 1972. Pattern of the Soil Cover. Academy of Sciences of the USSR. Geographical Institute and All-Union Academy of Agricultural Sciences, Dokuchaev Soil Institute. Translated from Russian, Israel Program for Scientific Translations, p. 274.

Glinka, K.D. 1914. Die Typen der Bodenbildung, ihre Klassifikation und geographische Verbreitung. Bornträger, Berlin. p. 365.

Hudson, B.D. 1992. The soil survey as paradigm-based science. *Soil Sci. Soc. Am. J.* 56:836–841.

Isbell, R.F. 1996. The Australian Soil Classification System. Australian Soil and Land Survey Handbook. CSIRO, Australia. p. 143.

Krogh, L. and Greve, M.H. 1999. Evaluation of World Reference Base for Soil Resources and FAO Soil Map of the World using nationwide grid soil data from Denmark. *Soil Use Manage.* 15.3:157–166.

Marbut, C.F. 1928. A scheme of soil classification. Proceedings First International Congress of Soil Science, Washington, DC, 1927. Vol. 4: 1–31.

Nachtergaele, F.O., Spaargaren, O., Deckers, J.A., and Ahrens, R. 2000. New developments in soil classification. World Reference Base for Soil Resources. *Geoderma* 96:345–357.

Purnell, M.F., Nachtergaele, F.O., Spaargaren, O.C., and Hebel, A. 1994. Practical topsoil classification – FAO proposal. Transactions 15th International Congress of Soil Science, Acapulco. Vol. 6b: 360.

Richardson, J.L., Wilding, L.P., and Daniels, R.B. 1992. Recharge and discharge of groundwater in the aguic moisture regime, 212–219, in J.M. Kimble, Ed., *Proceedings 8th Int. Soil Correlation Meeting (VIII ISCOM): Characterization, Classification and Utilization of Wet Soils.* USDA Soil Conservation Service/Soil Management Support Services, p. 327.

Ruellan, A. 1984. Les apports de la connaissance des sols intertropicaux au développement de la pédologie. Science du Sol, Bulletin de l'Association française pour l'Etude du Sol, No. 2.

Schlichting, E. 1970. Bodensystematik und Bodensoziologie., Zeitschrift für Pflanzenernährung und Bodenkunde, 127 Band, I. Heft.

Shishov, L.L., Tonkonogov, V.D., Lebedeva, I.I., and Gerasimova M.I. 2001. Russian Soil Classification System, V.V. Dokuchaev Soil Science Institute, Moscow, p. 221.

Simonson, R.W. 1997. Evolution of Soil Series and Type Concepts in the United States, in *History of Soil Science*. Yaalon, D.H. and Berkowicz, S., Eds., Catena Verlag, Advances in Geoecology 29:79–108.

Soil Classification Working Group, 1991. Soil classification, a taxonomic system for South Africa. Memoirs on Agricultural Natural Resources of South Africa, No. 15, Pretoria.

Soil Survey Staff, 1960. Soil classification: a comprehensive system. 7th Approximation. USDA, Soil Conservation Service. Washington, DC, p. 265.

Soil Survey Staff. 1999. Soil taxonomy, a basic classification for making and interpreting soil surveys. Agriculture Handbook 436, 2nd ed. USDA, Natural Resources Conservation Service. Washington, DC, p. 869.

Swanson, D.K. 1993. Comments on 'The Soil Survey as Paradigm-Based Science.' *Soil Sci. Soc. Am. J.* 57:1164.

United Nations Environmental Programme/ISSS/ISRIC/FAO. 1995. Global and National Soils and Terrain Digital Databases (SOTER). Procedures Manual (revised edition). World Soil Resources Report 74. FAO, Rome, p. 125.

CHAPTER 3

Soil Classification: Past and Present

Robert J. Ahrens, Thomas J. Rice, Jr., and Hari Eswaran

CONTENTS

BACKGROUND AND HISTORY

Although not recognized as disciplines until the nineteenth century, pedology and soil science in general have their rudimentary beginnings in attempts to group or classify soils based on productivity. Early agrarian civilizations must have had some way to communicate differences and similarities among soils. The earliest documented attempt at a formal classification of soils seems to have occurred in China about 40 centuries ago (Lee, 1921). The Chinese system included nine classes based on productivity. The yellow, soft soils (soils derived from loess) were considered the best, followed by the rich, red soils. The evidence suggests that the Chinese soil classification system was used to levy taxes based on soil productivity (Simonson, 1962).

Cato (234–149 BC), a Roman scientist, contrived a classification of soils based on farming utility. His system employed 9 classes and 21 subclasses, and it guided decisions about use and care of the land for production of food and fiber (Stremski, 1975). The decline of the Roman Empire coincided with a general stagnation in the field of soil science, as noted by the low number of major contributions in the discipline until the nineteenth century.

The nineteenth century saw renewed interest in studying soil characteristics, in order to relate tax assessment to soil productivity. In Russia, this effort helped establish the discipline of pedology. The Russian government in 1882 hired V. V. Dokuchaiev to guide a program that would map and classify soils as a basis for tax assessment (Simonson, 1962).

Dokuchaiev and his students launched a new era in pedology that promoted the description and characterization of soils as natural bodies with a degree of natural organization, rather than viewing soils simply as mantles of weathered rock. This important notion fostered the concept of the pedon, from which data could be collected and compared. Even after the concept of the pedon took hold among pedologists to facilitate data collection, soil science still lacked standards to classify soils

Table 3.1 Families of Reddish Prairie Soils in the Southern Great Plains Correlation Area

Family	Stage	Texture Class	Drainage	Degree of Weathering	Size of Solum
Craig	Maximal	Medium	Good	Strong	Medium
Dennis	Medial	Medium to moderately fine	Good to moderately good	Strong	Medium
Hockley	Maximal	Moderately coarse	Good to moderately good	Strong	Medium
Kirkland	Medial	Moderately fine	Good to moderately good	Medium	Medium
La Bette	Medial	Loamy	Good	Medium	Medium
Pratt	Minimal	Coarse	Good	Weak	Medium
Teller	Minimal	Loamy	Good	Weak	Medium
Tishomingo	Medial	Moderately coarse	Good	Strong	Thin
Wilson	Maximal	Loamy	Moderately good	Strong	Medium

and describe the morphology and properties of soil profiles. This lack of standards hampered pedology and resulted in classification schemes shrouded with cloudy concepts that lacked operational definitions.

As an example, the U.S. 1938 classification system (USDA, 1938) followed the concepts of zonal and azonal soils, lacked operational definitions, and consequently failed to meet all the needs of the soil science community. In the 1938 system, one of the zonal soils, Reddish Prairie Soils, is described as dark-brown or reddish-brown soil grading through reddish-brown heavier subsoil, medium acid. This description is very vague, and without the knowledge that these soils occur in the southern Great Plains of the United States, the soil scientist might believe that these soils are in several parts of the world. Aside from the indistinct categories within the 1938 scheme, the system did not offer a means to differentiate soils both among taxa and within the same taxa. For example, Table 3.1 illustrates the families from a card dated November 26, 1951, used presumably by the correlators and field soil scientists to differentiate among the Reddish Prairie Soils. Obvious deficiencies include a lack of definitions for column headings such as stage, degree of weathering, and size of solum. Additionally, there are no operational definitions to differentiate any of the classes within the columns. This means that the differences among the differentiae, such as the degrees of weathering, are based on judgment and experience. The terms may have valid meaning to the local soil scientists. Soil scientists from different parts of the world, however, converging on the southern Great Plains, could engage in interesting discussions, but likely not reach agreement on whether a given soil exhibits medium or strong weathering. Furthermore, the differentiae are defined in neither the *Soil Survey Manual* (Soil Survey Staff, 1951) nor anywhere else.

The information in Table 3.1 is useful only to those who already have a familiarity with these soils. The differentiae provide little value in distinguishing these soils, even for the most experienced soil scientist.

Table 3.2 is also a card dated November 25, 1951, that attempts to provide facts about the Craig soils. Again, the information is scant and provides little value for a soil scientist unfamiliar with these soils or the area in general. Table 3.3 is a modern description of the same soil series.

MODERN SOIL CLASSIFICATION

After World War II, agriculture felt the effects of economic reconstruction and the expansion of global markets, and there was a renewed interest in soil conservation and alternative land uses, which helped invigorate soil survey activities. Soil scientists began identifying many new soils, and classification systems needed to track all the newly recognized soils. The United States Soil Conservation Service (now the Natural Resources Conservation Service), under the leadership of Guy Smith, accepted the challenge and made giant strides in improving soil classification. Work to develop a new U.S. soil classification system commenced in 1951.

Table 3.2 Description of Craig Soil (circa 1952)

Great Soil Group	Reddish Prairie (maximal)
Family	Craig
Series included	Craig
Drainage class	Good
Texture class	Loamy (medium)
Horizons	**Degree of Development**
A1	Strong
A3 & B1	Medium
B2	Strong
C	
Degree of weathering	Strong (moderately strong)
Size of profile	Medium
Kind of phases	Depth, slope, erosion
Parent material	Residuum from interbedded cherty limestone and shale
Climate	Moderately humid, temperate

Table 3.3 A Modern Description of the Craig Series

LOCATION CRAIG OK + MO

The Craig series is a member of the clayey-skeletal, mixed, thermic family of Mollic Paleudalfs. These soils have very dark brown and very dark grayish brown silt loam A horizons, dark grayish brown silt loam E horizons, brown silt loam BE horizons, dark yellowish brown and yellowish red very cherty clay loam Bt horizons, and BC horizons.

TAXONOMIC CLASS: Clayey-skeletal, mixed, thermic Mollic Paleudalfs

TYPICAL PEDON: Craig silt loam — rangeland.

(Colors are for moist soil unless otherwise stated.)

A1 — 0–7 in., very dark brown (10YR 2/2) silt loam, dark gray (10YR 4/1) dry; moderate medium and fine granular structure; hard, friable; many fine roots; common fine pores; medium acid; gradual smooth boundary (6 to 12 in. thick)

A2 — 7–12 in., very dark grayish brown (10YR 3/2) silt loam, grayish brown (10YR 5/2) dry; weak fine granular structure; hard, friable; common fine roots and pores; few medium fragments of chert; strongly acid; gradual smooth boundary (0 to 10 in. thick)

E — 12–16 in., dark grayish brown (10YR 4/2) silt loam, light brownish gray (10YR 6/2) dry; weak medium granular structure; hard, friable; common fine roots and pores; few medium fragments of chert; strongly acid; gradual wavy boundary (3 to 5 in. thick)

BE — 16–21 in., brown (10YR 5/3) silt loam, pale brown (10YR 6/3) dry; weak medium subangular blocky structure; hard, friable; few fine roots and pores; 10% medium fragments of chert; common 2 to 8 mm dark concretions; strongly acid; gradual wavy boundary (3 to 12 in. thick)

Bt1 — 21–25 in., dark yellowish brown (10YR 4/4) very cherty clay loam; yellowish brown (10YR 5/4) dry; moderate very fine blocky structure; hard, friable; few fine roots and pores; 60 to 70% by volume of chert fragments from 2 mm to 100 mm in diameter; thin patchy clay films on faces of peds and chert fragments; common 2 to 5 mm dark concretions; strongly acid; gradual wavy boundary (4 to 16 in. thick)

Bt2 — 25–42 in., yellowish red (5YR 5/6) very cherty clay loam; common fine, medium, and coarse reddish and brownish mottles on the chert fragments; weak very fine blocky structure; hard, friable; few fine roots and pores; 75 to 85% by volume chert fragments from 2 to 100 mm; thin patchy clay films on faces of peds, on chert fragments, and in pores; strongly acid; gradual wavy boundary (10 to 30 in. thick)

BC — 42–60 in., yellowish red (5YR 5/6) very cherty clay loam; common reddish and brownish mottles; structure is obscured by the chert; hard, friable; fractured chert ranges from 2 to 100 mm in diameter and occupies about 85% of the volume; strongly acid

TYPE LOCATION: Craig County, Oklahoma; about 5 miles southeast of Vinita; about 3150 ft south and 50 ft east of the northwest corner of sec. l2, T. 24 N., R. 20 E.

RANGE IN CHARACTERISTICS: Solum thickness ranges from 60 to more than 80 in. The depth to horizons containing more than 35% chert by volume ranges from 15 to 30 in. The soil ranges from medium acid through very strongly acidic throughout.

The A horizon is black (10YR 2/1), very dark brown (10YR 2/2), very dark gray (10YR 3/1), very dark grayish brown (10YR 3/2), or dark brown (10YR 3/3). It is loam, silt loam, cherty loam, or cherty silt loam. Coarse fragments more than 3 in. diameter range from 0 to 5% of the volume and coarse fragments less than 3 in. diameter range from 0 to 35% of the volume.

continued

Table 3.3 (continued) A Modern Description of the Craig Series

LOCATION CRAIG OK + MO
The E horizon is dark gray (10YR 4/1), dark grayish brown (10YR 4/2), brown (10YR 4/3, 5/3), gray (10YR 5/1), or grayish brown (10YR 5/2). Texture and coarse fragments are similar to those in the A horizon.
The BE horizon is very dark grayish brown (10YR 3/2), dark brown (10YR 3/3; 7.5YR 3/2), dark yellowish brown (10YR 3/4, 4/4), dark grayish brown (10YR 4/2), brown (10YR 4/3, 5/3; 7.5YR 4/2, 4/4, 5/2, 5/4), grayish brown (10YR 5/2), or yellowish brown (10YR 5/4). It is loam, silt loam, clay loam, silty clay loam, cherty loam, cherty silt loam, cherty clay loam, or cherty silty clay loam. Coarse fragments more than 3 in. in diameter range from 0 to 5% of the volume and coarse fragments less than 3 in. diameter make up 1 to 50% of the volume.
The Bt horizon is brown (10YR 4/3, 5/3; 7.5YR 4/4, 5/4), dark yellowish brown (10YR 4/4), yellowish brown (10YR 5/4, 5/6), strong brown (7.5YR 5/6), reddish brown (5YR 4/3, 4/4, 5/3, 5/4), or yellowish red (5YR 4/6, 5/6). The lower Bt horizon also includes yellowish brown (10YR 5/8), pale brown (10YR 6/3); light yellowish brown (10YR 6/4), brownish yellow (10YR 6/6, 6/8), strong brown (7.5YR 5/8), light brown (7.5YR 6/4), reddish yellow (7.5YR 6/6, 6/8), or yellowish red (5YR 4/8, 5/8). The Bt horizon is cherty silty clay loam, cherty clay, cherty silty clay, very cherty silty clay loam, very cherty clay loam, very cherty clay, or very cherty silty clay. The upper 20 in. clay percentage ranges from 35 to 45. Coarse fragments more than 3 in. in diameter range from 5 to 10% of the volume and coarse fragments less than 3 in. in diameter range from 35 to 90% of the volume.
The BC horizon is strong brown (7.5YR 5/6, 5/8), reddish yellow (7.5YR 6/6, 6/8; 5YR 6/6, 6/8), yellowish red (5YR 4/6, 4/8, 5/6, 5/8), red (2.5YR 4/6, 4/8, 5/6, 5/8), or light red (2.5YR 6/6, 6/8). It is very cherty clay loam or very cherty clay. Coarse fragments more than 3 in. diameter range from 5 to 10% of the volume and coarse fragments less than 3 in. diameter range from 65 to 90% of the volume.
An R layer of cherty limestone occurs at depths ranging from 5 ft to 30 ft below the surface.
COMPETING SERIES: These are the Boxville, Braxton, Eldon, Eldorado, and Riverton series. Boxville and Braxton soils have clayey control sections. Eldon soils have mesic temperatures. Eldorado and Riverton soils have loamy-skeletal control sections.
GEOGRAPHIC SETTING: The Craig soils are on uplands. Slope gradients range from 0 to 5%, mainly less than 3%. The Craig soils are formed in residuum weathered from cherty limestones. The average annual precipitation ranges from about 37 to 47 in., the annual Thornthwaite P-E indices from 64 to about 80, and the average annual air temperature ranges from 57° to about 62° F.
GEOGRAPHICALLY ASSOCIATED SOILS: These are the competing Eldorado series, and the Bates, Dennis, and Parsons series. Bates, Dennis, and Parsons soils contain little or no chert.
DRAINAGE AND PERMEABILITY: Well drained; medium runoff; moderately slow permeability.
USE AND VEGETATION: Some areas cultivated to small grains and sorghums. Some areas are in native range of tall prairie grasses or in improved pasture. The native vegetation is tall grass prairie.
DISTRIBUTION AND EXTENT: Northeastern Oklahoma and possibly in southwestern Missouri, northwestern Arkansas, and southeastern Kansas. The series is minor in extent.
MLRA OFFICE RESPONSIBLE: Salina, Kansas.
SERIES ESTABLISHED: Craig County, Oklahoma; 1931.

During the same period, there were intensive activities under way in Europe to develop national systems. A most notable contribution was that of the French pedologists who had begun to develop their system in the early fifties, and published it in 1967 (CPCS, 1967). The U.S. System saw its debut in 1960 as the 7th Approximation, which was the first operational version of Soil Taxonomy. In the meantime, there were other groups developing concepts and terminology for specific uses. Two outstanding contributions include the Soil Map of the World Project, for which a legend was developed by the Food and Agricultural Organization of the United Nations (FAO, 1971–1981). Another group published the Soil Map of Africa (D'hoore, 1964). Later, the first effort toward a Soil Map of Europe was initiated (Dudal et al., 1970). Although legends were developed for these small-scale maps, the process also helped to develop units at the higher levels of classification. The maps then became a technique for validating the higher levels.

Similar discussions occurred in Europe. FAO organized several working meetings to develop the legend for the world map. Field trips during such meetings were critical in testing concepts and developing criteria. Commission V of the International Society of Soil Science (ISSS) also played an important role in this process through conferences and symposia. Each national, regional, or international group had an opportunity to report on its progress and obtain critical evaluation of its efforts. The universities and research communities developed methods for soil characterization and testing of the theoretical concepts. Thus the sixties and seventies were a period of intense activity in the development of soil classification systems; the activities were spurred both by national needs and a soft competition.

Perhaps the greatest modern breakthrough in soil classification is the recognition that the soil-forming processes frequently leave markers in the forms of diagnostic horizons and features. In turn, diagnostic horizons and features can be defined in terms of observable and measurable properties. One of the most difficult considerations in establishing concise definitions is that soils are not discontinuous natural units. Gradual transitions of soil properties and soil bodies occur on any landscape. The choice for the differentiating criteria becomes of paramount importance in applying the definitions of the diagnostic horizons or features in the field.

When the definitions are written using well-defined differentiating criteria and are applied consistently, soil scientists with different backgrounds and experiences should arrive at the same conclusions, regardless of any differing views on the genetic aspects of the soil. The genesis of the soil is important to the classification because it permits us to place similar soils in the same or similar taxa. Additionally, the genesis plays a major role in mapping soils because it helps us develop our predictive model of soil-landscape segments that can be delineated into usable soil maps with viable interpretations.

In summary, the diagnostic horizons represent the genetic aspects of soils, but genesis does not appear in the definitions. Well-defined diagnostic horizons and features allow soil scientists with different views and experiences to describe the same horizons and features, even though all the genetic processes that produced the horizons and features are not known or fully understood.

The diagnostic horizons and features form the building blocks of the various taxa of a soil classification system, and provide a powerful tool for communicating information about the soil and for differentiating among soils. According to Soil Taxonomy (Soil Survey Staff, 1999), the Craig series from Table 3.1 is in the family of clayey-skeletal, mixed, active, thermic Mollic Paleudalfs. Soil scientists, who are familiar with Soil Taxonomy, will know that the Craig soil has a thick argillic horizon with at least 35% clay and 35% rock fragments, and adequate bases. The Craig soil occurs in a warm, humid, or semihumid climate on stable landscapes. The surface layer is dark, likely from the accumulation of organic matter. The classification of the soil provides significant information about the properties of the soil.

The classification also provides a way to compare the soils quantitatively. The Dennis series from Table 3.1 is a fine, mixed, active, thermic Aquic Argiudolls (Soil Survey Staff, 1999). This series has more bases and fewer rock fragments than does the Craig series. The Craig series is also better drained than the Dennis series. The differences between the two series can be quantified: The Dennis series has a mollic epipedon that is 25 cm or more thick and less than 35% rock fragments. The Craig series has an umbric epipedon and more than 35% rock fragments in the upper 50 cm of the argillic horizon.

Soil classification systems have come a long way from their humble beginnings as a means of levying taxes based on production, and have progressed through various stages, including the descriptive stage illustrated above to rather sophisticated, quantitative systems. Most modern soil classification systems are developed to complement and support soil survey activities. Many countries have curtailed soil survey activities, thus the long-term value to pedologists is as a means of communicating important properties about the soil, and helping to differentiate among soils in a consistent manner.

Cline (1949) indicated that classifications are not truths that are discovered, but contrivances made by humans to suit their purposes. It is quite apparent from this book that many countries have developed sophisticated soil classification systems that fit their needs. Although Soil Taxonomy and the World Resource Base have been adopted by several nations, one of the lingering criticisms is that there is no universal soil taxonomic system, as there is for plants and animals. The Australian (Isbell, 1996), New Zealand (Hewitt, 1998), and Canadian (Agriculture Canada Expert Committee on Soil Survey, 1987) soil classification systems, to name a few, are directed toward a national effort. Although many countries have developed national soil classification systems, there are common features among them. Most national soil classification systems have shifted toward more quantitative definitions and criteria for diagnostic horizons and features, permitting the formation of mutually exclusive taxa. Concepts and models of soil genesis have guided the selection of diagnostic horizons and features, and it is no surprise that many national soil classification systems share common or roughly equivalent diagnostic horizons and features that provide a means of communication among soil scientists from various countries.

IMPROVEMENTS NEEDED

The diagnostic horizons and features represent a major innovation in soil classification that has been embraced by most pedologists, but there remain issues that have not been resolved to everyone's satisfaction. Soils, unlike discrete plants or animals, form a continuum over the earth's surface. Soil delineations are represented by one or more soils as a map unit; in reality, however, the map units contain many soils, not just the few designated in the map unit name. The confusion lies in classifying the pedon and its dimensions, and using the classification to represent the map unit. The concept of the pedon has been scrutinized (Holmgren, 1988), but not resolved. The map unit and pedon appear simple and straightforward at first, but continue to be sources of confusion or discomfort for many pedologists.

Anthropogenic soils pose another challenge to the soil science community. Humans have influenced and drastically changed the soil for centuries. At what point does the human influence change the classification of a soil? We have altered the surface by plowing and adding fertilizers, but when is the soil significantly changed to warrant different taxa from the native soil? Are there markers in the soil that capture the impact of humans on the soil resource? Or must we rely on sources outside the soil, such as the history of the area?

The World Resource Base (WRB, 1998) and other classification systems have made bold attempts to capture the human influences. The Anthrosol order in WRB tries to group together all the agricultural soils that are significantly impacted by humans. The Anthrosol order is required to have diagnostic horizons that are human-influenced. For example, the terric horizon is one of the diagnostic horizons used to key to the Anthrosol order. According to WRB (1998), "The terric horizon develops through additions of earthy manure, compost or mud over a long period of time. It has a non-uniform textural differentiation with depth. Its color is related to the source material or the underlying substrate. Base saturation is more than 50%." The requirement of base saturation is quantitative. The criterion of nonuniform textures requires some judgment on the part of the pedologist and may not be applied uniformly by all. "Additions of earthy manure, compost or mud" refers to the mode of deposition, and may be difficult to differentiate from nonhuman eolian and alluvial deposition. Does the mode of deposition affect use or management of the soil? Should soils like this have separate taxa because of the anthropogenic influences? These questions will be debated, and the answers depend largely on the purposes of the classification system. The soil science community is discussing the issues raised above, but will not likely reach agreement.

SUMMARY

Soil classification systems have evolved into sophisticated communication tools. Probably the greatest contribution in the last 50 years is the development of diagnostic horizons and features with associated quantitative definitions, which allow pedologists with different experiences to classify soils the same way.

Many countries have developed their own classification systems, whose features depend on the needs and soils of the country. Although there is no one-soil classification system that is used by all countries, most pedologists are familiar with diagnostic horizons and features, and have used them as an international means of communication.

Even with all the advances in soil classification, there are still classification difficulties between the soils that we classify and the soils that we map. Soils influenced and forever modified by humans present one of the greatest classification challenges. Although some classification systems have developed taxa for these soils, there are still questions as to their utility.

REFERENCES

Agriculture Canada Expert Committee on Soil Survey. 1987. The Canadian system of soil classification. 2nd. Edition. *Agric. Can.* Publ. 1646.

Cline, M.G. 1949. Basic principles of soil classification. *Soil Sci.* 67:81.

Commission de Pédologie et de Cartographie des Sols. 1967. Classification des sols. Lab. Geol. Pedol., Ecole Nat. Supr. Agron. (Grignon, France).

D'Hoore, J.L. 1964. Soil Map of Africa, Scale 1:5,000,000. Joint Project No. 11. Commission for Technical Cooperation in Africa Publ. Ceuterick, Louvain, Belgium.

Dudal, R., Tavernier, R., and Osmond, D. 1970. Soil Map of Europe, 1:2,500,000; explanatory text. FAO, Rome.

Food and Agricultural Organization of the United Nations. 1971–1981. FAO/UNESCO Soil Map of the World, 1:5 million. Vols. 1–10. FAO/UNESCO, Rome.

Hewitt, A.E. 1998. New Zealand Soil Classification, Manaaki Whenua Press, p. 133.

Holmgren, G.G.S. 1988. The point representation of soil. *Soil Sci. Soc. Am. J.* 52:712–716.

Isbell, R.F. 1996. *The Australian Soil Classification*, CSIRO Publishing, Collingwood, p. 143.

Lee, M. Ping-Hua. 1921. Columbia University. Studies in History, Economics, and Public Law 99.

Simonson, R.W. 1962. Soil classification in the United States. *Science.* 137(3535):1027–1034.

Soil Survey Staff. 1951. Soil Survey Manual. U.S. Dept. Agrie. Handbook No. 18. U.S. Government Printing Office, Washington, DC.

Soil Survey Staff. 1999. Soil Taxonomy: A Basic System of Soil Classification for Making and Interpreting Soil Surveys. 2nd ed. Agric. Handbook No. 436, U.S. Government Printing Office, Washington, DC, p. 869.

Stremski, M. 1975. Ideas underlying soil systematics. Polish Edition. TT73–54013, Foreign Scientific Publ. Dept., National Center for Scientific, Technical and Economic Information, Warsaw.

United States Department of Agriculture. 1938. Soils and Men. Yearbook of Agriculture. U.S. Government Printing Office, Washington, DC.

World Reference Base for Soil Resources. 1998. FAO, Rome.

CHAPTER 4

Conceptual Basis for Soil Classification: Lessons from the Past

Richard W. Arnold and Hari Eswaran

CONTENTS

ABSTRACT

The publication of a second edition of Soil Taxonomy in 1999 was a benchmark event that, among other things, enabled us to evaluate the progress of soil classification over five decades. The quest for a single international system will continue to be elusive for another few decades, and Soil Taxonomy will still have imperfections. However, it is a living system in the sense that it serves a function. Because of this and the fact that it has been amenable to changes without major alterations to the basic structure, the system will continue to serve in the future, needing refining only when knowledge and applications warrant it. Facts determined by precise observation, experimentation, and measurement minimize prejudicing the future, and are the building blocks of a classification system. Systems built solely on empirical relationships among facts or genetic theories falter quickly. As there are relics of classification systems around the world, we considered that it would be useful to elucidate some of the lessons learned during this glorious period of Pedology. These lessons will be useful when further refinements are attempted. They will also be useful to others when they embark on a similar exercise, and to some countries that are developing national systems.

INTRODUCTION

"Science is a process of discovery, not confirmation. Let us allow for the occasional, delicious surprise that makes us rethink all we thought we knew."

P. Shipman

With the publication of the second edition of Soil Taxonomy (Soil Survey Staff, 1999), a new era in the development of the system was put in place. In the 25 years since the first edition (Soil Survey Staff, 1975), significant changes in our understanding of soils have taken place as a result of systematic investigations accompanying the progress in soil survey. The progress has been worldwide, with developments in other classification systems and with the contributions of many foreign scientists to Soil Taxonomy. Many countries are in the process of or contemplating revisions of their respective systems. A notable event is the publication of the World Reference Base for soil classification (Deckers et al., 1998), an effort by the International Union of Soil Science supported by the Food and Agriculture Organization of the United Nations.

Historical developments leading to the publication of the first edition of Soil Taxonomy (Soil Survey Staff, 1975) were reviewed by Smith (1983). Changes in the second edition (Soil Survey Staff, 1999) have been mainly in the addition of new classes at different categoric levels, requiring reorganization of the categories and definitions. The Orders of Andisols and Gelisols were introduced, and Aridisols and Oxisols were significantly modified. Appropriate changes in the lower categoric levels were made with new definitions or new classes. The basic structure of the system has not changed since its 1975 debut (Soil Survey Staff, 1975). The basic principles that guided the development of the system have been explained in several publications (Arnold, 1990; Arnold et al., 1997; Cline, 1949, 1963, 1971; Simonson, 1963; Smith, 1963; Witty and Eswaran, 1990). Most of the accepted notions of Soil Taxonomy and its relationship to soil genesis were brought together in two volumes of a book edited by Wilding, Smeck, and Hall (1983).

Despite the voluminous amount of information on Soil Taxonomy, students of the system and knowledgeable users still have questions that are not adequately addressed or elaborated on. Guy Smith, the prime coordinator for development of the system, provided answers for some of the major decisions made in designing the system, and Forbes (1986) summarized these. However, Guy Smith stated at the outset of his interviews, "I rather concealed the reason for doing this when I wrote Soil Taxonomy. If I had explained why we did this or that, the reader would be more apt to pay attention to the reason than to the actual definition. We wanted a test of the definition, not of the reason." The reasoning for the structure of the system would perhaps remain speculative. Some of the rationale, however, was explained later by Guy Smith and others for various reasons: (a) to better appreciate the system, (b) to compare it with other systems to test for potential weaknesses in Soil Taxonomy, and (c) to use it as an exercise for developing an ideal classification system.

The purpose of this paper is to evaluate Soil Taxonomy in a retrospective manner, and to enunciate some principles related to the architecture of the system. Such an analysis could help others develop national systems, or at least suggest areas for their consideration. We have organized the paper around questions that lead us to concepts that are the foundation of soil classification throughout the world. The questions are these: Why soils? What is soil? Why classification? and What objectives? These points are followed by consideration of the consequences of the answers to these questions. We conclude with some ideas on minimizing prejudice for the future.

WHY SOILS?

Throughout history, there have been numerous soil classifications. Our concern is why people were interested in soils. What was their purpose? Below are four different perspectives, among many options.

1. Soil survey. Want to understand soils well enough to predict the consequences of alternative uses and where they will occur. Because many soil survey programs were initiated in Agriculture Ministries, there was major emphasis on agriculture and forestry. Those in Ministries of Natural Resources dealt more with soil as geologic deposits and engineering materials. Today there is a new emphasis on environmental aspects.
2. Genesis. Want to understand soils well enough to reveal the genetic relationships among them. This entails the history of evolution of soils and the pedological development of soil features. Emphasis has often been on how soil-forming factors and processes have influenced the evolution and development of the pedosphere.
3. Use interpretation. Want to understand soils well enough to predict their behavior and response to management. Emphasis is on the dynamics of current interactions in the environment, with special emphasis on human-induced changes and expectations. In developing countries, managing soils is linked to processes of desertification and the goal of food security.
4. Correlation. Want to provide a uniform way to link equivalent and near-equivalent soil classes in different systems together to provide an integrated global overview. Emphasis is on providing a reference base that reveals the commonality of soils, regardless of their original classification.

WHAT IS SOIL?

Above are some good reasons to study soils, but what is the user's perception of soil, and will that concept satisfy his or her purpose? What is the universe of soils, what is a population, and what kind of an individual is needed? As we shall see, there are choices.

1. Geographic body. Soil is accepted as a function of climate, organisms, relief, parent material, and time. Because these factors are distributed geographically, so are the functional processes by which soils occur in landscapes. If observed at points on the landscape, they seem to be individuals; but if observed in a trench, they are a continuum. A soil has many properties: physical, chemical, and biological. The evolution of soils and their distribution patterns are of interest: where and why do the soils occur where they do? Soils, as such, are little geographic bodies; they are polypedons or their equivalent (Knox, 1965).
2. Small representative volume. Soil (Jenny, 1963) is conceived as S = f(cl, o, r, p, t). These factors influence processes that produce properties. There is interest in (a) the factors and how they interact; (b) the processes that occur and have occurred; (c) the many properties that result from the soil-forming processes; (d) the sets of properties, how they formed, and how they influence or are being influenced now. The details of the pedosphere are less important for the primary purpose. It may be critical, however, for the genetic study of soils as landscapes. Soils are small volumes with many features; they are pedons or their equivalent.
3. Profile data sets. A soil is a set of properties—physical, chemical, biological—that behave or interact in unique ways with the ambient environment. Based on relationships of properties, environmental conditions, use, and management, it is possible to predict what will happen and why. Extrapolation to landscapes is, at some time, necessary, and can build on the knowledge of others. Profiles and data sets are the entities of interest.

There is general agreement that all soils (regardless of size or shape) have morphology. This body or individual requires definition to recognize it either as a member of a population or as a part of a continuum. Two attributes of interest are the central concept of the individual, and the boundary limits of the individual. There are options and there are choices.

WHY CLASSIFICATION?

Classification (Smith, 1963) is a system contrived by humans to organize ideas, as they know them, according to their concepts of order (relationships). There is uncertainty associated with the facts that are obtained, in the degree of association among the facts, and in the ideas and concepts whose relationships to each other indicate patterns of order. Classification is not a truth that can be discovered—it is only a set of relationships, an abstraction of knowledge, a scientific model of the subject of interest.

As an abstraction of knowledge, a classification relies on facts. The facts are data that have been derived from observations and measurements that are known in terms of the operations by which they were acquired. As a set of relationships, a classification relies on the correlations (relationships) among various facts that are determined empirically, without the necessity of reasons for them. And as a model, it deals with concepts that describe and explain the elements and relationships of our current understanding. Insofar as the relationships are believed to be cause-and-effect ones, and not merely empirical ones, they form the basis of explanations.

We need classification because there are too many objects (soils) to remember and too many relationships that may exist (Cline, 1963). We need a way to organize information into patterns that we can recognize and use more efficiently and effectively. This means there will be groupings and separations in such a system. The adequacy of a classification system is evaluated against its purpose and not the expectations of the evaluator. If a purpose is implied or assumed but not stated, then any evaluation becomes somewhat tenuous. There are options; there are choices.

WHAT OBJECTIVES?

We expect a classification to achieve our objectives, which are conveniently grouped as applied and scientific objectives.

Applied objectives commonly require direct interpretations. These have been called technical groupings (Bartelli, 1978), and each one is a type of classification. Some are based on a single property, such as stony soils, wet soils, or organic matter content. Others are based on simple interpretations, such as capability groupings or land-use potentials. There are as many technical groupings as there are objectives (reasons) for grouping soils. Most soil information systems, such as USDA-NRCS's National Soil Information System (NASIS), contain information and classes of numerous technical groupings. The general public and users of soil-related information are more familiar with these objectives, because the groupings can be applied directly to answer specific questions of interest to them.

Scientific objectives are expectations that concepts of order (in nature) will be revealed and, as such, a system needs to (a) give identity to otherwise unidentified individuals in the population, (b) provide identified groups of individuals, and (c) organize classes in such a manner that they will abstract concepts of order. Most hierarchical systems (Allen and Starr, 1982) provide identity to the objects of interest, however, they require at least one step of reasoning to reveal either genetic relationships or applications to practical problems. There are many options and many choices.

THE ARCHITECTURE OF THE SYSTEM

The decade of the 1950s was a remarkable period in the development of soil science. Many national soil classification systems were being conceived, and the Food and Agriculture Organization of the United Nations was embarking on a major effort to produce the soil map of the world. National and global soil survey programs required an acceptable classification system, but soil databases were sparse. The path of least resistance (for acceptance by peers) was to enhance existing systems. In the United States, Dr. Guy Smith was charged with the task of developing a system, and he decided to be guided by principles and concepts that deviated from existing molds.

Hierarchies

Soil Taxonomy is structured as a nested hierarchy, in which classes of the lower levels are an integral part of and confined by the defining properties of the classes at higher levels. With the established linkages between the levels or categories in a hierarchy, the user can appreciate and remember the functional relations of the classes. Use of a hierarchy is facilitated by appropriate design of the nomenclature of the categories and the classes therein, in which the name signifies the position of the category and its membership in the system, including some defining properties.

The process of building up hierarchies is generally one of aggregation. Entities with similar behavior or that result from similar sets of processes are clustered together; unique objects may deserve a separate cluster, or another rationale may be used in the aggregation process. The rationale of this aggregation process leads to the principles upon which the system is developed.

Hierarchies are conceptual. However, the system that is developed has a defined purpose, which is the making and interpreting of soil surveys. Thus it is desirable that each of the categoric levels has a meaningful relationship to the range of scales to which the information could be applied. This linkage to scale adds value to the system and assists in defining the functions of the category. It is also a check on the information content (categoric relevance) in the context of the scale (cartographic accuracy) as conceived by Orvedal and Edwards (1941). Unlike the hierarchy, which is a conceptual construct, scale is real and quantifiable. In using agro-ecological information, Fresco (1995) indicates that there is an optimal scale at which each process can and must be studied, so that scale-sensitivity is built into the system, helping to link one category to another.

A fundamental problem in developing hierarchies of natural systems is that they are continua, rather than discrete objects. Breaking a continuum using defined limits of selected properties introduces subjectivity into the system, and is frequently the source of disagreement. This is particularly the case with soils for which not only the absolute value of the property has to be considered, but also the position in the soil where that feature is exhibited. The case for discussions on limits is reduced with the use of scientific rationale, the selection of properties that have a number of co-related properties, or a property that exhibits an extreme value from the population. As the information content increases in the lower categories, it is in the categories that properties with narrow limits are introduced to define the categories. In the higher categoric levels, concepts about processes and factors, or conditions that determine the processes, can be used to provide aggregation of the lower level units. Table 4.1 lists the major defining process(es) employed at the Order categoric level in Soil Taxonomy. Definition of the processes is given by Buol et al. (1980). All classification systems employ such sets of concepts, and Table 4.2 attempts to show how other classification systems (Russian by Shishov and Sokolov, 1990; WRB by Deckers et al., 1998; and Brasilian by SPI, 1999) define their higher categories using similar concepts. These are more abstract concepts based on the understanding of the nature of the population (Arnold, 1990). Selection of "controls of processes," or sets of conditions, also needs good rationale.

A number of soil classification systems have evolved with time. Soil Taxonomy has six categories and the World Reference Base (WRB) (Deckers et al., 1998), which is being considered as a classification system, has only two. The objectives of a system obviously determine the number of categories. The question, however, remains: Having decided on the number of categories, have the objectives been satisfied? Authors of systems usually indicate that the number is adequate for their purpose. It would be useful to have some guidelines for making this determination. The decision relates to the purpose of the classification and the function of the categories, and to the scale at which the use of the category is relevant. Recently, Triantafilis et al. (2001) developed a procedure to classify soil information. With the use of appropriate algorithms, they attempted to reduce the subjectivity of conventional soil classifications in deciding the number of classes and

Table 4.1 Assumed Processes and Soil Properties Characteristics of Soil Orders

Order	Dominant Processes	Description	Dominant Properties
Gelisols	• Gelification • Cryoturbation	Extended periods of freezing conditions followed by thawing above the permafrost result in volume changes, dislocation of material, and unique conditions.	Freeze-thaw reworked soils with or without high amounts of organic matter, with high proportion of raw humus.
Histosols	• Paludization • Humification • Mineralization	Slow accumulation of organic matter in waterlogged conditions with or without low temperatures; mineralization related to hydrologic conditions.	Characterized by organic soil materials; mineral soil materials may be present as strata in the deposit or as minor component of the deposit.
Spodosols	• Podzolization • Eluviation, Illuviation	Acid complexation and removal of Fe and Al with organic matter from upper part of soil leading to a bleached horizon and deposition of the complexes in a subhorizon called spodic horizon.	A coarse textured soil with a spodic horizon, with or without an albic horizon.
Andisols	• Cumulization • Mineral alteration • Melanization	Alteration of material with dominance of weatherable minerals, formation of organo-mineral complexes, and darkening of material due to humus enrichment.	Very high variable charge, low permanent charge colloids, and a net charge close to zero. Layering of material indicating periodicity of deposition.
Oxisols	• Ferralitization • Allitization	Alteration of primary weatherable minerals to their oxyhydrate state through extreme weathering; clay colloids enriched with iron and/or aluminum-reducing ability to be translocated.	Low weatherable mineral content in the sand fraction, low permanent charge and proportionately high pH-dependent charge, yellowish to reddish colors, and deep zones of alteration.
Vertisols	• Pedoturbation	High shrink/swell potential due to high clay content, with a dominance of smectites in the clay, resulting in slickenside formation and shrinkage cracks.	Negation of any translocation process, and homogenization of material through pedoturbation accompanying shrink/swell processes.
Aridisols	• Calcification • Salinization • Alkalization	Accumulation of carbonates and salts more soluble due to net upward movement of moisture in the soil; minimal expression of products of translocation or alteration due to low intensity of processes.	Accumulation of solubles. Translocation of mineral particles or alteration of minerals takes place at the margins of the areas with aridic moisture regimes. Products of previous processes, preserved.
Ultisols	• Leaching • Eluviation, Illuviation	Translocation and subsequent accumulation of clay in a lower horizon in an acid environment.	A coarse-textured surface horizon overlying a horizon with a clay augmentation.
Mollisols	• Melanization • Humification • Mineralization	Humus enrichment in a cool climate in base-rich environments.	Dark, organic matter-rich surface horizon in a base-rich soil.
Alfisols	• Eluviation, Illuviation • Calcification	Translocation and subsequent accumulation of clay in a lower horizon in a base-rich environment.	A coarse-textured surface horizon overlying a horizon with a clay augmentation.
Inceptisols	• Mineral alteration • Decomposition	Minimal expression of development due to low intensity of accumulation of material, alteration of minerals, or other processes.	Weak subsurface structure development or evidence of alteration and hydrolysis to release iron resulting in some reddening of soil material.
Entisols	• Cumulization	Absence of profile development except for some accumulation of organic matter due to rapid accretion or removal of soil material or very low intensity of processes.	Evidence of deposition of soil material not erased by soil-forming processes, or only ochric epipedon present as evidence of soil formation.

Definition of processes given by Buol et al., 1980

Table 4.2 Determinants of Processes and Their Use in Some National and International Classification Systems

Dominant Determinants of Processes	Major Processes	Soil Taxonomy	WRB Soil Classes	Russian Classification	Brazilian Classification
		Nature of material			
Organic	Paludization, Humification, Mineralization	Histosols	Histosols	Peat	Latossolos
Sesquioxidic	Ferralitization, Allitization	Oxisols	Ferralsols	Residual peat	
Vitric volcanic	Cumulization, Mineral alteration,	Andisols	Andosols	Soddy organo-accumulative	
Smectitic	Melanization	Vertisols	Vertisols	Volcanic	
Rock at shallow depth	Pedoturbation		Leptosols	Lithozems	Vertissolos
Extremes of climate					
Low temperature	Gelification, Cryoturbation	Gelisols	Cryosols	Cryozems	
Low moisture availability	Calcification, Salinization, Alkalization	Aridisols			
		Flux of water in the soil			
Accumulation of discrete particles	Eluviation, Illuviation, Calcification	Alfisols	Albeluvisols Alisols Nitisols Acrisols	Texture differentiated	Alissolos
	Leaching, Eluviation, Illuviation	Ultisols	Luvisols Lixisols		
	Podzolization, Eluviation, Illuviation	Spodosols	Podzols	Alfe-humus	Nitossolos
Accumulation as organo-mineral complexes Accumulation through dissolution and crystallization	Calcification, Gypsification, Salinization, Alkalization		Solonetz Solonchaks Gypsisols Durisols Calcisols	Alkaline clay-differentiated Halomorphic Crusty	Argissolos Luvissolos Esposossolos

continued

Table 4.2 (continued) Determinants of Processes and Their Use in Some National and International Classification Systems

Dominant Determinants of Processes	Major Processes	Soil Taxonomy	WRB Soil Classes	Russian Classification	Brazilian Classification
	Favorable Moisture and Temperature Conditions or Anthropic Modifications				
Accumulation of organic matter	Melanization, Humification, Mineralization	Mollisols	Chernozems	Metamorphic Low-humus accumulative calcareous	Chernossolos
Accumulation of anthropic products	Cumulization, Eutrophication		Kastanozems Phaeozems Umbrisols Anthrosols	Anthropogenic-accumulative	
Favorable aerobic subsurface conditions	Mineral alteration, Decomposition	Inceptisols	Cambisols	Fersiallitic	Cambissolos
Periodic water saturation leading to redoximorphic conditions	Gleyfication, Paludization, Humification		Gleysols	Gleyzems	Gleissolos
	Ferrification		Plinthosols		Plintossolos
Rates of erosional or depositional processes exceed pedogenesis	Cumulization	Entisols	Fluvisols Planosols Arenosols Regosols	Weakly developed Alluvial	Neossolos Planossolos

how individuals are allocated to them. This, or a modified approach, will enable a better rationalization for the number and selection of categories in a classification system.

Although aggregation of field-level units to higher and more abstract groups is probably the most appropriate procedure in natural systems, some soil classification systems have begun by defining the highest category, usually based on traditional systems or preconceived notions. Many of the early systems, including WRB, have adopted this procedure. Lack of detailed information on the soil individual is usually the reason. The Russian system (Shishov and Sokolov, 1990) adopts both procedures. Having defined the Types, higher units (Order and Trunk) of the Russian system are formed by aggregation of the Type, whereas lower units are formed by desegregation of the Types. In Soil Taxonomy, the soil series is the lowest category, and these are aggregated in five higher levels. Conceptually, the soil individual is the polypedon, which is constituted by the pedon and acceptable variability.

Principles

The principles of Soil Taxonomy set forth by Cline (1949) and others provide the basis for the architecture of the classification system that has been developed. They have served to guide the modification of the system since its publication in 1975. Many listed below are re-enforcement of the principles existing in the first edition (Soil Survey Staff, 1975). A few reflect the changes that have evolved with the systematic application of the system in the United States and globally. The fundamental principles were guided by the system of logic enunciated by John Stuart Mill (Cline, 1963) and do not change.

Soil Taxonomy is designed for the making and interpreting of soil surveys, which is the most important use of Soil Taxonomy. It was and is designed as a tool for correlation in soil survey programs. It was also designed as the technical language of communication between soil scientists and those knowledgeable about soils. The categories of the system (Ahrens and Arnold, 1999) enable one to make interpretations regarding use and performance of soils; the accuracy of such interpretations is a function of the categoric level, with specificity increasing at lower categories. For site-specific predictions and management, information contained in the lowest category must be supplemented with information about those properties that show spatial (short distance) and temporal variability. Basic considerations for the principles include notions of the following:

1. Equity. Each category in the system should provide a place for all soils in a landscape, or have the capability to provide such a place without distortion of the system.
2. Transparency. The categories of the system must have unambiguous definitions and clearly stated functions; knowledgeable persons equipped with the same information must arrive at the same classification.
3. Science-based. The system should not prejudice future applications. This is ensured by adhering to logical scientific rules and by minimizing the use of biases, popular feelings, and traditional approaches; historical information must be continually validated for its relevance and acceptance.
4. Architecture and linkage. The design of a classification system should facilitate its use in digital processing systems, and where possible, with easy linkage to other natural resource classification systems.
5. Ecosystem links. As soils are a component of an ecosystem, ecosystem parameters are valid as surrogates for defining categories, particularly at the high categories.
6. Flexibility. To serve its intended purpose and to reduce subjectivity in application of the system, flexibility should not result in distortion of system integrity.

The Categories of the System

The categories of the hierarchical system are designed to serve specific functions, and the use of any categoric level has restrictions imposed by its information content. Conceptually, categories follow these rules:

A category has completeness, because it permits all members of the population with similar attributes to be included.

A category is defined by an abstract concept that characterizes the population.

The definition is the only basis by which the individuals of the population can be separated; different properties or marks may be used to recognize or satisfy that basis.

The usefulness of a category is judged in terms of its definition, its function(s), and its potential uses. In Soil Taxonomy, the categories are fixed, but other evolving classification systems usually rely on their previous system to determine the number of categories. Determining or developing a rationale for each of the categories in the three terms could augment this approach. Table 4.3 summarizes the rationale for each of the categories in the Soil Taxonomy.

The *definition* is a statement of the conceptual framework that permits the construction of classes in that category providing the boundary conditions for lower categories within each class. It provides the "raison d'etre" for that category.

The *function* determines the need for the category and is a statement of the purpose for which it was constructed. If a category does not perform its function, or if it is believed that the designated functions could be achieved by other means, then the role of the category must be examined.

Potential uses provide examples of the application of the categoric level. It also suggests appropriate scale and the smallest land area where the information content of the category is applicable.

Table 4.1 provides a basis for refining current systems or developing new national systems.

ACCOMPLISHING THE OBJECTIVE

The questions we set out with were these: Why soils, what is soil, why classification, and what objectives? Herein lies the foundation of systems and the biases that they promulgate.

Purpose

Purpose is the driving force of any classification, and it is the only real basis for judging the adequacy of the system. Does the system satisfy its own purpose and objectives?

1. Soil survey is a complex process because it is designed to allow predictions of behavior of soils (applied objectives) and to support the recognition and mapping of soils (scientific objectives). The geography of factors and processes implies that soils are pieces of landscapes, and that genesis based on static property correlations will be needed. Interpreting behavior implies current interactions of properties and environments (dynamics), including manipulations by humans. Its purpose is to do both, mainly by identifying bodies and kinds of soils. Some have argued that because of the contrasting (though interrelated) nature of objectives, one system will not accomplish both objectives.
2. Genesis is less of a mixed bag, as genetic relationships can be separated into facets that emphasize special factors, features, or concepts. Particular attention can be given to what, where, and why parent materials are the initial stage of evolution. Deciphering pathways of processes that seem to result in specific kinds of soils may be a main purpose. Development of morphology has always been important, and so the static-historical properties are given more emphasis. Geographic distributions of kinds of soils as a consequence of evolution and pedogenetic development are valued purposes as well.

Table 4.3 Defining Characteristics of the Categories in Soil Taxonomy

Category	Definition	Functions	Potential Uses
Order	Soils having properties (marks) or conditions resulting from major soil-forming processes that are sufficiently stable in a pedologic sense, and that help to delineate broad zonal groups of soils.	Depict zones where similar soil conditions have occurred for general understanding of global patterns of soil resources Establish global geographic areas within which more specific factors and processes result in the diversity of soils	General global/continental/regional assessment global climate change studies AMS: < 1:10,000,000 MSD: > 40,000
Suborder	Soils within an order having additional properties or conditions that are major controls, or reflect such controls on the current set of soil-forming processes, and delineate broad ecosystem regions.	Demarcate broad areas where dominant soil moisture conditions generally result from global atmospheric conditions Delineate contiguous areas with similar natural resource endowments Ecosystems with distinct vegetation affinities usually determined by limiting factor/s of soil moisture or conditions	Demarcate areas in regions or large countries for assessment and implementation of economic development Analysis of international production and trade patterns Priority setting for multipurpose uses of land resources AMS: 1:1,000,000 to 1:10,000,000 MSD: 4,000 to 40,000
Great Group	Soils within a suborder having additional properties that constitute subordinate or additional controls, or reflect such controls, on the current set of soil-forming processes, including landscape-forming processes.	To demarcate contiguous areas with similar production systems or performance potentials	Development of strategic plans for regional development Basis for coordinating national resource assessment and monitoring programs Infrastructure development to assure equity in development AMS: 1:250,000 to 1:1,000,000 MSD: 250 to 4,000
Subgroup	Soils within a great group having additional properties resulting from a blending or overlapping of sets of processes in space or time that causes one kind of soil to develop from, or toward, another kind of soil: *Intergrades* show the linkage to the great group, suborder, or order level; *Extragrades* have sets of processes or conditions that have not been recognized as criteria for any class at a higher level, including non-soil features; The soil is considered to be the '*typic*' member of the class if the set of properties does not define intergrades or extragrades.	To demarcate production land units with similar land use and management requirements	Targeting research and development for specific land use or cropping systems Implementing conservation practices and ecosystems-based assistance Community development projects and monitoring sustainability Basis for diversification of agriculture and uses of land resources Basis for implementing environmental management programs and modeling ecosystem performance AMS = 1:100,000 to 1:250,000 MSD: 40 to 250

continued

Table 4.3 (continued) Defining Characteristics of the Categories in Soil Taxonomy

Category	Definition	Functions	Potential Uses
Family	Soils within a subgroup having additional properties that characterize the parent material and ambient conditions; the three most important properties are particle size, mineralogy, and the soil temperature regime.	Demarcate resource management domains characterized by similar management technology and production capabilities	Basic units for extension/technology transfer Modeling cropping systems performance Addressing socioeconomic concerns AMS: 1:25,000 to 1:100,000 MSD: 2.5 to 40
Series	Soils within a family having additional properties that reflect relatively narrow ranges of soil-forming factors and of processes, determined by small variations in local physiographic conditions, that transform parent material into soil.	To delineate land units for site-specific management of farms.	Implementing soil-specific farming Designing farm-level conservation practices AMS: >1:25,000 MSD: <2.5

AMS = Appropriate map scale.

MSD = Minimum size delineation; smallest area that can be delineated on a map with a legible identification; in hectares.

3. Use interpretations have technical groupings to provide estimates of behavior and responses when soils are used or managed. The set of properties, their environmental interactions, and expected changes when managed do not rely on why or how the properties came into being; rather, they require empirical relationships of properties and their measured behavior.
4. Equivalency is a means of comparing units or classes in different systems of classification and indicating their similarity. Traditionally, it has not itself been a classification. It has, however, many similarities regarding definitions and their applications.

With careful attention to details, each system can be adjusted so that it satisfies its own purpose and objectives. There will not be a truth discovered, nor will there be a perfect fit—only a satisfactory agreement of a purpose and a system. It is absolutely essential that a carefully designed and worded purpose be stated for a proposed classification. Without it there can be no testing or evaluation of the suitability of a system of classification. A system may be very adequate for its stated purpose, but not serve other purposes or desires well. It seems that very few soil classification schemes have been independently evaluated, although opinions about their usefulness are abundant.

Each purpose implies that soils exist and may be recognized, identified, and their characteristics used to group or segregate individuals (or components of a continuum). Most scientists agree that soils have morphology with sets of associated chemical and biological features that uniquely specify entities that may be treated as individuals of a population of interest.

Soil survey and genetic soil geography require that members of a population be bodies or segments of landscapes. Many genetic schemes need a small arbitrary volume and associated features, but do not require geographic attributes to organize them in ways that reveal concepts of order (especially in nature).

Interpretations do not require genetically determined bodies, or profiles, and mainly rely on data sets and their behavior patterns. Why and how a profile evolved may be interesting but not necessary for valid predictions of behavior and response. Correlation tools likely work best with profiles or small volumes and data sets for the comparisons that are made to determine membership in new classes.

What does matter is whether the basic unit is a higher level one, such as a Russian Soil Type or American Great Group, or a member of the lowest level, such as the Series in Soil Taxonomy. In the former, extra information is used to subdivide the unit; in the latter, features and properties are generalized or removed from consideration, as new higher level units and groups are formed and considered. If soils are individuals as well as bodies, or represent bodies in landscapes, when are they scale dependent?

The question of soil classification is less of an issue, because everyone recognizes the improbability of remembering all the kinds of soils, their myriad properties, and their relationships with each other. Whether dealing with profiles, pedons, or polypedons, the number is overwhelming—thus a multilayered or nested scheme seems appropriate. Kellogg (1936) remarked that soil classification helps us to remember the significant characteristics of soils, to synthesize our knowledge about them, to see the relationships to one another and to their environments, and to develop predictions of their behavior and responses to management and manipulation. This exemplifies the functions or objectives of a classification.

The sheer numbers commonly dictate a hierarchical system. It can be a word/text document or a mathematical expression of a multicategorical system with dendrograms and similarity indices or their equivalent. Each is constructed to abstract characteristics of the population, and to reveal patterns associated with concepts of order. In addition, there is a need to have inclusive classes that generalize and group the individuals. This is especially needed when developing generalized maps. It does not imply, however, that generalized classes provide large areas on generalized maps. It means that large areas may be named in terms of a few classes. There is a need for an orderly organization of understanding. Classes at successively higher levels imply segregation of successively more important qualities, and it is the relationships among the categorical levels that depict concepts of order.

As interpretations are important, there should be devices or means that can be adapted to different organizational schemes that have hierarchies of classes that are different from those of the system being devised. No one system can emphasize attributes or concepts that are equally important for all perspectives. Classes (groups of soils) can be combined or separated in many ways outside of the system, and a multicategorical system makes this possible. There can be a group of "all soils having an aquic moisture regime" in a separate scheme, even though they do not exist as such in the present one.

We note the following stated objectives of a soil classification system:

- Remember significant characteristics of soils; mnemonics may be helpful.
- Use relationships to help synthesize knowledge.
- Show relationships among classes and categories to each other and to their environments.
- Enable predictions of behavior and responses to management.
- Group soils of similar genesis, or by other desired concepts.
- Give identity to otherwise unidentified individuals of a population.
- Provide groups that can be regrouped or subdivided for applied objectives.
- Abstract concepts of order in nature.
- Provide classes that have real counterparts in nature (mainly as mappable soil bodies).
- Component bodies in nature must be pertinent to applied objectives.
- A system that can be applied consistently by competent scientists working independently needs to be objective in the sense that the system uses the properties of soils, and not the beliefs of the classifier.
- Group soils according to a maximum number of common properties that reflect a common genesis.

There are so many variables in definitions, purposes, and in concepts that chaos can be the only result. Nevertheless, there have been many classifications that have worked well for their intended purpose, and have met their objectives with varying degrees of success. In many places, conscious and unconscious biases creep in and prejudice the future.

MINIMIZING PREJUDICE FOR THE FUTURE

A classification scheme is based on experiences that obtained facts and developed relationships. Consequently, it is subject to inadequacies of that experience. The base of specialized knowledge now involves some 20,000 scientific journals (Yaalon, 1999). A soil classification system is a human contrivance—thus it is subject to inadequacies of the human mind. It is said that geniuses have habits of thought that constantly involve alternative solutions, employ innovative and continual questioning, and have more dead ends than most synthesizers of soil knowledge. Thus there are many possibilities for future information and the obsolescence of what exists today.

As we draw closer to accepting a classification as a truth, the concepts on which it is constructed are accepted as facts, and then we resist change. Research in soil science may be molded in the same patterns as previous research efforts. Small incremental confirmations of the same kind of information obtained in the same or similar ways lull us into complacence. Observations of soils, landscapes, interactions, and relationships tend to be limited to the features that were known previously. Many years passed before fragipans were recognized as possible pedological features. The analysis of empirical relationships is often limited to those noted in the past. Even our explanations of new facts and hypotheses may be channeled into the patterns of the past.

The probabilities of accepting or rejecting facts and relationships suggest that both classification and understanding are not truths. New experiences and facts are a certainty, and a classification scheme attempts to organize current understanding. Such understanding is largely composed of explanations that relate unfamiliar things with familiar things. Genetic concepts and hypotheses are based mainly on what we believe are cause-and-effect correlations; however, such relationships can be supported by empirical correlations that do not require reasons. That is, a soil may be placed in a taxon without knowing its genesis if the theory was not a criterion of recognition. Facts determined by precise observation, experimentation, and measurement minimize prejudicing the future, as do empirical relationships among facts.

REFINING THE SYSTEM: "RULES OF ENGAGEMENT"

As we venture into the new millennium, there will be constant new demands for soil survey information. As science progresses, our understanding of soils will change. Soil classification, which is a scientific reflection of the state of the science, must change and even adapt to the new technologies of processing information. Owners of systems must be alert and must respond to changing needs. Some countries are in the process of developing national systems or revising their systems. The following is a set of ideas that may be useful:

1. Clearly state the purpose and objectives of the system.
2. Describe and define a basic unit to be classified.
3. Be clear about whether or not geographic attributes are required.
4. Quantify and use measurable soil properties. (If temperature and moisture patterns in soils are not soil attributes, know why not.)
5. Understand the scientific method and use it.
6. Understand and use categories and classes correctly in a hierarchical system. There are accepted protocols for such systems. Do not violate them without providing viable alternatives (e.g., mutually exclusive classes vs. fuzzy set classes).
7. Constantly evaluate classes for appropriate central concepts (within-class attributes) and boundary limits (among class attributes).
8. If showing concepts of order in nature, critically evaluate the bases of differentiation for categories as meaningful cause-and-effect relationships.
9. If showing interpretive groupings for applied objectives, define the classes in terms of the interpretation, and not soils per se.
10. Do not use concepts or theories as criteria of recognition.
11. Evaluate the criteria that are deliberately chosen to see if they show what is desired.
12. Accept new facts and relationships as approximations of the moment.
13. Search for failure to accommodate new facts and relationships in a classification system.
14. Let each experience test the theory behind the system.
15. Establish a mechanism to constantly ensure that testing of past experience will occur.

REFERENCES

Ahrens, R.J. and Arnold, R.W. 1999. Soil Taxonomy, in *Handbook of Soil Science*. M.E. Sumner, Ed. CRC Press, Boca Raton, E-117–E-135.

Allen, T.F.H. and Starr, T.B. 1982. *Hierarchy: Perspectives for Ecological Complexity*. The University of Chicago Press, Chicago.

Arnold, R.W. 1990. Soil taxonomy, a tool of soil survey, in Soil Classification. B.G. Rozanov, Ed. Publ. Centre for International Projects, USSR State Committee for Environmental Protection, Moscow, 94–111.

Arnold, R.W., Ahrens, R.J. and Engel, R.J. 1997. Trends in soil taxonomy—a shared heritage. *Comm. Austrian Soil Sci. Soc. Heft.* 55:167–170.

Bartelli, L.J. 1978. Technical classification system for soil survey interpretation. *Adv. Agron.*, 30:247–289.

Buol, S.W., Hole, F.D. and McCracken, R.J. 1980. *Soil Genesis and Classification*. Iowa State University Press, Ames.

Cline, M.G. 1949. Basic principles of soil classification. *Soil Sci.* 67:81–91.

Cline, M.G. 1963. Logic of the new system of soil classification. *Soil Sci.* 96:17–22.

Cline, M.G. 1971. Historical highlights in soil genesis, morphology, and classification. *Soil Sci. Soc. Am. J.* 41:250–254.

Deckers, J.A., Nachtergaele, F.O. and Spaargaren, O.C., Eds. 1998. World Reference Base for Soil Resources: Introduction. International Society of Soil Science (ISSS), International Soil Reference and Information Centre (ISRIC), and Food and Agriculture Organization of the United Nations (FAO). Leuven, Belgium.

Forbes, T.R. 1986. The Guy D. Smith Interviews. Rationale for Concepts in Soil Taxonomy. Soil Management Support Services, Tech. Mono. 11. Publ. Dept. of Agronomy, Cornell University, Ithaca, New York.

Jenny, H. 1941. *Factors of Soil Formation*. McGraw-Hill, New York.

Kellogg, C.E. 1936. Development and significance of the Great Soil Groups of the United States. USDA Misc. Publ. 229:1–40.

Knox, E.G. 1965. Soil individuals and soil classification. *Soil Sci. Soc. Am. Proc.* 29:79–84.

Orvedal, A.C. and Edwards, M.J. 1941. General principles of technical grouping of soils. *Soil Sci. Soc. Am. Proc.* 6:386–391.

Shishov, L.L. and Sokolov, I.A. 1990. Genetic classification of soils in the USSR, in Soil Classification. B.G. Rozanov, Ed. Publ. Centre for International Projects, USSR State Committee for Environmental Protection, Moscow. 77–93.

Simonson, R.W. 1963. Soil correlation and the new classification system. *Soil Sci.* 96:23–30.

Smith, G.D. 1963. Objectives and basic assumptions of the new classification system. *Soil Sci.* 96:6–16.

Smith, G.D. 1983. Historical developments of Soil Taxonomy: Background, in L.P. Wilding, N.E. Smeck, and G.F. Hall, Eds. *Pedogenesis and Soil Taxonomy I. Concepts and Interactions*. Elsevier, New York, 23–49.

Soil Survey Staff. 1975. Soil Taxonomy: A Basic System of Soil Classification for Making and Interpreting Soil Surveys. U.S. Dept. Agric. Handbook 436. U.S. Government Printing Office, Washington, DC.

Soil Survey Staff. 1999. Soil Taxonomy: A Basic System of Soil Classification for Making and Interpreting Soil Surveys, 2nd edition. U.S. Dept. Agric. Handbook 436. U.S. Government Printing Office, Washington, DC.

SPI. 1999. Sistema Brasileiro de Classificacao de Solos. Publ. Serviço de Produção de Informação (SPI), EMBRAPA, Brasilia, Brazil.

Triantafilis, J., Ward, W.T., Odeh, I.O.A., and McBratney, A.B. 2001. Creation and interpolation of continuous soil layer classes in the Lower Namoi Valley. *Soil Sci. Soc. Am. J.* 65:403–413.

Wilding, L.P., Smeck, N.E., and Hall, G.F., Eds. 1983. *Pedogenesis and Soil Taxonomy I. Concepts and Interactions*. Elsevier, Amsterdam.

Wilding, L.P., Smeck, N.E., and Hall, G.F., Eds. 1983. *Pedogenesis and Soil Taxonomy II. The Soil Orders*. Elsevier, Amsterdam.

Witty, J., and Eswaran, H. 1990. Recent changes in Soil Taxonomy, in Soil Classification. B.G. Rosanov, Ed. Report of Int. Conf. Alma-Ata. UNEPCOM, Moscow. 112–122.

Yaalon, D.H. 1999. On the importance of international communication in Soil Science. *Eurasian Soil Sci.* 32:22–24.

CHAPTER 5

Soil Classification and Soil Research

Winfried E.H. Blum and Michiel C. Laker

CONTENTS

INTRODUCTION

Since the middle of the 20th century, society has been generous in funding research (Lubchenco, 1998), and this was a period when almost every field of scientific endeavor saw great advances. The development of new research techniques and technologies greatly boosted research outputs during this period. One of the most important results of these endeavors was to raise of the world average life expectancy at birth from about 45 years in 1970 to about 65 years in the year 2000. As lifetime expectancy, for hundreds of years, was only about 45 years, the increase of about 20 years over three decades can be considered as one of the most outstanding achievements of science, especially the medical sciences, but also the natural sciences, such as biology, chemistry, hydrology, agricultural sciences, nutritional sciences, environmental sciences in general, and soil sciences.

In this context, soils play a prominent role in two aspects: One is in a positive sense, because soil influences the lifetime of humans through adequate nutrition, including clean water, clean air, and the maintenance of biodiversity. It is also in a negative sense, however, in the case of contamination and pollution of soil and the food chain, of drinking water resources, and of the space in which we live. Therefore soil research is of great importance for sustainable development—but in

0-8493-1339-2/03/$0.00+$1.50

which sense? How important is soil classification? What is the relationship between soil research and soil classification?

SOIL CLASSIFICATION

"Classification is a basic requirement of all science and needs to be revised periodically as knowledge increases" (Isbell, 1996). This is the case for all natural systems. For natural sciences, classification "serves as a framework for organizing our knowledge of natural systems including soils and provides a means of communication among scientists, and between scientists and users of the land" (Isbell, 1996).

Unlike other natural systems, soils do not occur as discrete entities but as a continuum over the landscape. This has led to much controversy regarding what should and can be classified in terms of soils: Does it make sense to classify individual soil profiles, or should larger soil bodies, even soil landscapes ("soilscapes") be classified? Much of this controversy stems from failure to distinguish between soil classification *senso stricto* and soil surveying (soil mapping), of which soil classification is a component, at different levels of detail. Individual soil profiles can be classified perfectly, but profiles that are very similar can seldom be mapped out separately, and units such as soil associations need to be used in mapping. Unfortunately, there is a tendency to see this as something unique to soils, which is not the case. A similar situation is found with natural vegetation; for example, individual plants can be classified perfectly, but very seldom, if ever, will a pure stand of a single plant species be found that can be mapped separately. However, for aspects such as determining the broad vegetation patterns of a country, it is necessary to map different vegetation units. They also form a continuum over the landscape. These are (like soils) defined and mapped at different levels of detail from biomes (at national level) to plant communities (at detailed level).

Unlike for plants and animals, a universal soil classification system does not yet exist that is used globally in all countries, and a wide array of classification systems exists. Soil classification has historically been linked to agricultural uses of soils. Because dominant soils and agricultural systems differ widely among different countries, a broad range of systems exists today. As the demands for soil information are becoming broader and include nonagricultural uses, such as those concerning environmental issues, soil classification systems are expected to be modified. There is an inherent danger in the development of purely technical or utilitarian classification systems, because their actual utility is often low, and with new developments they tend to become quickly obsolete (Kellogg, 1961; Laker, 1978). Laker (1981) listed the following three criteria for a well-designed soil classification system (at national level):

1. It must be comprehensive—that is, it must be capable of accommodating all of the soils found in the country. Class definitions must be clear, rigorous, mutually exclusive, and based on factual statements of soil properties.
2. Only soil properties that are easy to measure and to comprehend should be used as criteria for higher classification levels.
3. Classification systems must be well-structured, so that the similarities and differences among soils can be easily understood.

Several decades ago, Manil (1959) was "hoping, however, that in the not too distant future pedologists would agree on a universal classification, i.e., universally accepted after discussion and mutual concessions. It might simply be a system universally accepted as a system of reference without necessarily being universally applied." In 1998 the IUSS adopted a motion requesting that all soil scientists adopt the World Reference Base for Soil Resources (WRB), developed under the auspices of the IUSS, as a reference soil classification system. In the WRB publication, it is stated that "WRB is designed as an easy means of communication amongst scientists to identify, characterize and name major groups of soils. It is not meant to replace national soil classification systems, but to be a tool for better correlation between national systems" (WRB Working Group, 1998).

TYPES OF SCIENTIFIC RESEARCH AND TYPES OF SOIL CLASSIFICATION SYSTEMS

Before discussing the interrelationships between soil classification and soil research, it is necessary to define the kinds of research that are relevant. Essentially, two kinds of scientific research can be distinguished:

- "Basic" research, which is curiosity driven, has no specific targets except to understand nature through rational, analytical approaches, e.g., the morphology of a soil body, surface reactions of soil constituents, soil organisms, and their function within soils and others.
- "Targeted" research, which responds to problem-oriented questions, mostly so-called "applied" empirical/practical research with a background within soil sciences or in other natural, social, economic, and technical sciences, e.g., soil quality for specific uses, etc.

Classification systems are conceptual frameworks for dividing a continuum into units. Analogous to the above two kinds of research, most soil classification systems can also be divided into two main groups:

- "Taxonomic" systems that are developed essentially for scientific communication. These are systems, which deal "only with the higher categories, in which case its chief interest lies in the scientific side" (Manil, 1959).
- "Pragmatic" (utilitarian, technical) systems, which deal with the "lower categories which are essential for more immediate practical purposes" (Manil, 1959). These are developed for technical communications in regard to aspects such as land suitability, etc. Since these classifications are developed for specific purposes, the purpose of the system determines the manner in which it is structured and the emphasis given to diagnostic properties and attributes.

During the 1950s and 1960s, a large debate was waged in the *Journal of Soil Science* regarding "natural" and "artificial" soil classification systems, the latter supposedly being the "practical" system (Leeper, 1956; Kubiëna, 1958; Manil, 1959; Jones, 1959; and Muir, 1962).

INTERRELATIONSHIPS BETWEEN SOIL CLASSIFICATION AND SOIL RESEARCH

Information on properties of the soil is used to structure a soil classification system. Scientific research from other fields within Soil Sciences, such as Soil Physics, Soil Chemistry, and Soil Mineralogy, provides the basis for the identification and selection of appropriate diagnostic soil properties and attributes for classification, especially for the development of meaningful class limits. Without the necessary input from specialists in these fields, no proper classification system can be constructed. But to make useful contributions, these specialists must, like the pedologists, know what the "actual soils" in the field are like, and not try to mold classifications in terms of "pure ideas, abstractions" (Manil, 1959). Just as the specialists from other fields must realize their responsibility for the development of a soil classification system, the pedologists must realize that they cannot "do it alone," and must interact with their colleagues. "If one part isolates itself it can become sterile because soil biology, chemistry, physics, mineralogy, and pedology interact with each other. In this narrow sense we should all be holists at heart even though developments in science require us to specialize" (Wild, 1989).

An ideal classification system should be able to deliver the appropriate kind of information, or at least infer the major attributes of the soil through the names of the taxa, thus enabling researchers in other fields of soil science to select appropriate soils for their (basic or applied) research.

From basic research as well as from applied (targeted) research, the following questions may be directed to soil classification:

- What type of information can be delivered?
  - Conceptual (e.g., type of classification, type of hierarchical order, etc.)
  - Morphological/micromorphological information (e.g., pedality—types, grades, and sizes of aggregates—within diagnostic horizons, etc.)
  - Physical information (e.g., color, texture, pore space distribution, shearing resistance, etc.)
  - Chemical information (e.g., pH, CEC, base saturation, etc.)
  - Mineralogical information (e.g., type of minerals etc.)
  - Biological information (e.g., habitat functions, biological activities, etc.)
- At what scale can this information be delivered?
  - In space: macro- to microscale?
  - In time: information about short-, medium-, long-term processes

Most of these questions have to be answered in order to meet the demands from soil scientists and other scientists with regard to the range of uses of the soil. The tools to generate the data or information (including concepts and principles) are provided by soil research. The classification system is the tool that enables the assimilation of the information and the delivery of the information to the user.

It must be kept in mind that not all of the types of information listed above or other important information can be built into a formal classification system. These (usually more practical) criteria are handled as soil phases during soil surveying (Manil, 1959; Van Wambeke and Forbes, 1986). This is, unfortunately, often seen as something that is unique to soil classification, although that is by no means the case. Manil (1959) makes it clear that "the best system of classification will never be able to give all the information necessary for all practical purposes. In botany, for instance, it is not possible to include in one and the same systemization the classification by anatomical and physiological characteristics and the utilitarian properties of plants."

Within soil sciences, basic and applied (targeted) research is differentiated as shown in Table 5.1, which shows the new scientific structure of the International Union of Soil Sciences. This new scientific structure for determining, in a systematic manner, the range of inter-relationships between soil science research and soil classification (Blum, 2000).

D1: Division 1 – Soil in Space and Time

The commissions under this division are the following:

Commission 1.1: Soil Morphology
Commission 1.2: Soil Geography
Commission 1.3: Soil Genesis
Commission 1.4: Soil Classification

All commissions within this division are focusing on soil classification issues in their respective research fields, looking into soil morphology, the geographical distribution of soil morphological features, and processes of soil genesis before they come to soil classification itself. Research under Commissions 1.1 to 1.3 is indispensable for improving soil classification.

Table 5.1 Soil Classification and Scientific Research in Soil Science

D1	**Soil in Space and Time**
C1.1	Soil Morphology
C1.2	Soil Geography
C1.3	Soil Genesis
C1.4	Soil Classification
D2	**Soil Properties and Processes**
C2.1	Soil Physics
C2.2	Soil Chemistry
C2.3	Soil Biology
C2.4	Soil Mineralogy
D3	**Soil Use and Management**
C3.1	Soil Evaluation and Land Use Planning
C3.2	Soil and Water Conservation
C3.3	Soil Fertility and Plant Nutrition
C3.4	Soil Engineering and Technology
C3.5	Soil Degradation Control, Remediation, and Reclamation
D4	**The Role of Soils in Sustaining Society and the Environment**
C4.1	Soils and the Environment
C4.2	Soils, Food Security and Human Health
C4.3	Soils and Land Use Change
C4.4	Soil Education and Public Awareness
C4.5	History, Philosophy, and Sociology of Soil Science

The basic research in soil morphology can of course be done without reflecting on soil classification issues; the same is true for soil genetic approaches and soil geography, which may aim only at the distribution of specific soil types. But it will run into problems if soil classification is not respected as a basis of soil geographical distribution.

D2: Division 2 – Soil Properties and Processes

The commissions under this division are the following:

Commission 2.1: Soil Physics
Commission 2.2: Soil Chemistry
Commission 2.3: Soil Biology
Commission 2.4: Soil Mineralogy

The different Commissions dealing with soil physics, soil chemistry, soil biology, and soil mineralogy are less interested in soil classification, as long as they are focusing on basic research in their respective fields. Generally, the smaller the size of the research topic, especially at micro- and sub-micro-scale, the less important are soil classification issues.

This is, for example, true in research on specific soil physical parameters, such as pressure resistance, shearing resistance, solute movement in soil pores and others, as well as in soil chemistry, looking at element distribution in different soil particles, or exchange and precipitation processes at inner soil surfaces and others. Also, research in basic soil biology, looking, for example, at different distribution patterns of soil microorganisms such as bacteria and fungi, their relation to mineral surfaces, or their reactions in relation to pH, organic matter distribution, and others, must not be linked to soil classification. This is also true for basic soil mineralogical research, as far as it is focused on specific soil minerals and their characterization.

In general, it can be said that the classical, disciplinary fields in soil sciences, such as physics, chemistry, biology, and mineralogy, are not very much focused on soil classification issues. Frequently scientists working in these fields have very little or no soil classification information, because of a lack of scientific interest. Only in cases in which these soil disciplines are asking about the geographical distribution of specific soil properties and processes are they obliged to look into soil classification issues in order to understand their geographical distribution in space and time. This is, then, more related to applied research than to basic research.

This frequent apathy, and in the worst cases even antagonism, of specialists in these fields toward soil classification is very unfortunate, because they should play a very important role in constructive interaction with pedologists during the development of classification systems. After all, soil physical (texture; soil water relations), chemical (CEC; base content; etc.), and mineralogical (clay minerals; sesquioxides) criteria are the most important parameters for determination within a classification system.

D3: Division 3 – Soil Use and Management

The commissions under this division are the following:

Commission 3.1: Soil Evaluation and Land Use Planning
Commission 3.2: Soil and Water Conservation
Commission 3.3: Soil Fertility and Plant Nutrition
Commission 3.4: Soil Engineering and Technology
Commission 3.5: Soil Degradation and Control, Remediation, and Reclamation

All commissions within this division depend very strongly on high quality soil classification, because it is the basis for efficient land suitability evaluation, planning, and management. On the other hand, soil classifiers should carefully note the requirements of the experts in these fields regarding soil classification.

D4: Division 4 – The Role of Soils in Sustaining Society and the Environment

The commissions in this division are the following:

Commission 4.1: Soils and the Environment
Commission 4.2: Soils, Food Security and Human Health
Commission 4.3: Soils and Land Use Change
Commission 4.4: Soil Education and Public Awareness
Commission 4.5: History, Philosophy, and Sociology of Soil Science

All the commissions under this division have an interest in soil classification, because they need information about the distribution of specific soil characteristics or clusters of different soil characteristics relevant in space and time, for their specific research needs.

INTERRELATIONSHIP BETWEEN SOIL CLASSIFICATION AND RESEARCH IN OTHER FIELDS OF SCIENCE

Soil classification also has a role in other natural, social, economic, and technical sciences. In Table 5.2, fields other than soil sciences are listed, as well as possible questions regarding soil classification. In this table, it is clear that soil classification is important for many scientific fields: agriculture, horticulture, forestry, many fields within biology (such as botany, plant sociology, and zoology), ecology, hydrology, geography, geology, land use planning, urban planning and architecture, archaeology, social sciences (such as ethno-sociology and economic sciences), and technical sciences, in which soils and their distribution may play an important role, especially in the foundation work for technical infrastructures, such as roads and buildings.

Table 5.2 Soil Classification and Scientific Research in Other Natural, Technical, Social, and Economic Sciences

Field of Science	Possible Questions to Soil Classification
Agriculture:	soil fertility, precision agriculture, etc.
Horticulture:	selection of species according to soil type, etc.
Forestry:	forest growth, reforestation, etc.
Biology (e.g., Botany, Zoology):	plant sociology, species distribution, etc.
Ecology:	resilience, etc.
Hydrology:	ground water protection against contamination, etc.
Geography:	soil distribution and land use, etc.
Geology:	soil distribution and weathering processes, etc.
Land use planning:	soil distribution, etc.
Urban planning and architecture:	soil distribution and rainwater infiltration, etc.
Archaeology:	soil distribution and conservation status of remnants, etc.
Social Sciences (e.g., ethno-sociology):	survival strategies of indigenous people, etc.
Economic Sciences:	life (health) insurance risks and environmental conditions, etc.

CONCLUSIONS

The following are general conclusions:

- Soil classification is important not only for scientific research in soil sciences themselves, but also for many other natural, social, economic, cultural, and technical sciences.
- The main contribution of soil classification is to allow for the defining of clusters of different soil characteristics, relevant in space and time to the specific research needs in the respective fields of sciences.
- The access of nonsoil scientists to soil classification is very limited because there is not enough information available in a written form to understand soil classification, which must be urgently improved.
- A user-friendly key to soil classification must be developed as soon as possible.

REFERENCES

Blum, W.E.H. 2000. Challenge for Soil Science at the Dawn of the 21st Century, in Soil 2000: New Horizons for a New Century. Australian and New Zealand Second Joint Soils Conference, Volume 1: Plenary Papers. J.A. Adams and A.K. Metherell, Eds. 3–8 December 2000, Lincoln University, New Zealand Society of Soil Science, 35–42.

Isbell, R.F. 1996. *The Australian Soil Classification.* CSIRO Publishing, Collingwood.

Jones, T.A. 1959. Soil Classification – A destructive criticism. *J. Soil Sci.* 10:196–200.

Kellogg, C.E. 1961. Soil interpretation in the soil survey. U.S. Government Printing Office, Washington, DC.

Kubiëna, W.L. 1958. The classification of soils. *J. Soil Sci.* 9:9–19.

Laker, M.C. 1978. Soil science in relation to development processes in less developed areas. Proc. 8th Nat. Congr. Soil Sci. Soc. S. Afr., 182–187.

Laker, M.C. 1981. The value of soil resource inventories in national research planning. SMSS Tech. Monogr. 1, 149–157. SMSS, USDA, Washington, DC.

Leeper, G.W. 1956. The classification of soils. *J. Soil Sci.* 7:59–64.

Lubchenco, J. 1998. Entering the century of the environment: A new social contract for science. *Science.* 279:491–496.

Manil, G. 1959. General considerations on the problem of soil classification. *J. Soil Sci.* 10:5–13.

Muir, J.W. 1962. The general principles of classification with reference to soils. *J. Soil Sci.* 13:22–30.

Van Wambeke, A. and Forbes, T. 1986. Guidelines for using Soil Taxonomy in the names of soil map units. SMSS Tech. Monoger. No. 10. SCS, USDA, Washington, DC.

Wild, A. 1989. Soil scientists as members of the scientific community. *J. Soil Sci.* 40:209–221.

WRB Working Group, 1998. World Reference Base for Soil Resources. FAO World Soil Resources Report 84. FAO, Rome.

CHAPTER 6

Back to the Old Paradigms of Soil Classification

J. Bouma

CONTENTS

ABSTRACT

Though soil taxonomies will further evolve as our knowledge grows, we should now primarily focus on application of the existing systems, and use insights gained to further improve taxonomies in the future. The analogy with plant and animal taxonomy shows that they have moved on to process-oriented studies and, lately, to biotechnology. The latter is, as yet, no option for soil science, but a strong focus on soil processes and the effects of various forms of management is a realistic parallel development, considering the demands of society. As taxonomy has evolved into a *de facto* two-dimensional pedon-oriented activity, its natural links with dynamic three dimensional landscape processes and soil survey have been weakened. Those links should be re-established in terms of considering soils as living bodies in living landscapes, and reviving intense interaction with land users, which has always been a key element of soil survey. Soil taxonomy provides an excellent way to stratify soil information. A plea is made to extend soil survey interpretations beyond the current empirical level, and to add dynamic characterizations using computer simulation models. This includes the effects of management on any given soil series, to be observed in the field at many locations that are identified through soil maps. Two examples are presented in which effects of past management on soil organic matter content, within a given soil series, could be defined by regression analysis. Any type of soil, as defined by soil classification, has a unique "window of opportunity." Such windows should be communicated much more effectively to our many known and as yet unknown customers.

0-8493-1339-2/03/$0.00+$1.50

INTRODUCTION

Soil classification has played a central role in soil science over the last century. As soil surveys were made, the understanding of relationships between soils and the soil-forming factors has increased, and this is reflected in the classification systems. Distinction among different types of soil helped to categorize and stratify the large body of information being assembled during soil survey, and this was certainly important in the pre-computer period—but it remains important now. The earlier notion that the instant availability of basic data retrieved using information systems would make soil classification less relevant has proven to be unrealistic. Huge volumes of unstratified data tend to confuse, rather than enlighten, the prospective user. As time went by, soil classification systems developed using different categories, ranging from orders to soil series. Their practical application under field conditions was facilitated by emphasis on soil morphological features, allowing classifications to be made in the field without having to take and analyze many samples. The problem, of course, has always been the fact that soils are not separate individuals, as are animals and plants. Soils form continuous bodies in landscapes, thus soil taxonomic systems have a different character from that of other taxonomic systems. But several taxonomic systems have been completed, and are now in use, such as Soil Taxonomy (Soil Survey Staff, 1998) and the FAO Classification system (FAO, 1998). The time is right to ask whether these systems have now found their final form, or whether new developments are to be expected or desirable. Also, it is of interest to see what the impact of soil classification has been on pedology, which is described here as the study of soils in the broadest sense, and how, in turn, pedological studies have shaped soil classification.

In this paper we argue that soil classification appears to have lost its original links with soil survey, as expressed by relations between soils and landscapes, and by independent field observations by surveyors who were in touch with a wide variety of land users. This observation is relevant as we face developments in regional land-use studies involving many new forms of interaction with different types of stakeholders (e.g., Bouma, 2001). Given such developments, in what way can soil classification still contribute to the effective communication of soil expertise to a wide variety of soil users, including not only farmers, contractors, and sanitarians, but also planners, politicians, and involved citizens?

Developments in plant and animal taxonomy are relevant to soil science. Attention from these professions has moved from static taxonomy to studies of dynamic processes and, lately, to genetic engineering, which is now providing a major thrust to biological work, making it the hottest area in science. This happened because funding for taxonomic studies was difficult to obtain even though taxonomies were far from complete. In soil science, there is a corresponding analogy in terms of a shift from static classification to process studies, while genetic modification of soils has, of course, as yet no practical significance. The closest analogy is the effect of soil management on soil properties, which is characteristically different for different soils.

Soil classification systems will never be completed, because our increasing knowledge about soil processes is reflected in these systems. There is good reason, however, to temporarily stop modifying systems at this time, because classification activities tend to become rather self-centered and inward-looking. Emphasis on dynamic process-studies of soils in a landscape context and on assessing the effects of soil management deserves priority now to broaden our insights which, in time, can and should be the basis for further development of taxonomic systems.

SOIL CLASSIFICATION AS A PURPOSE IN ITSELF

For many decades, soil classification has been a separate object of study within soil science. Development of a universally accepted classification system was a worthwhile objective requiring a lot of data, particularly that derived from chemical and mineralogical studies. All scientists who were involved with pedological studies that had a focus on soil classification remember field excursions where long discussions were held on Soil Taxonomy in and next to large pits, testing keys of taxonomic systems. To some, this emphasis on classification was problematic, as little attention was paid to quantitative process-studies, to soil-landscape analysis, and to land use and its effect on soils.

Still, as long as classification systems were being developed, one could support such a focus. Now, however, systems have reached a certain degree of maturity, and it is wise to first significantly advance our understanding of dynamic soil processes before focusing on the development of yet another revision of existing classification systems.

SOIL SURVEY INTERPRETATION, OLD AND NEW

Soil survey interpretations have been a prime vehicle for introducing soil classification to a wide variety of soil users. Particularly in the United States, interpretations of soil maps for uses ranging from agricultural suitability to suitability for on-site application of septic tank effluent have been quite succesful over the years. They demonstrated the implications for land use of having a wide variety of soils occurring in landscapes. Standard soil survey interpretations were made in terms of defining relative suitabilities or limitations for a wide variety of land uses. Increasingly, more detailed and quantitative information was needed to answer more specific questions, and better monitoring methods and computer simulation models have been used to provide such information (e.g., Bouma and Jones, 2001). At first, questions associated with water movement in soils received the most emphasis, but applications were later extended to plant growth and movement of solutes (e.g., Bouma et al., 1980). Unfortunately, use of simulation modeling in the context of soil survey interpretation has remained limited in the United States. Elsewhere, simulation has been widely used in the context of land evaluation, which was often based on soil maps defining regional soil patterns. As models became more advanced, lack of appropriate data to feed the models formed an increasing problem. The advance of pedotransferfunctions, which relate available soil survey data (that cannot be used as such) to data that can be used in modeling has been an important justification for using soil survey data (e.g., Wosten, 1997; Wosten et al., 1999). Also, simulations could be improved by a functional characterization of a given soil pedon. In other words, pedological soil horizons were compared to see whether they were different in terms of physical (Wosten et al., 1985) or chemical (Breeuwsma et al., 1986) behavior. When different pedological horizons showed comparable behavior, they were lumped into thicker functional horizons. Thus fewer functional horizons could be used in the simulations, making the work more efficient. Functional characterization has also been applied recently to farmers' fields, defining management units for precision agriculture (Van Alphen and Stoorvogel, 2001).

Soil survey interpretations or land evaluation, whether qualitative or quantitative, are provided for soil taxa, often at the soil series level. This has remained attractive over the years because the occurrence of soil taxa in the field is expressed on soil maps through the map legend. As of old, interpretations relate to areas of land which is crucial for all applied work. Of course, much variability is involved here, which explains all the work that has been done on spatial and temporal variability in soil science, to be discussed next.

VARIABILITY IN SPACE AND TIME: DISTINGUISHING MEANS AND PURPOSES

Classical soil survey interpretations stratify landscapes in mapping units that are each represented by a "representative profile," for which assessments are made that are implicitly expected to apply to the entire units, ignoring spatial variability within the units and assuming sharp boundaries among them. Geostatistics, as recently reviewed by Heuvelink and Webster (2001), has been developed and perfected as a tool during the last decades to deal with variations in space and time. These authors conclude that a combination of soil classification and geostatistics is advantageous, as this can deal with the occurrence of sharp boundaries, while internal variation within mapping units can be expressed in quantitative terms. Both approaches were compared for the first time by van Kuilenburg et al. (1982). Merging the discrete and continuous models of spatial variation, however, is more promising for practical applications. Stein et al. (1988) applied the continuous model in each stratum, which excludes spatial autocorrelation across boundaries. An alternative technique was presented by Heuvelink and Huisman (2000). Another approach that aims to escape from the abrupt boundary assumptions of soil maps is based on fuzzy set theory. Objects do not belong to one class only, but can be members of multiple classes, with class-specific membership values that add up to 1 (e.g., McBratney, 1992). Geostatistics and fuzzy sets allow us to express variability within mapping units on soil maps. This variability can be expressed in terms of occurrence of different soil series (at detailed levels), which provides a link with interpretations as discussed above.

BACK TO LIVING SOILS IN LIVING LANDSCAPES

Soil classification is essentially an expression for a two-dimensional feature—the soil profile, or pedon. Functioning of soil is, of course, a three-dimensional process, which is highly affected by landscape processes: Infiltration higher up the slope is often associated with seepage at lower parts of the slope, for example. Computer techniques are currently available to characterize such landscape processes in quantitative terms using Digital Terrain Models and simulation models of various forms (e.g., Schoorl and Veldkamp, 2001). Remote sensing techniques can be used to fine-tune such models to real conditions (e.g., Cloutis et al., 1999; Baban and Luke, 2000). Landforms and environmental attributes, such as local relief, upstream area, and vegetation, serve as additional information to the soil mapping process, which may include interpolation of point data (Lagacherie and Voltz, 2000; McKenzie and Ryan, 1999). Modern applications of pedology call for dynamic characterization of soil behavior in a landscape context, and modern techniques add new dimensions to the possibilities available to soil scientists.

The Effect of Soil Management

Soil classification is focused on natural soil profiles that are, in principle, unaffected by recent human management. As is stated in the introduction to Soil Taxonomy, ploughing a soil does not change its classification. Of course, severe erosion may change soil classifications dramatically, but this is different from the effects of soil management. Soil management may, however, strongly change soil properties of a given soil series in a manner that is quite significant for soil use. This does not necessarily correspond with a change in classification. For instance, Droogers and Bouma (1997) showed that about 40 years of different types of management in a prime agricultural soil in the Netherlands (a loamy, mixed, mesic Fluventic Eutrudept) resulted in significantly different soil properties, even though the classification did not change. Biological farming resulted in a higher soil quality, expressed by higher organic matter contents, while traditional arable farming led to depletion of organic matter and a lower soil quality. Computer simulations showed that nitrogen transformations were different as well. Pulleman et al. (2000) followed up these studies by documenting the land-use history of 40 sites where this type of soil occurred in the region, using the soil map to identify the sites. They derived the following equation by regression analysis:

$$\text{SOM} = 20.7 + 29.7\ \text{C-I} + 7.5\ \text{C-V} + 7.5\ \text{M-IV}\ (\text{r square} = 0.74)$$

where SOM = soil organic matter content; C-I and C-V = crop types in periods I (1–0 years from now) and V (3–1 years from now) (C = 1 for grass and 0 for arable land). M-IV = management type in period IV (7–3 years from now) (M = 1 for organic and 0 for conventional farming).

This study shows that soils in the landscape reflect past management practices and can thus provide a treasure of information to the soil scientist. Using soil classification and soil maps as a means of stratification, one can obtain more specific results, compared to a hypothetical situation in which the landscape is sampled at random, without benefiting from stratification made possible by soil classification. This procedure clearly presents an alternative to complex generic deterministic simulation models, and can also be used to predict organic matter contents. We recommend that more studies be done on different conditions in a given soil type, with the objective of relating these conditions to different types of documented management or land-use practices of the past. Once such relations are established, they allow predictions to be made about the effects of different types of management. This is important information for many users, and represents a significant step beyond defining relative suitabilities of a given soil type for a given type of land use, which is the classic mode of soil survey interpretation.

The Importance of Communication

Soil classification has a somewhat dusty image both inside and outside soil science. This can be improved by demonstrating the significance of using soil classification as a means to stratify soil information. Rather than talk about soils in general, we talk about soil units, which occur in characteristic spots in the landscape and which are constituted by a unique combination of soil properties. So far, such properties have been expressed mostly in terms of classic soil survey interpretations, which list relative suitabilities of soils for a wide variety of uses. The modern user of soil information wants more. He or she does not look for a judgment, but for indications as to which options are realistic when dealing with various types of land use (e.g., Bouma, 2000; 2001). Such options can be listed for any soil series, providing a window of opportunity that is characteristic for each soil series. Modern land-use problems indicate that the old paradigm that "anything can be done anywhere" does not apply anymore. Conflicting demands have to be balanced, and soil information showing the options for different soil series can be very important, because non-soil scientists tend to lump all soils together, considering areas to be developed as blank sheets of paper. Communication of soil information, perhaps in the format of "The Story of [the Miami silt loam]" could be a means to freshen up the image of soil science to a broad audience, certainly when it includes specific examples of the effects of different types of management on soil properties as described above.

CONCLUSIONS

1. Even though classification systems will evolve further as more information becomes available, we should temporarily stop further modifications of these systems, and focus more strongly on using the available classifications, applying them as a medium to stratify soil information for a wide range of modern applications.
2. Plant and animal scientists have moved on from taxonomy to process-studies and biotechnology. We should follow this example by emphasizing process-studies in soils in a landscape context, and by studying real and potential effects of different types of management within a given type of soil.
3. Taxonomy has increasingly become a two-dimensional pedon-oriented activity. We should go back to our roots and consider dynamic three-dimensional landscape processes, and recall the intense interaction with stakeholders that was an essential ingredient of soil survey. The latter is crucial for communicating effectively with our modern stakeholders.
4. Effects of different management practices on any given type of soil are out there to be observed in the field. Such observations should be made more often. Together with process-studies supported by simulation modeling, they can provide unique and characteristic windows of opportunity for every soil type, and will be effective in communicating our expertise to our many customers.

REFERENCES

Baban, S.M.J. and Luke, C. 2000. Mapping agricultural land use using retrospective ground reflective data. Satellite sensor imagery and GIS. *Int. J. Remote Sensing*. 21:1757–1762.

Bouma, J. 2000. Land evaluation for landscape units. in *Handbook of Soil Science*, M.E. Summer, Ed. CRC Press. Boca Raton. E393–411.

Bouma, J. 2001. The role of soil science in the land use negotiation process. *Soil Use Manage*. 17:1–6.

Bouma, J. and Jones, J.W. 2001. An international collaborative network for agricultural systems applications (ICASA). *Agricultural Systems*, CRC Press, Boca Raton.

Bouma, J., de Laat, P.J.M., van Holst, A.F., and van den Nes, Th.J. 1980. Predicting the effects of changing water table levels and associated moisture regimes for soil survey interpretations. *Soil Sci. Soc. Am. J.* 44:797–802.

Breeuwsma, A., Wosten, J.H.M., Vleeshouwer, J.J., van Slobbe, A.M. and Bouma, J. 1986. Derivation of land qualities to assess environmental problems from soil surveys. *Soil Sci. Soc. Am. J.* 50:186–190.

Cloutis, E.A., Connery, D.R. and Dover, F.J. 1999. Agricultural crop monitoring using airborne multi-spectoral imagery and C-band synthetic aperture radar. *Int. J. Remote Sensing*. 20:767–788.

Droogers, P. and Bouma, J. 1997. Soil survey input in exploratory modeling of sustainable soil management practices. *Soil Sci. Soc. Amer. J.* 61:1704–1710.

Food and Agricultural Organization of the United Nations. 1998. World Reference Base for Soil Resources. World Soil Resources Report 84. FAO. Rome, Italy.

Heuvelink, G.B.M. and Webster, R. 2001. Modeling soil variation: Past, present and future. *Geoderma*. 100:269–303.

Heuvelink, G.B.M. and Huisman, J.A. 2000. Choosing between abrupt and gradual spatial variation? in *Quantifying Spatial Uncertainty in Natural Resources: Theory and Applications for GIS and Remote Sensing*. H.T. Mowrer and R.G. Congalton, Eds. Ann Arbor Press. Chelsea, MI. pp. 111–117.

Lagacherie, P. and Voltz, M. 2000. Predicting soil properties over a region using sample information from a mapped reference area and digital elevation data: A conditional probability approach. *Geoderma* 97:187–208.

McBratney, A.B. 1992. On variation, uncertainty and informatics in environmental soil management. *Aust. J. Soil Res*. 30:913–935.

McKenzie, N.J. and Ryan, P.J. 1999. Spatial prediction of soil properties using environmental correlation. *Geoderma* 89:67–94.

Pulleman, M.M., Bouma, J., van Essen, E.A. and Meijles, E.W. 2000. Soil organic matter content as a function of different land use history. *Soil Sci. Soc. Am. J.* 64:689–693.

Schoorl, J.M. and Veldkamp, A. 2001. Linking land use and landscape process modeling: A case study for the Alora region (south Spain). *Agric. Ecosyst. Environ*. 85:281–292.

Soil Survey Staff. 1998. Keys to Soil Taxonomy. 8th Edition. U.S. Government Printing Office, Washington, DC.

Stein, A., Hoogerwerf, M. and Bouma, J. 1988. Use of soil-map delineations to improve (co)kriging of point data on moisture deficits. *Geoderma*. 43:163–177.

Van Alphen, B.J. and Stoorvogel, J.J. 2001. A functional approach to soil characterization: Support of precision agriculture. *Soil Sci. Soc. Am. J.* 64:1706–1713.

Van Kuilenburg, J., de Gruijter, J.J., Marsman, B.A., and Bouma, J. 1982. Accuracy of spatial interpolation between point data on moisture supply capacity, compared with estimates from mapping units. *Geoderma*. 27:311–325.

Wosten, J.H.M., Bouma, J., and Stoffelsen, G.H. 1985. The use of soil survey data for regional soil water simulation models. *Soil Sci. Soc. Am. J.* 49:1238–1245.

Wosten, J.H.M. 1997. Pedotransfer functions to evaluate soil quality, in E.G. Gregorich, and M.R. Carter, Eds., *Soil Quality for Crop Production and Ecosystem Health. Developments in Soil Science*. Vol. 25. Elsevier, Amsterdam. pp. 221–245.

Wosten, J.H.M., Lilly, A., Nemes, A. and le Bas, C. 1999. Development and use of a database of hydraulic properties of European soils. *Geoderma*. 90:169–185.

CHAPTER 7

Incorporating Anthropogenic Processes in Soil Classification

Ray B. Bryant and John M. Galbraith

CONTENTS

ABSTRACT

This chapter considers the need for incorporating anthropogenic processes in soil classification systems, describes fundamental approaches to soil classification and their underlying concepts that affect the incorporation of anthropogenic processes, and proposes possible approaches to incorporating anthropogenic processes in Soil Taxonomy and like systems. Soil classification schemes serve to facilitate communication about soils by organizing the tremendous number of individual soils into groups of soils according to similarity. A problem facing Soil Taxonomy is that soils that may contain garbage, coal ash, or construction debris are grouped with Entisols or Inceptisols, and this results in a loss of credibility in the system. Additionally, wide ranges of properties within classes that include anthropogenic soils result in large numbers of series that are unwieldy when one attempts to correlate series and interpret map units. There are major problems with incorporating anthropogenic processes in systems such as Soil Taxonomy that use morphology-based criteria, because not all anthropogenic soils contain morphological evidence of anthropogenic processes. Three approaches to incorporating anthropogenic processes are proposed for use, singularly or in combination, to minimize impacts on Soil Taxonomy and like systems that rely on morphological criteria for defining classes. Previously unused observations, such as landforms, could be used as criteria for classification; relational observations and data (to include historical records) could be used as criteria for classification; and the knowledge of process of formation could be allowed for use in a single order. These approaches do not present a complete solution to the problem, but do provide a means of separating soils in which anthropogenic processes are the primary process of formation.

INTRODUCTION

Anthropogenic (also appears as Anthrogenic in some publications) soil-forming processes are defined as those actions by humans that modify and control soil-forming processes. Humans themselves are not a process, but they act as a factor of soil formation (Pouyat and Effland, 1999). These anthropogenic processes can be generally grouped into physical alterations, hydrologic alterations, and geochemical alterations. Examples of specific processes are excavation and deposition of soil, sediment, and rock; deposition of artifacts (human-manufactured or modified objects and materials); mechanical mixing or compaction; long-term increase in phosphorous and organic matter through agriculture and waste disposal; long-term protection from natural flooding, artificial flooding and drainage; accelerated erosion; alteration of soil reaction by chemical additions; and contamination with pollutants. Some of these processes may lead to formation of anthropogenic soils, defined as those soils that have been profoundly affected by anthropogenic soil processes and have morphological characteristics and properties that are irreversible, or very slowly reversible (Galbraith and Bryant, 1999).

Anthropogenic soils may fall into two distinct categories, based on their content of artifacts (Galbraith et al., 1999). There is no set of morphological characteristics or properties that uniquely typify anthropogenic soils that have few artifacts. In fact, most physically altered anthropogenic soils exhibit soil profiles that, in terms of morphology, are indistinguishable from stratified or weakly developed soils in natural landscapes. Although many processes related to long-term agricultural use have profoundly affected soil properties over a time scale of a few thousand years, these anthropogenic processes are relatively fast-acting, compared to natural processes of soil formation, and are usually documented in historical records (Fanning and Fanning, 1989).

The morphological characteristics and properties of anthropogenic soils are usually not congruent with neighboring soils that formed under natural soil-forming processes. For example, the anthropic epipedon (Soil Survey Staff, 1999) that formed under long-term irrigation has high organic matter content relative to the expected levels in nonirrigated soils, and in comparison to surrounding non-irrigated soils. Soils forming in transported material, such as mine spoil, occupy landforms that can usually be identified as artificial or out of sync with the surrounding naturally occurring landforms. Although the soils that form on these artificial landforms often have morphology that can be explained by anthropogenic processes, their morphology ranges from strikingly dissimilar to indistinguishable, in comparison to soils formed under natural processes. Identical classification of anthropogenic soils and surrounding natural soils, in some systems, is the cause of much frustration for those attempting to identify the major reason for their existence, behavior, properties, and location (Sencindiver and Ammons, 2000).

This chapter considers the need for incorporating anthropogenic processes in soil classification systems, describes fundamental approaches to soil classification and their underlying concepts that affect the incorporation of anthropogenic processes, and proposes possible approaches to incorporating anthropogenic processes in Soil Taxonomy and similar systems.

THE NEED FOR INCORPORATING ANTHROPOGENIC PROCESSES

Anthropogenic processes should be incorporated into existing soil classification systems because they are widely found, impact large numbers of people and extensive land areas, and may produce soils that are distinctly different from existing soils. Additionally, in some current systems such as Soil Taxonomy, the lack of sufficient classes for anthropogenic soils and processes results in areas that are often not mapped correctly or are not mapped at all. Anthropogenic soils are found on all continents, but are most likely to be found in areas of long-term human habitation, where population density is high, and in countries where large mechanical equipment is readily available. For example, three major land types with anthropogenic soils are high-density residential, transportation corridors, and mining areas. There are estimated to be about 150 Mha of urban land, about 1% of the world's land surface area (Grubler, 1994). The amount of urban land and urban population is expected to increase rapidly (National Institute of Urban Affairs, 1994). Countries in Europe and Asia that have long-term habitation have felt the need to distinguish urban soils from natural soils, especially those with pollutants, garbage, and artifacts (Burghardt, 1994; Kimble et al., 1999). High-density residential areas include less than 20% of natural soils (Smith, 1999) and include a high percentage of impervious materials such as buildings and pavements that cover the soil (Hernandez and Galbraith, 1997). These residential areas are also the home to several million metric tons of municipal wastes each year. In addition, there are probably over 30 MKm of roads in the world, and most of these have a corridor of 15 to 125 m where the soil has been highly altered. This equates to anywhere from 28.1 to 233.5 Mha of anthropogenic soils in the world (United States Central Intelligence Agency, 2000). Although the global geographical extent is difficult to estimate, major and minor surface mining of soil, rocks, and minerals is by far the most extensive anthropogenic process.

Anthropogenic processes often have profound and permanent effects on soils. For example, human activity may lead to destruction or removal of soil horizons; burial of soil with pavements, other soil, rock, garbage, or debris; contamination of the soil by air- or water-borne pollutants; removal of the soil; or permanent alteration of the soil hydrology (Galbraith and Bryant, 1999; Stroganova, 1999). Upland soils that contain garbage and other artifacts at depth are certain to have been mixed or transported by humans. Human-altered upland soils frequently lack highly developed genetic soil horizons, as compared to surrounding natural soils, and may be easily distinguished from natural soils by virtue of their occurrence on anthropogenic landforms. Humans that live around anthropogenic soils are usually aware of their history, and do not consider them to be natural soils. Resource managers prefer that anthropogenic soils be classed separately from natural soils because of the important management and interpretive differences (Craul, 1999). In particular, Craul cites problems in sustainably vegetating urban soils because of improper respect paid to soil properties and microclimates by planners. Urban soils are unique, thus the USDA-NRCS has established an urban soils program (Scheyer, 1999), and urban soils are recognized and classified in numerous taxonomies around the world (Burghardt, 1994; Kimble et al., 1999).

Soil Taxonomy was designed to be a tool for making and interpreting soil surveys of agricultural land and other areas where soils formed dominantly from natural soil-forming processes. Disturbances in the surface or plow layer were specifically addressed by the system to prevent separation of agricultural soils from their undisturbed equivalents. There are a few categories in Soil Taxonomy that distinguish anthropogenic soils (Ahrens and Engel, 1999; Galbraith and Bryant, 1999), and some mapping of urban landscapes in the United States has begun (Levin, 1999; Hernandez, 1999; Smith, 1999). However, complete acceptance of these maps into the existing National Cooperative Soil Survey system of mapping has not been achieved, and challenges to research (Pouyat and Effland, 1999) and mapping of anthropogenic soils remain (Southard, 1999), especially with regard to soils polluted with heavy metals (Russell-Anelli et al., 1999) and toxic organic compounds. European countries have longer experience with industrial development and have developed several categories of toxic or polluted urban soils (Gerasimova, 1999; Sobocká, 1999; Tonkonogov and Lebedeva, 1999).

The need for incorporating anthropogenic processes into soil classification systems may be evaluated in terms of the purpose of existing classification schemes. All soil classification schemes facilitate communications by organizing the tremendous number of individual soils (those soils having similar characteristics that are significantly different from other soils) into groups that have more similarity and less differentiation within the group than in comparison with other soils. A classification system loses credibility as a tool of communication when members of a single group have vastly different morphological characteristics, as is often the case with soils in urban environments (Burghardt, 1994). For example, human-transported soils that contain garbage, coal ash, or construction debris are currently grouped with naturally formed Entisols or Inceptisols at the subgroup and family levels in Soil Taxonomy, even though their morphology and behavior are easily recognized as being distinct (Hernandez and Galbraith, 1997).

A recognized purpose of many classification systems is to group like soils for purposes of correlation and interpretation. Correlating and interpreting soils forming in mine spoil is a major concern on large acreages in the U.S. and other countries (Sencindiver and Ammons, 2000). Soil scientists working with these soils would prefer a taxonomic class that excludes soils that are not derived from mine spoil, to focus on distinguishing characteristics that are important to the use and management of these anthropogenic landscapes. At the very least, those anthropogenic soils that have been modified to the extent that they would be considered dissimilar to their parent soil and have limiting interpretations should be placed into a separate class.

Some classification schemes, such as Soil Taxonomy, were purposely designed to provide a framework for research on soil-forming processes. To meet that objective, classes were constructed to imply processes of genesis, but criteria are based on observed morphology or measurable properties. Although the major anthropogenic process responsible for the formation of an anthropogenic soil may be a matter of historical record, the processes of future change and development in response to the active factors of soil formation, climate, and organisms is of interest to researchers. Incorporating anthropogenic processes into soil classification would provide a framework for these research efforts.

FUNDAMENTAL APPROACHES TO SOIL CLASSIFICATION

Concepts valid during the development of a system necessarily prejudice the future modifications of the system, and may or may not complicate the incorporation of anthropogenic processes. Fundamental concepts that affect the incorporation of anthropogenic processes can be discussed in terms of three broad approaches to soil classification. The ecologically based soil classification system, typified by the 1938 U.S. Soil Classification System and the Russian Soil Classification System contemporary to that time, used the ecological zone in which soils formed as criteria for classification. For example, the *Mountain Meadow soils* included those soils that formed in a mountain meadow. In accordance with that concept, anthropogenic soils could easily be grouped according to the anthropogenic environment in which they formed. Mine soils could be defined as those formed on mined lands; urban soils could be defined as those formed in urban environments.

Two other more recent approaches to soil classification can be described as "process-based with descriptive morphology" and "morphology-based with implied process of formation." The Chinese Soil Taxonomic Classification (Chinese Soil Taxonomy Research Group, 1995; Zitong et al., 1999), the Revised Legend of the Soil Map of the World (Food and Agriculture Organization, 1996), and the World Reference Base for Soil Resources (International Soil Science Society Working Group RB, 1998) are examples of systems that use processes of soil formation as criteria for identifying soil class, and describe the central concept of the class in terms of morphological characteristics that are typical of soils that belong in the class. For example, as defined in the World Reference Base, a "Luvisol" is a soil in which clay has been washed out of the upper part of the soil and accumulated in the lower part of the soil. Moderate to high activity clays and low aluminum saturation further characterize Luvisols. These specifically named systems and other systems of this kind do include anthropogenic processes. In the World Reference Base, Anthrosols are defined as those soils that have been transformed by anthropogenic processes to the extent that the original soil is no longer recognizable or remains only as a buried soil. In these systems, the process of formation is the essential criteria for classification—thus including anthropogenic processes is not problematic. Similarly, soils in China that formed over centuries of sedimentation resulting from long-term irrigation have morphologies identical to naturally occurring Fluvents that formed on floodplains. Although the physical process of sedimentation is the same, Chinese Soil Taxonomy separates these soils where anthropogenic activities controlled the processes of sedimentation, and recognizes the anthropogenic nature of their origin.

Soil Taxonomy (Soil Survey Staff, 1999) and the Brazilian System of Soil Classification (EMBRAPA, 1999) are examples of systems that use morphology-based criteria for defining classes, but processes of soil formation are implied in the construct of the classes. The use of perceived knowledge of the processes of soil formation is expressly avoided. Neither of these systems completely incorporates anthropogenic processes, although some horizons and conditions that developed as a result of anthropogenic processes are recognized. These systems share concepts and structural characteristics that complicate the incorporation of anthropogenic processes; the problems associated with incorporating anthropogenic processes are discussed in terms of Soil Taxonomy. As previously described, anthropogenic soils may or may not contain morphological evidence of anthropogenic processes. Although soils forming in mine spoil do contain some morphological features that can be specifically and uniquely related to processes of their origination, those features are not always present in all polypedons. The same is true of artifacts that may or may not be present in deeply mixed soils that supported ancient civilizations. As such, these features do not provide a means of consistently distinguishing these soils from naturally occurring Entisols, Inceptisols, or other orders. Additionally, reclamation practices can and do result in reconstructed morphologies that fit the technical description of Mollisols or other soil orders, further complicating any effort to group anthropogenic soils and separate them from naturally occurring soils with similar morphology. The heart of the problem lies in the desire to clearly and consistently identify those soils that formed primarily as a result of anthropogenic processes, although soil morphology does not always provide clues to these processes. For example, destructive processes, such as excavation and accelerated erosion, may have profound impacts on pedon and landform morphology, but leave little or no morphological evidence behind that can be tied conclusively to human activities. Entisols and many Oxisols and Andisols that have been deeply mixed (>1 m) would still meet the criteria for inclusion in their respective orders, as would some Histosols and Inceptisols (Soil Survey Staff, 1999), and many soils that have been thoroughly mixed to between 50 and 100 cm deep would still qualify in the same soil order as their undisturbed analogs. Credibility with users will be lost if anthropogenic soils are grouped with soils that formed under natural processes because the morphological evidence to separate them is lacking, especially when the true process of soil formation is a matter of historical record, the soils contain abundant artifacts, or they occur on evident anthropogenic landforms.

INCORPORATING CLASSES OF ANTHROPOGENIC SOILS

Additions to Soil Taxonomy have been faced and dealt with many times in the last 25 years, although there has been a reluctance to add new soil orders. There is a question of whether anthropogenic soils resemble recent additions. In the 1960s, the authors of Soil Taxonomy were considering whether to have a separate soil order for soils with aquic conditions, or whether to add an aquic suborder to each order. The unifying character of these soils was their aquic soil moisture regime that gives them similarities of behavior and interpretation. They have representatives that meet the requirements of all other soil orders. They are distributed throughout the world, although they are more common in level, low-lying areas, and less common in arid areas. It was decided to add an aquic suborder to each order, rather than classify them as a separate order (Smith, 1986).

The two most recent soil orders that were added to Soil Taxonomy are the Andisols and Gelisols. It was decided that soils that had andic soil properties (or high potential to develop them) had unique andic or vitric properties that governed their behavior and management. Soils with andic soil properties existed previously in the system (mainly as Anthrepts), but were not represented in all orders. Andisols have some similarity of occurrence in areas of geologically recent volcanism, and are not found across all continents. The Gelisols were separated because of their current or former permafrost content, which gives them similar behavior and interpretation. These frozen soils were found in a few soil orders, mainly Histosols, Inceptisols, and Spodsols. Gelisols have a predictable zone of occurrence and are not ubiquitous (Soil Survey Staff, 1999).

In accordance with the logic used to classify aquic conditions, Andisols and Gelisols, we should first consider whether anthropogenic soils have more in common with each other than they do with the soils from which they originated (their parents). As noted earlier, there is a wide variety of distinct processes that control formation of anthropogenic soils. Although they formed under anthropogenic processes, they may or may not share common, unique properties or behaviors that are distinct from existing natural soils. Those anthropogenic soils that have had significant additions of artifacts potentially have unique properties that govern their behavior and interpretation, and cause them to be distinctly different from any existing soil classes. Conversely, other anthropogenic soils do not have a unique set of characteristics that are distinctly different from existing soils. For example, reclamation practices can result in soil profiles that have the morphological characteristics of Mollisols, and mechanically disturbed Oxisols, Vertisols, and Gelisols might very well still meet the criteria for classification within the order of their respective parent soils. If classified with their natural soil counterparts, it can be assumed that anthropogenic soils would be represented in most, if not all, soil orders, because they occur in all climates and land areas.

The categorical structure of Soil Taxonomy and existing class structure poses additional challenges for incorporating anthropogenic soil processes in an efficient and logical manner. The most efficient method of incorporation would be to add a new soil order at the very beginning of the Keys, to avoid proliferation of classes throughout Soil Taxonomy and disruption of existing classes. All other classification systems that recognize anthropogenic soil processes do include a separate order of Anthrosols. Suborders and great groups could be structured to group soils according to dominant anthropogenic process, anthropogenic materials, and properties important to interpretation.

Alternatively, the formation of new suborders of Entisols based on anthropogenic materials, as proposed by Fanning and Fanning (1989), would be an expedient approach if classes of soils were added only after proof of their existence. This would temporarily capture most of the soils with drastic physical or hydrologic alteration, and would keep the number of new classes at a minimum. However, there would eventually be similar suborders (or exclusion statements) added to all other orders, or the addition of a great number of suborders to Entisols. Logic would demand that aquic characteristics of Entisols be recognized after anthropogenic characteristics, thereby disrupting existing classes and leading to constraints in recognizing other characteristics, because of a limited number of categories within the system.

Given the problems of classifying soils formed by anthropogenic processes within Soil Taxonomy and other systems that use morphology-based criteria for defining classes, the challenge becomes one of modifying the system, but with minimal impact on existing classes, procedures of identification, and underlying concepts and principles that guided the original development of the system. Addition of a new soil order at the beginning of the Keys appears to be the most efficient means of incorporating anthropogenic processes in the long run.

DIAGNOSTIC CRITERIA FOR ANTHROPOGENIC SOILS

There are two major approaches to developing diagnostic criteria for anthropogenic processes in Soil Taxonomy and like systems that rely on morphological criteria for defining classes. The first approach would involve working within existing Soil Taxonomy procedures and protocols by defining new anthropogenic soil diagnostic horizons and materials, along with new rules and classes.

For example, diagnostic horizons with high content of specified kinds of artifacts and an anthroturbic horizon of highly mixed soil materials that contain remnants of diagnostic horizons could be defined. However, if limited to morphological evidence in the pedon, this approach would fail to capture many of the anthropogenic soils, because of their lack of morphological evidence of anthropogenesis. As a result, soils of similar process of origin would occur both in classes of anthropogenic soils and soils formed under natural conditions.

The second approach to incorporating anthropogenic processes is to allow previously unused observations, such as landform type (anthropogenic vs. natural), relational observations and data (to include historical records), and the knowledge of process of formation, to be used to identify the anthropogenic soil order at the beginning of the key to soil orders. These new types of evidence could be used to distinguish different types of anthropogenic soils, but would be restricted from use in other parts of the keys. As precedent for restricting the use of these criteria from other parts of the key, separate control sections, epipedons, diagnostic horizons, and diagnostic materials already exist for organic soils in Soil Taxonomy (Soil Survey Staff, 1999), and most of them cannot be used in mineral soil orders. Additionally, the family level is full of examples where classes and categories are applied only to a restricted set of mineral soils. Alternative particle-size classes, mineralogy classes, and reaction classes are used and defined based on the higher classification of the soil in question.

Anthropogenic soils occur on anthropogenic landforms, such as rice paddiess, old raised fields, landfills, mine spoils, excavation areas, fill areas, disturbed construction sites, and others. Experienced soil classifiers are also experienced in landform identification, which is an integral component of soil survey. The identification of an anthropogenic landform could be used as criteria or partial criteria for identifying an anthropogenic soil and placing it in a class apart from soils that formed in natural, undisturbed landscapes. This approach would work well for those soils that have experienced significant physical alteration, but it does not address anthropogenic processes that result only in chemical alteration. Decisions will have to be made regarding the kind and degree of anthropogenic alteration to be recognized in Soil Taxonomy at various levels.

Relational observations in combination with data could be used to address issues such as anthropogenic alteration of soil chemical properties. The use of relational data is not unprecedented in Soil Taxonomy, although it may only be used with respect to the identification of anthropogenic processes as originally included in the system. Anthropic epipedons found in soils of kitchen middens are described as having high base saturation, compared to adjacent soils. Data alone do not address the source of an observed property, but if used in conjunction with relational data, such as comparisons between limed soils and soils that have not been limed, criteria could be developed for identifying the effects of these kinds of anthropogenic processes. Similarly, eroded Mollisols could be identified in comparison to uneroded counterparts to assess the degree of accelerated erosion. However, a proposal to use relational data for this purpose was recently considered and rejected by the Soil Taxonomy Committee. The review expressed the concern that consistency among soil scientists in interpreting the definition of taxa would be threatened if the classification of soils based on properties that may have been present were to be allowed. This decision may restrict the use of relational data as criteria for classes of naturally formed soils, but it may not be viewed negatively if restricted for use within classes of anthropogenic soils.

Although strictly avoided for classifying soils formed under natural soil-forming conditions, the use of knowledge of process of formation could be allowed for use in classifying anthropogenic soils. Classifying soils developed in mine spoil as Spolanths could be simply based on knowledge that the soil in question is formed in material that originated as a result of the mining process (Sencindiver and Ammons, 2000). Diagnostic horizons may not be required to identify a soil to the order or suborder level, but they may be very helpful in defining classes in the lower categories. Meaningful separations within suborders (Spolanths, for example) should be based more on observable morphological characteristics or chemical properties that affect the use and management of these soils, or that imply the direction of genesis of these soils in response to the active factors of soil formation under ambient conditions. If this approach is used, an Anthrosol order should be placed at the beginning of the key to allow use of these restricted criteria for first identifying a soil as a member of this order. This approach is used in the Australian Soil Taxonomy, in which knowledge of process is allowed in classifying anthropogenic soils, but subsequent orders are defined using morphology-based criteria. Their experience should be evaluated prior to adopting this approach. It seems probable that the most common debatable classification would arise from the inclusion of a soil that some might argue has not been sufficiently modified by anthropogenic processes to warrant inclusion in Anthrosols. Restricting modified rules for defining criteria to a single Anthrosols order is not a perfect solution. Difficulties in defining extragrades within other orders would continue to be a problem. For example, most taxonomists would probably prefer that a moderately eroded Mollisol be classified as an anthropogenic extragrade of Mollisols, rather than an Anthrosol. Nevertheless, a separate soil order of Anthrosols that allows special rules for identifying anthropogenic soils and the broad spectrum of anthropogenic properties seems to be the best option for incorporating anthropogenic processes in Soil Taxonomy, with minimal disturbance to the rest of the system.

CONCLUSIONS

To those taxonomists who are purists with respect to the original fundamental principles of Soil Taxonomy, the proposed approaches to incorporating anthropogenic processes in soil classification systems such as Soil Taxonomy that utilize morphology-based criteria are sure to be unacceptable. It could also be argued that adopting different rules of classification within a single soil order in Soil Taxonomy is little more than publishing two separate classification systems within the same book. However, it is our opinion that incorporating all anthropogenic soil processes in Soil Taxonomy under the present set of principles of classification is not possible, because some anthropogenic soil processes do not leave morphological evidence. The evidence of anthropogenic activity is often expressed in the lack of horizonation, altered chemistry, or difference in landform relative to surrounding parent soils.

Soil Taxonomy was designed in part for the purpose of classifying, in an objective manner, soils that formed by natural processes of soil formation. It provides a framework for assessing our knowledge of soil formation. By exempting all or part of the system from the principles that provide for the objective classification of soils without the use of knowledge of process of their formation, the system may be weakened with respect to this purpose of classification. However, the injury may be minimized if the new rules are restricted for use within a single soil order that falls out first in the key to soil orders. Taxonomists should decide just how important this aspect of the system is for future applications. In our opinion, weakening this aspect of Soil Taxonomy in one soil order is a reasonable alternative to restructuring classes throughout Soil Taxonomy, and to being unable to incorporate anthropogenic soils into our classification (and mapping) system at all.

REFERENCES

Ahrens, R.J. and Engel, R.J. 1999. Soil Taxonomy and anthropogenic soils, in J.M. Kimble, R.J. Ahrens, and R.B. Bryant, Eds. Classification, Correlation, and Management of Anthropogenic Soils. Proceedings—Nevada and California, Sept. 21-Oct. 2, 1998. USDA-NRCS, Nat. Soil Survey Center, Lincoln, NE.

Burghardt, W. 1994. Soils in urban and industrial environments. *Z. Pflanzenern Bodenkunde* 157:205–214.

Chinese Soil Taxonomy Research Group. 1995. Chinese Soil Taxonomy (Revised Proposal). Chinese Agri Sci. Tech. Press, Beijing, China (in Chinese).

Craul, P.J. 1999. *Urban Soils: Applications and Practices.* John Wiley & Sons, New York.

EMBRAPA. 1999. Brazilian Soil Classification System. EMBRAPA, National Center for Soil Research, Rio de Janeiro, Brazil.

Fanning, D.S. and Fanning, M.C. 1989. Soil Morphology, Genesis, and Classification. John Wiley & Sons, New York. Food and Agriculture Organization of the United Nations. 1996. Digital Soil Map of the World and Derived Soil Properties. Land and Water Development Division. UNESCO. Computer optical disc (CD-ROM) Version 3.5.

Galbraith, J.M. and Bryant, R.B. 1999. ICOMANTH Circular Letter Number 2, in J.M. Kimble, R.J. Ahrens, and R.B. Bryant, Eds. Classification, Correlation, and Management of Anthropogenic Soils. Proceedings—Nevada and California, Sept. 21-Oct. 2, 1998. USDA-NRCS, Nat. Soil Surv. Center, Lincoln, NE.

Galbraith, J.M., Russell-Anelli, J.M., and Bryant, R.B. 1999. Major kinds of humanly altered soils, in J.M. Kimble, R.J. Ahrens, and R.B. Bryant, Eds. Classification, Correlation, and Management of Anthropogenic Soils. Proceedings—Nevada and California, Sept. 21-Oct. 2, 1998. USDA-NRCS, Nat. Soil Surv. Center, Lincoln, NE.

Gerasimova, M. 1999. Heavy metals in Soddy-podzolic soils and Chernozems of European Russia (an overview of the literature), in J.M. Kimble, R.J. Ahrens, and R.B. Bryant, Eds. Classification, Correlation, and Management of Anthropogenic Soils. Proceedings—Nevada and California, Sept. 21-Oct. 2, 1998. USDA-NRCS, Nat. Soil Surv. Center, Lincoln, NE.

Grubler, A. 1994. Technology, in W.B. Meyer and B.L. Turner II, Eds. *Changes in Land Use and Land Cover: A Global Perspective.* Cambridge University Press, Cambridge, UK.

Hernandez, L.A. 1999. New York City Soil Survey Program, in J.M. Kimble, R.J. Ahrens, and R.B. Bryant, Eds. Classification, Correlation, and Management of Anthropogenic Soils. Proceedings—Nevada and California, Sept. 21-Oct. 2, 1998. USDA-NRCS, Nat. Soil Surv. Center, Lincoln, NE. 226 pp.

Hernandez, L.A. and Galbraith, J.M. 1997. Soil Survey of South Latourette Park, Staten Island, New York City, NY. Dept. of Soil, Crop, and Atm. Sci., Cornell University, Ithaca, NY.

International Society of Soil Science Working Group RB. 1998. *World Reference Base for Soil Resources: Introduction,* J.A. Deckers, F.O. Nachtergaele and O.C. Spaargaren, Eds. First Edition. International Society of Soil Science (ISSS), International Soil Reference and Information Centre (ISRIC) and Food and Agriculture Organization of the United Nations (FAO). Wageningen/Rome.

Kimble, J.M., Ahrens, R.J. and Bryant, R.B. Eds. 1999. Classification, Correlation, and Management of Anthropogenic Soils, Proceedings—Nevada and California, Sept. 21-Oct. 2, 1998. USDA-NRCS, Nat. Soil Surv. Center, Lincoln, NE. 226 pp.

Levin, M.J. 1999. Soil mapping in an urban environment—the city of Baltimore, Maryland, in Kimble, J.M., R.J. Ahrens, and R.B. Bryant, Eds. Classification, Correlation, and Management of Anthropogenic Soils, Proceedings—Nevada and California, Sept. 21-Oct. 2, 1998. USDA-NRCS, Nat. Soil Surv. Center, Lincoln, NE.

National Institute of Urban Affairs. 1994. Urban Environmental Maps: Delhi, Bombay, Vadodara, Ahmedabad. NIUA, New Delhi, India.

Pouyat, R.V. and Effland, W.R. 1999. The investigation and classification of humanly modified soils in the Baltimore Ecosystem Study, in J.M. Kimble, R.J. Ahrens, and R.B. Bryant, Eds. Classification, Correlation, and Management of Anthropogenic Soils. Proceedings—Nevada and California, Sept. 21-Oct. 2, 1998. USDA-NRCS, Nat. Soil Surv. Center, Lincoln, NE.

Russell-Anelli, J.M., Bryant, R.B., and Galbraith, J.M. 1999. Evaluating the predictive properties of soil survey—soil characteristics, land practices, and concentration of elements, in J. M. Kimble, R.J. Ahrens, and R.B. Bryant, Eds. Classification, Correlation, and Management of Anthropogenic Soils. Proceedings. USDA-NRCS-NSSC, Lincoln, NE, pp. 155-168.

Scheyer, J.M. 1999. Overview of the USDA-NRCS Urban Soils Program, in J.M. Kimble, R.J. Ahrens, and R.B. Bryant, Eds. Classification, Correlation, and Management of Anthropogenic Soils. Proceedings—Nevada and California, Sept. 21-Oct. 2, 1998. USDA-NRCS, Nat. Soil Surv. Center, Lincoln, NE.

Sencindiver, J.C. and Ammons, J.T. 2000. Minesoil Genesis and Classification, in R.I. Barnhisel, R.G. Darmody, and W.L. Daniels, Eds. Reclamation of Drastically Disturbed Lands. Ch. 23, Agronomy No. 41. Am. Soc. Agronomy, Madison, WI.

Smith, H. 1999. Soil Survey of the District of Columbia. 1976, in J.M. Kimble, R.J. Ahrens, and R.B. Bryant, Eds. Classification, Correlation, and Management of Anthropogenic Soils. Proceedings—Nevada and California, Sept. 21-Oct. 2, 1998. USDA-NRCS, Nat. Soil Surv. Center, Lincoln, NE.

Smith, G.D. 1986. The Guy Smith Interviews: Rationale for Concepts in Soil Taxonomy, in T.R. Forbes, Ed. SMSS Tech. Monogr. No. 11. USDA-NRCS. U.S. Government Printing Office, Washington, DC.

Sobocká, J. 1999. Anthropogenic soils and problems of their classification in Slovokia, in J.M. Kimble, R.J. Ahrens, and R.B. Bryant, Eds. Classification, Correlation, and Management of Anthropogenic Soils. Proceedings—Nevada and California, Sept. 21-Oct. 2, 1998. USDA-NRCS, Nat. Soil Surv. Center, Lincoln, NE.

Soil Survey Staff. 1999. *Soil Taxonomy:* A Basic System of Soil Classification for Making and Interpreting *Soil Surveys,* 2nd ed. U.S. Dept. Agric. Handbook 436. U.S. Government Printing Office, Washington, DC.

Southard, S. 1999. Work Group 3—mapping altered landscapes, in J.M. Kimble, R.J. Ahrens, and R.B. Bryant, Eds. Classification, Correlation, and Management of Anthropogenic Soils. Proceedings—Nevada and California, Sept. 21-Oct. 2, 1998. USDA-NRCS, Nat. Soil Surv. Center, Lincoln, NE.

Stroganova, M.N. 1999. Urban soils—their concept, classification, and origin, in J.M. Kimble, R.J. Ahrens, and R.B. Bryant, Eds. Classification, Correlation, and Management of Anthropogenic Soils. Proceedings—Nevada and California, Sept. 21-Oct. 2, 1998. USDA-NRCS, Nat. Soil Surv. Center, Lincoln, NE.

Tonkonogov, V. and Lebedeva, I. 1999. A system for categorizing technogenic surface formations (humanly modified soils), in J.M. Kimble, R.J. Ahrens, and R.B. Bryant, Eds. Classification, Correlation, and Management of Anthropogenic Soils. Proceedings—Nevada and California, Sept. 21-Oct. 2, 1998. USDA-NRCS, Nat. Soil Surv. Center, Lincoln, NE.

United States Central Intelligence Agency. 2000. National basic intelligence factbook. Supt. of Documents, U.S. Government Printing Office, Washington, DC. Also: The World Factbook [computer file]. Online version latest edition: http://www.odci.gov/cia/publications/factbook/index.html

Zitong, G., Ganlin, Z., and Guobao, L. 1999. The Anthrosols in Chinese Soil Taxonomy, in J.M. Kimble, R.J. Ahrens, and R.B. Bryant, Eds. Classification, Correlation, and Management of Anthropogenic Soils. Proceedings—Nevada and California, Sept. 21-Oct. 2, 1998. USDA-NRCS, Nat. Soil Surv. Center, Lincoln, NE.

CHAPTER 8

Developments in Soil Chemistry and Soil Classification

Goro Uehara

CONTENTS

ABSTRACT

The best predictors of soil chemical behavior are mineralogy and specific surface. Mineralogy yields information on the physicochemical nature of surfaces, and specific surface provides information on the extent of the surface. Recent advances in quantitative mineralogy enable soil scientists to apply knowledge of mineral properties to predict soil behavior and performance. If, in addition to this, quantitative mineralogy is supplemented with total chemical analysis, the elements can be allocated to minerals and the amorphous fraction. Accessory properties that can be predicted from knowledge of mineralogy, specific surface, and elemental composition include surface charge characteristics, ion exchange capacity, noncoulombic adsorption-desorption reactions, aggregation, aggregate stability, pore size distribution, and water, energy, and gas transport coefficients. A classification system that includes mineralogy, including the amount and composition of the amorphous fraction, and specific surface as differentiating criteria should enable users to make better predictions of soil behavior and performance.

The goal of advancing soil chemistry and soil classification is to enable individuals who are responsible for the care and management of land to make better long-term predictions of how soils will behave and perform when used for specified purposes. Two ways to achieve this goal are to develop better prediction models and to improve the characterization of soil constituents. The latter is the more critical, because a model can only be as good as the knowledge we have about a soil's characteristics.

While soil surveys and classification enable users to find useful information about how a soil will behave and perform, users are increasingly asked to make more precise predictions about the

0-8493-1339-2/03/$0.00+$1.50

long-term consequences of land use practices. These include, for example, predictions of agronomic crop performance, the fate and whereabouts of agricultural and industrial chemicals in the vadose zone, and rheologic behavior of materials for engineering interpretation and use of soils. The irony is that soil characterization data are primarily used for soil survey and classification, and are rarely used as inputs for prediction models. Ideally, soil characterization data should be the same as input data for dynamic, process-based simulation models. If this were the case, it would result in greater demand and use of soil surveys and soil characterization data. But soil science is far from being ready for use as identical parameters for classifying soils and for use as input in prediction models.

Soil texture and clay content are still the most commonly used predictors of soil behavior, but too often fail when used alone for this purpose. We know, for example, that a very fine Vertisol and very fine Oxisol will behave and perform differently. For this reason, the family category of Soil Taxonomy (Soil Survey Staff, 1999) specifies particle size as well as mineralogy for grouping soils into performance classes. But there are instances in which samples with nearly identical texture and mineralogy produce unexpectedly different results in the field and when analyzed in the laboratory for accessory properties, such as absorption isotherms, potentiometric titration curves, and zero points of charge.

The failure of particle size and mineralogy to predict soil behavior stems from the fact that both are surrogates of two other more fundamental soil properties, namely, specific surface and surface charge density. Although particle size substitutes for specific surface, we expect a very fine kaolinitic sample to have a lower surface area than a very fine montmorillonitic sample. Mineralogy also provides additional information about the sample's surface charge characteristics. This implies that if specific surface and quantitative mineralogy were used as differentiating criteria in place of particle size and qualitative mineralogy, more consistent results would follow. I believe there is still some unfinished business in soil science that needs to be resolved before better prediction will be possible. This unfinished business is the quantitative analysis and characterization of noncrystalline or amorphous materials in soils.

QUANTIFYING AMORPHOUS MATERIALS

Amorphous materials are generally thought to be confined to Andisols and related soils formed from volcanic rock. Well-known minerals with short-range-order, such as allophane and imogolite, are often grouped with, and treated as, amorphous materials. Amorphous materials have properties normally associated with high specific surface, but do not lend themselves to particle size analysis or mineralogical analysis. Allophane and imogolite exceed smectites in specific surface, and posses high water retention capacity. But unlike smectitic soils, allophanic soils are endowed with very high saturated hydraulic conductives. The surface charge on amorphous materials, including allophane and imogolite, are pH dependent, and the net surface charge can be net negative, net zero, or net positive, depending on pH and the silica-sequioxide ratio of the material. Characterization data for a range of materials of this kind can be found in SCS, USDA (1976).

Amorphous materials would not be a cause for concern if they were confined to Andisols. Tenma (1965) applied a method proposed by Hashimoto and Jackson (1965) to measure amorphous silica and alumina content of Andisols, Oxisols, and Ultisols of Hawaii. Hashimoto's and Jackson's (1958) method involves boiling samples in 0.5 NaOH for 2.5 min, and was designed to dissolve amorphous silica and alumina without affecting crystalline minerals. Tenma (1965) performed the dissolution analysis on the clay fraction separated from deferrated samples previously treated with H_2O_2 to remove organic carbon. His results confirmed the high amorphous silica and alumina levels in Andisols, but showed unexpectedly high levels in the Oxisols and Ultisols. The amorphous silica and alumina content in the clay fraction ranged from 3.9% to 11.6% and 3.2% to 12.7%, respectively, in the Oxisols and 1.3% to 7.4% and zero to 6.7% in the Ultisols. The results of Tenma's (1965) Master's degree thesis were not published because of the uncertainty of the dissolution procedure to discriminate between amorphous materials and crystalline minerals. But signs of large quantities of amorphous materials in non-Andisols would not go away. Electron microscope examination of clays consistently showed what appeared to be coatings on all particles (Jones and Uehara, 1973). The coatings were in fact ubiquitous and the problem was not in finding them, but in locating uncoated mineral surfaces. Uehara and Jones (1974) published electron micrographs of finely ground quartz, albite, and obsidian, all showing coatings as thick as 400 Å, and speculated on the role such coatings might play in crust cementation. In the same article they showed coatings on gibbsite and goethite particles from a highly weathered Oxisol, and guessed that the coating was mainly alumina. The electron micrographs seem to show the coatings occupying a large fraction of the soil volume. The reviewers of the paper wondered if the coatings, which we assumed were amorphous, were artifacts produced during sample preparation.

In Soil Taxonomy (Soil Survey Staff, 1999), acid oxalate extractable Al and Fe are used as criteria for andic soil properties. This method is used to selectively extract organically complexed Fe and Al, noncrystalline hydrous oxides of Fe and Al, allophane, and amorphous aluminosilicate (Wada, 1989). What is needed, however, is not simply a criterion for andic properties, but a method to quantify and characterize noncrystalline components of soils.

The most recent work by Jones et al. (2000) overcomes many of the uncertainties connected with earlier effort to study amorphous soil materials. The method depends on spiking a sample with a known amount of crystalline mineral, and comparing the measured with the added amount. If the sample is free of amorphous materials, the measured and the added amount of spike would be the same. If not, the measured amount would exceed the added value, because the x-rays upon which this method depends cannot detect the amorphous fraction, and in the normalization process, the "missing" fraction is allocated to the spike.

What is remarkable about the Jones et al. (2000) study is the high amorphous mass fractions measured in the clay fraction of non-Andisols. In eight samples collected from 0–15 cm depth of five Oxisols, two Mollisols, and one Ultisol, they measured mass fraction of amorphous materials between 29% and 40%. These amorphous contents would be even higher if they were expressed in terms of volume fraction, owing to higher water content and lower density of amorphous materials. The average loss on ignition of the crystalline phases for ten samples, including two Andisols, was 30%, whereas the water and other volatiles associated with the amorphous fraction was 70% by weight, indicating that the latter has a low specific gravity.

The above raises new questions about amorphous materials in soils. Are the samples studied by Jones et al. unique because of their volcanic origin? Would the clay fraction of soils developed from nonvolcanic parent rock also contain significant amounts of amorphous materials?

We know from experience that quartz, when crushed to near clay-size particles, produces large quantities of amorphous silica (Uehara and Jones, 1994), and that amorphous silica can be seen by electron microscopy to be associated with montmorillonite (Uehara and Jones, 1994). But do soils formed from glass-free parent material produce amorphous coatings? For answers to this question, samples were obtained from the International Institute of Tropical Agriculture field station in Ibadan, Nigeria. The samples were taken from the plow layer of the upper (Oxic Paleustalf) and middle (Typic Plinthusalf) section of a toposequence in alley cropping plots, established by Dr. B.T. Kang many years earlier. Clay fractions from the two soil samples were analyzed by R.C. Jones using the method described by Jones et al. (2000). The amorphous mass fraction for the Alfisols was 19.1% for the upper and 15.7% for the middle member of the toposequence. These numbers are sufficiently high to make me believe that amorphous materials are more common in soil clays than is currently believed.

IMPLICATIONS

If amorphous materials occur as coating on mineral surfaces, the chemical properties of the soil should be dominated by the chemistry of coating. In electron micrographs (see Figure 8.1),

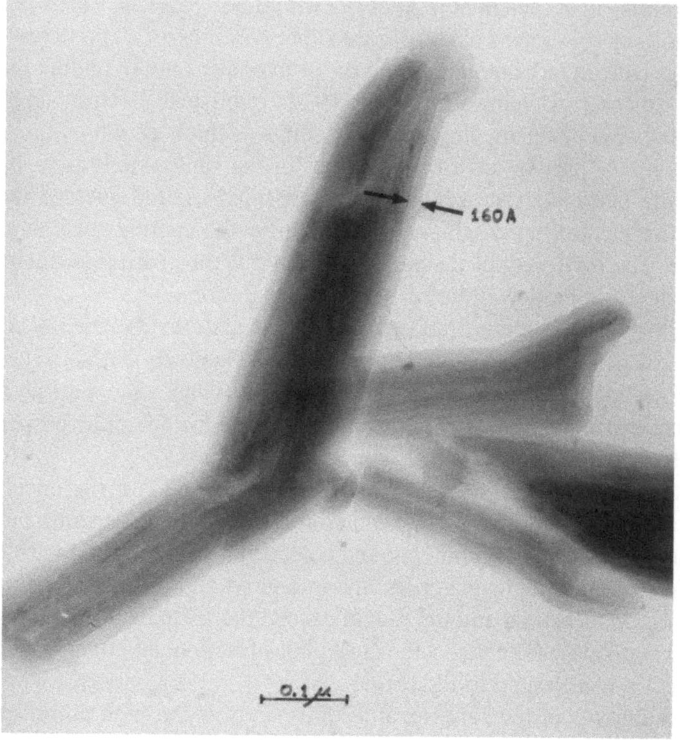

Figure 8.1 Halloysite particle from an Oxisol coated with a substance that appears to be amorphous. (EM by R.C. Jones.)

the amorphous fraction appears as a coat of paint on particles. The coatings on adjoining particles seem to coalesce when they come in contact, and appear to serve as a binding or aggregating agent. The picture changes when the volume fraction of amorphous materials exceeds 25%.

In Andisols, particles are few and far between. There is no meaning to particle size distribution or even to specific surface. Amorphous materials behave more like liquids than solids when moist, but turn into elastic solids when dried. Uehara and Jones's (1974) speculation about the role of amorphous silica on crust formation in arid regions was based on this model. Crusts that form in irrigation furrows lose their strength when rewetted, and Uehara and Jones (1974) speculated that high silica amorphous materials tended to soften and harden with wetting and drying.

On the other extreme, we have low silica, high iron, and aluminum materials, such as in plithite, that dry irreversibly. If we return to the earlier example of the very fine Oxisol and Vertisol, we can visualize not only particle size and mineralogical differences, but differences in soil structure and the resistance of structural units to the slaking action of water. What causes the self-mulching behavior of Vertisols, and why do soil aggregates of Oxisols so often resist slaking in water? It is easy to explain these characteristics in terms of the swelling or nonswelling properties of clay materials, but what cements the particles in a dry, self-mulching aggregate of a Vertisol, or water-resistant aggregate of an Oxisol? Are they the clay minerals themselves that act as cementing agents, or does a ubiquitous amorphous fraction that varies in composition and reaction with water play some role in soil cementation? Many more questions come to mind when one views surface coating on soil particles, as shown in Figures 8.2 to 8.5.

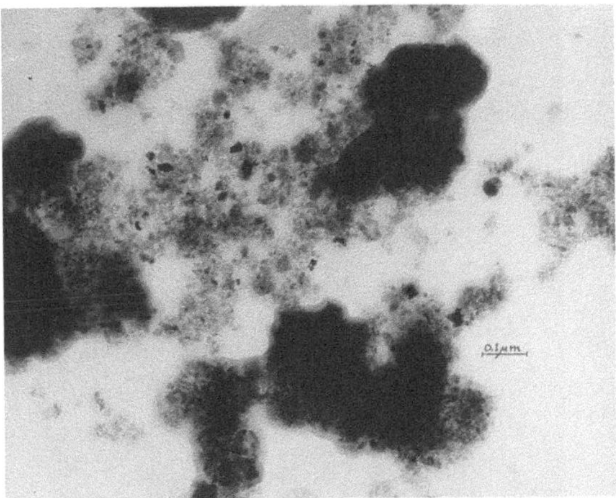

Figure 8.2 Microaggregate from an Oxisol coated with a substance that appears to be amorphous. (EM by R.C. Jones.)

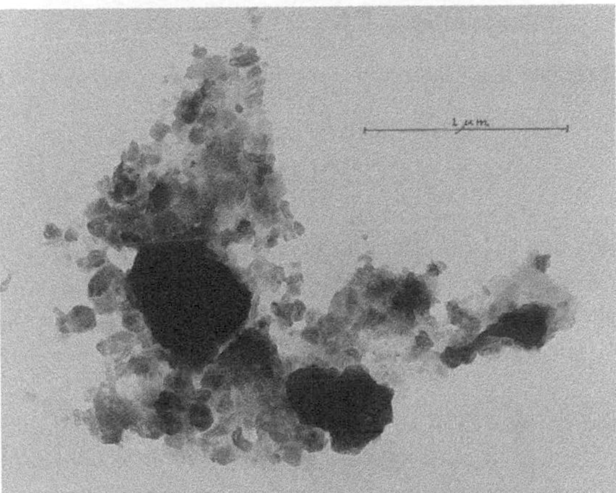

Figure 8.3 Halloysite particles are still recognizable in the amorphous matrix from a Haplustand. (EM by R.C. Jones.)

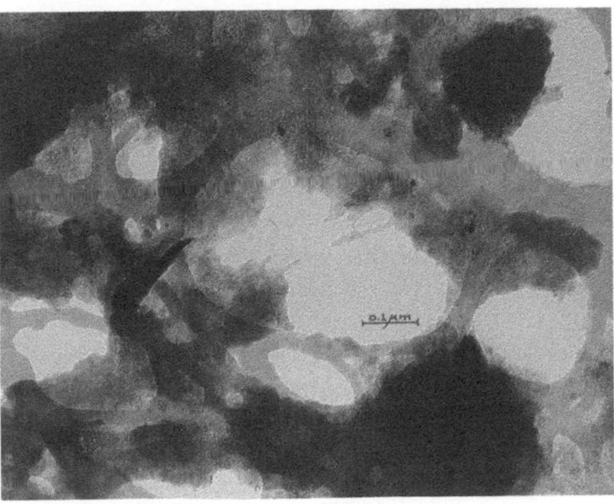

Figure 8.4 The solid matrix from a Hydrudand is virtually all amorphous. (EM by R.C. Jones.)

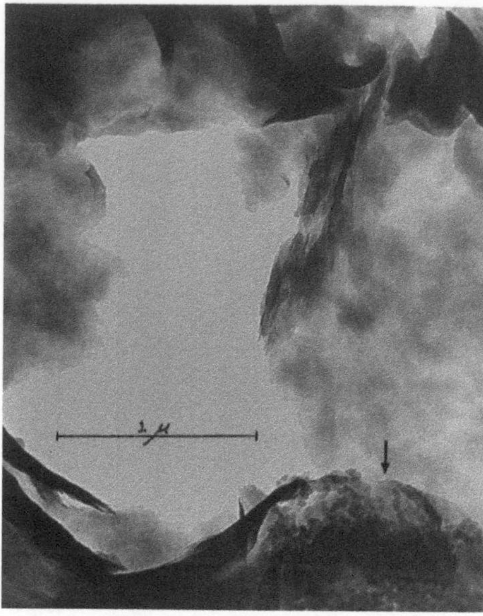

Figure 8.5 Cheto montmorillonite associated with amorphous silica (arrow). (EM by R.C. Jones.)

CONCLUSION

The chemistry and physics of soils depend so much on the nature and extent of surfaces that renewed research in this area may be worthwhile. Jones et al.'s (2000) method for quantifying and determining the chemical composition of amorphous materials provides that basis for new research to answer the following questions:

1. How extensive is the occurrence of amorphous materials in non-Andisols?
2. Will, for example, 5% to 10% by weight in the clay fraction be common for most soils?
3. Will knowledge of the amount and chemical composition of amorphous materials in clay fractions add significantly to explaining variances in soil behavior and performance?
4. How important are amorphous coatings on the behavior of coarse-textured soils?
5. Will the silica-sesquioxide ratio of the amorphous fraction influence its surface charge characteristics?
6. Does the amount and composition of the amorphous fraction influence the carbon sequestration potential of soils?

REFERENCES

Hashimoto, I. and Jackson, M.L. 1958. *Rapid Dissolution of Allophane and Kaolinite-halloysite after Dehydration.* Reprint from 7th National Conference on Clays and Clay Minerals. Pergomon Press, New York, 102–113.

Jones, R.C., Babcock, C.J., and Knowlton, W.B. 2000. Estimation of the total amorphous content of Hawaii soils by the Rietveld Method. *Soil Sci. Soc. Am. J.* 64:1100–1108.

Jones, R.C. and Uehara, G. 1973. Amorphous coatings on mineral surfaces. *Soil Sci. Soc. Am. Proc.* 37:792–798.

Soil Survey Staff. 1999. Soil taxonomy: A basic system of soil classification for making and interpreting soil surveys. 2nd Ed. USDA Agric. Handbook 436. U.S. Government Printing Office, Washington, DC.

SCS, USDA. 1976. Soil Survey Laboratory data and description for some soils of Hawaii. Soil survey investigation report No. 29. Soil Conservation Service, USDA in Cooperation with Hawaii Agric. Expt. Sta. and Hawaii Sugar Planters' Association. U.S. Government Printing Office, Washington, DC.

Tenma, H.H. 1965. Some characteristics of non-crystalline constituents in Hawaiian soils. M.S. Thesis. Department of Agronomy and Soil Science, University of Hawaii, Honolulu, HI.

Uehara, G., and R.C. Jones. 1974. Particle surfaces and cementing agents, in J.W. Cary and D.D. Evans, Ed. Soil Crusts. Chap. 3, Part 1, Agricultural Experiment Station, University of Arizona, Tucson, Arizona, 17-28.

Part 2

Developments in Classification Systems

CHAPTER 9

Demands on Soil Classification in Australia

R.W. Fitzpatrick, B. Powell, N.J. McKenzie, D.J. Maschmedt, N. Schoknecht, and D.W. Jacquier

CONTENTS

INTRODUCTION

Soil classification in Australia has had a checkered history with one of several contrasting systems acting as the national scheme at various times. Sometimes the use of these systems has been concurrent (see " the Early Systems of Classification) and also used alongside Soil Taxonomy

0-8493-1339-2/03/$0.00+$1.50

or FAO. There have been periods of aggressive debate over the role and application of general-purpose national schemes.

A persistent problem has been the limited appreciation of how the utility of a general-purpose soil classification is constrained by contrasting patterns of covariance in soil populations at local, regional, and national levels. For practical reasons, most Australian classification systems have depended heavily on morphological criteria. However, covariance between soil morphological properties and more relevant chemical, physical, and mineralogical properties is complex and sometimes poor (e.g., MacArthur et al., 1966; Webster and Butler, 1976; Butler and Hubble, 1977; McKenzie and MacLeod, 1989; Fitzpatrick, 1996; Fitzpatrick et al., 1999).

The substantial literature on spatial variability (e.g., Beckett and Webster, 1971; Wilding and Drees, 1983; Burrough, 1993) demonstrates that soil properties have varying levels of covariance. As a consequence, we should have modest expectations of the capacity of general-purpose classification systems to discriminate effectively against a broad range of criteria (e.g., soil fertility, hydraulic properties, engineering, genesis) across a wide range of soil conditions.

Several critics contend that general-purpose systems have inherent, and sometimes severe, limitations (e.g., Gibbons, 1961; Webster, 1968, 1977; FitzPatrick, 1971; Butler, 1980). A typical opinion is shown in the following statement by White (1996): "The effort put in to developing general-purpose soil classification systems in Australia seems to have been largely unrewarded in terms of the use made of this information by land managers." Basher (1997) in a paper titled "Is pedology dead and buried?" and others (e.g., Dudal, 1987; Zinck, 1993; White, 1993; Yaalon, 1996) suggest that the decline in the use of general-purpose soil classification systems can be ascribed to a range of factors, including the following:

- The specialized terminology used to name and classify soils in soil map legends and reports, and the range of systems in use for classifying soils
- The need for adjustments in soil survey techniques and soil classification to meet the requirements of potential users
- Insufficient attention given to presenting information in an accessible, purpose-orientated, user-friendly language and format
- Inadequate use of soil class criteria that are important to land use (i.e., physical properties such as porosity, infiltration rate, and permeability properties of the surface layer of soil) and an over-emphasis on taxonomic class criteria

In this chapter, our first objective will be to show that general-purpose classifications are being modified to remedy the limitations. The enduring value of good general-purpose classification systems will be as a tool to communicate soil information nationally. Australians have recognized this more clearly in recent times. Agencies undertaking soil and land resource surveys now provide predictions and interpretations of how a range of land qualities (e.g., salinity) vary across the landscape, rather than provide a single map of soil types. Various special-purpose classification systems have been developed for utilitarian ends, and some examples are considered in later sections. Descriptions of soils using national taxonomic schemes are used to aid communication, and this is increasingly being supported by good visual material to highlight differences between soils.

Some critics have also dismissed general-purpose soil classification systems as having limited value for soil fertility work (e.g., White, 1996). There are few who would expect general-purpose systems to provide good predictions of nutrient availability because of the covariance issues already mentioned and the strong dependence of these attributes on management history. Furthermore, most classification systems use less dynamic variables as differentiae to ensure stable allocations of profiles to classes over time. However, classification systems do have considerable value for stratifying behavior of soil groups in terms of nutrient dynamics. Such recognition (e.g., Fitzpatrick et al., 1999; NLWRA, 2001) has led to more constructive dialogue between the soil testing and pedological communities in Australia.

The general trend of soil and land resource survey programs away from pedological objectives toward practical land evaluation has also coincided with an improvement in the state of soil classification in Australia. This culminated in the publication of the Australian Soil Classification (Isbell, 1996), which officially replaced the Handbook of Australian Soils (Stace et al., 1968) and The Factual Key for the Recognition of Australian Soils (Northcote, 1979).

Needs of End-Users

Our focus here is on the provision of practical soil class information for a broad range of land management issues required by land holders, private enterprises, researchers, and government agencies.

There has been some excellent research leading to the development of numerical methods for soil classification (e.g., McBratney, 1994), but these are beyond the scope of the present paper. Our emphasis here is guided by four questions that are most frequently asked by users of land resource information:

- What soil properties are changing, vertically and laterally, in the landscape and with time?
- What are the most suitable approaches to characterize, monitor, predict, and manage the changes in soils?
- What soil measurements and user-friendly soil classifications are required to make suitable predictions about soil and landscape conditions and about sustainable land use?
- To what extent do soil processes and the management of soils, influence engineering infrastructure and water quality?

These four questions can be answered by using combinations of pedological data along with soil physical, chemical, hydrological, and mineralogical data relevant to a particular use. The combined information assists the understanding of how soils vary in landscapes, so that strategies can be developed for managing both spatial and temporal changes within them. This may often involve the development of a user-friendly specific purpose or technical soil classification systems.

Recently, pedologists in Australia have used existing information from soil maps in combination with data such as soil physics and mineralogy, to work in nonagricultural contexts. This chapter presents several case studies illustrating how applied soil science information has been classified and used to solve a wide range of practical problems for end-users.

APPRAISAL OF SOIL CLASSIFICATION SYSTEMS IN AUSTRALIA

The Early Systems of Soil Classification

The sophistication and effectiveness of classification can indicate the level of scientific maturity and understanding of a particular area of study, i.e., the level of knowledge and understanding of the entity under consideration. A major aim of classification is to usefully summarize the natural variability of forms the entity takes, and enhance communication about that entity. The first soil classification of Australian soils was devised by Prescott (1931, 1947) and incorporated concepts of the Russian system of soil classification, an approach that was followed broadly by others (Stephens, 1952; 1956; 1962; Stace et al., 1968). However, the Handbook of Australian Soils (Stace et al., 1968) also had many features in common with the American Great Group system with its 40 Great Groups, which were familiar to soil scientists throughout Australia (Moore et al., 1983). Leeper (1943; 1956) criticized these traditional classification schemes and emphasized "profile criteria such as marked texture contrast features" (Isbell, 1992). Although Leeper's classification was never used, many of its features were used by Northcote in his Factual Key to develop an objective system based solely on observed soil profile features. This classification became the basis for mapping of the continent through the Atlas of Australian Soils (Northcote, et al., 1960 to 1968).

Soil classification may spur or deter scientists with an interest in soils. If a classification system proves to be relevant and user-friendly, it stimulates and encourages further work because it is recognized for its inherent capacity to create order and enhance the useful understanding of soils. For example, The Factual Key for the Recognition of Australian Soils by Northcote (1979) provided a generation of Australian scientists with valuable conceptual understanding of soils in terms of textural-differentiation of profiles, the relative development of A2 (E) horizons, soil reaction trends, subsoil color and mottle differences. Many of the concepts in his classification also provided effective pedotransfer functionality (Bouma, 1989), particularly in terms of soil-water attributes (e.g., McKenzie et al., 2000).

If a classification is not useful, it hinders research. The Handbook of Australian Soils, by Stace et al. (1968), described 43 Great Soil Groups as central concepts, supported by representative profiles. Because of the lack of distinct separation between classes, many soils were inconsistently classified and distinguished, leading to conceptual confusion and pointless argument of subtle differences. While useful in many ways in a pedological context, it was significantly biased toward the population of agricultural soils, the subject of study for most of Australia's soil scientists. Many soils found in forests, rangelands, and nonagricultural contexts could not be allocated, as they did not match the central concepts.

The Australian Soil Classification System

The most recent national classification, the Australian Soil Classification (Isbell, 1996), is broad in its application, and its hierarchical structure generally allows for unambiguous allocation of unknowns to particular classes. It draws on concepts of the two previous systems mentioned and borrows from overseas classification schemes as well. A simplified outline is provided in Figure 9.1.

There are 14 Soil Orders at the highest level that reflect important features of the soil continuum in Australia. For example, as an arid and strongly weathered continent with an absence of glaciation, it has soil orders that reflect this. For instance, the Sodosol order is defined by a high subsoil exchangeable sodium percentage (ESP). It includes diverse soils that have been affected by salt during formation, often the consequence of past semi-arid or arid conditions. Other examples are the Kandosol and Ferrosol orders that usually have a history of significant weathering and leaching. The Australian Soil Classification not only deals with agricultural soils, but also all soils of the rangelands, the tidal zone, the arid zone, and human-made soils, which were less understood. Its strengths are derived from the large national and state databases used to develop the classification, its general-purpose orientation, and a set of clearly stated principles.

Many management-related properties of the scheme are revealed at the lowest level—the Family level. The Family level classifies soils based on soil depth, thickness, texture, and gravel content of A horizon and the maximum texture of the B horizon. For water-related devised functions (Bouma, 1989), the family criteria could well be the most technically useful part of the classification.

Application and Future Improvements

In the six years since its release, the Australian Soil Classification has been found to be very applicable to agricultural soils and adequate for many engineering purposes. However, it has attracted criticism regarding wet soils in general (Hydrosols soil order), tidal soils in particular, saline soils, arid zone soils, sands, and soils rich in ferruginous gravel. Its formulation has been an advance in that it focused attention on the limits of existing datasets and the inadequate knowledge of many soils, thus providing challenges for soil scientists and the direction of further useful research.

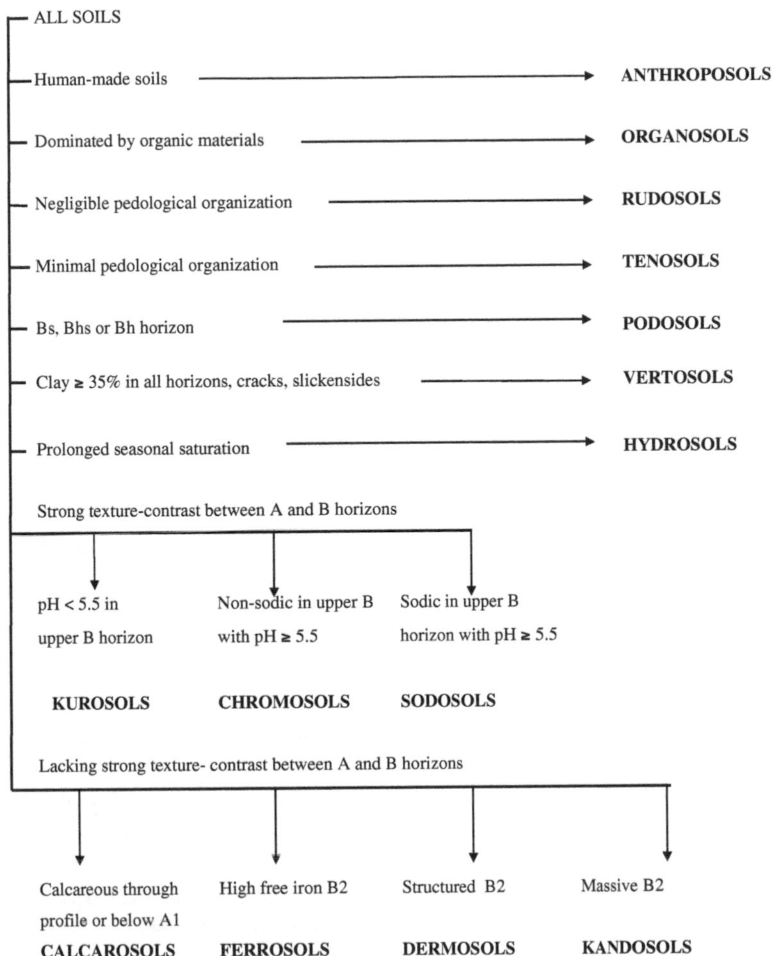

Figure 9.1 Structure of the Australian Soil Classification.

A major conceptual challenge relates to disturbed soils. At present they can be allocated to the suborder level of the Anthroposols soil order, or possibly as phases of other soil orders, such as drained, bunded, or artificially flooded soils. The properties of such soil materials are not found in current soil databases, but their inclusion should lead to further improvement as more information is accumulated (e.g., a proposed classification of minesoils by Fitzpatrick and Hollingsworth, 1995).

Other areas that require further development include the following:

- Better treatment of soil profiles with seasonal or more frequent changes in salinity.
- Improved characterization of sodic soils. The current dependence on exchangeable sodium percentage without reference to soil solution electrolyte concentration or composition is insufficient for predicting clay dispersion and related properties affecting soil behavior.
- Acid sulfate soils.
- Refinements to the Tenosol Order (this is being incorporated into a revised edition due for publication late in 2002).

Australian soil science and the classification schemes that reflect it may be considered to be "young science," as compared to the discipline of geology and its rock stratigraphic classification systems. Even when a tried and accepted general-purpose classification system is developed, there will always be a need for local or special-purpose schemes that provide distinctions and predictions not possible with a national, general-purpose scheme (e.g., Butler, 1980).

GENERAL-PURPOSE SOIL CLASSIFICATION SYSTEMS

The following case studies illustrate some of the new developments and the significant effort being devoted to public communication by developing better visual information relating to soils. A poster illustrating 48 common soils based on Isbell (1996) was published in 2001, and an illustrated guide to Australian Soils will be published in 2003 (similar to Deckers et al., 1998 and Bridges et al., 1998 for the World Reference Base). Several CDs have also been developed for teaching purposes.

Improved Communication Through a CD-based Interactive Key

The Australian Soil Classification has provided an excellent means for communication among scientists, and to a lesser extent among scientists and those in allied disciplines. However, training and access to the classification was initially restricted due to publication as a traditional key with a supporting volume on the underlying concepts and rationale (Isbell, 1996; Isbell et al., 1997). While this format suited pedologists, it was inappropriate for teaching and broader adoption. The Interactive Key to the Australian Soil Classification (Jacquier et al., 2001) aimed to change this state of affairs.

The Interactive Key is a CD-based system that uses innovative allocation and information retrieval software developed by CSIRO, originally for biological systematics (numerous interactive keys have been published for various groups of plants and animals). The Key allows allocation of soil profiles to the Order, Suborder, Great Group, or Subgroup levels within the Australian Soil Classification. A separate option within the Interactive Key allows allocation to the Family level. The Australian Soil Classification assumes soil profiles have been described according to the Australian Soil and Land Survey Field Handbook (McDonald et al., 1990).

The interactive method makes reliable allocation of soils to classes in the Australian Soil Classification more accessible to individuals without specialist knowledge. Use of the Interactive Key quickly provides users with an understanding of the Australian Soil Classification. The Interactive Key includes access to an abridged digital version of the Australian Soil Classification.

The Interactive Key uses a DELTA database (Dallwitz, 1980; Dallwitz et al., 1993 onwards). The software includes an interactive allocation and information retrieval system using the program Intkey (Dallwitz et al., 1995 onwards; 1998), descriptions, illustrations, references, and other subsidiary material. The database consists of a set of over 11,000 taxa (specified to Subgroup level) defined by 115 characters. A character is a feature or property that can be used to describe or compare taxa (e.g., clear or abrupt textural A/B horizon boundary). Characters include binary, ordinal, nominal, and numeric variables. In most cases, character states have been taken directly from the Australian Soil Classification (Isbell, 1996).

Most levels of the Australian Soil Classification require laboratory data from the profile of interest. The Interactive Key introduces surrogates for the characters requiring laboratory data, allowing allocation to be conducted to the Suborder level without laboratory data. Most steps during the allocation process have supporting graphical material to aid decisions, and images of typical profiles and their distribution are displayed when an allocation has been completed. The system allows for uncertainties and mistakes during the allocation process, with all possible options being displayed at each step. Copies of the Interactive Key are available from http://www.publish.csiro.au/

Improved Communication Through Use of Soil Profile Class and Soil Group Taxonomic Classifications

A Soil Profile Class is a group, or class, of soil profiles, not necessarily contiguous, which may be grouped on the similarity of their morphological characteristics and possibly some laboratory-determined properties. In Australia, the Soil Profile Class is defined at various levels of generalization (e.g., series, family, great soil group or phase, depending on the information available and the purpose of the survey and map scale (Gunn et al., 1988; Chapman and Atkinson, 2000). Soil classes and mapping units are not identical. Butler (1980) discusses the distinction between soil classification units and soil mapping units:

> The search for a merging between soil taxonomic entities (soil classes) with landscape area units (mapping units) has not been very successful—either we adopt a soil classification from the landscape itself in which case it fits in the landscape but not the rest of a soil taxonomic system designed to cover a wider area, or we impose the classes of a general taxonomic system on the landscape, and then it is unlikely to fit local soil modes or the individual character of the local landscape scene.

Butler (1980) supports the concept to "develop and use the soil classification that arises from the landscape itself," and contrasts this approach with national classifications, which may in some instances coincide with soil classes, but may even cross boundaries in the national scheme (Butler, 1980; McKenzie and Austin, 1989). Soil classes are usually based to some extent on a national classification scheme (e.g., defining them by incorporating their local uniqueness but also recognizing affiliations with a national scheme). For example, this has been done in various ways in each state by combining a local name with the taxonomic name, e.g., Dorrigo Red Ferrosols.

In several states and regions in Australia, a soil classification system or key has been devised especially to assist with the communication of soils information at a general level and to be suitable for use at the mapping scales required (i.e., regional). Each classification system has taken into account the most important features of that state or region. For example, in the western region of New South Wales, the devised soil classification system (Murphy and Murphy, 2000) uses soil physical properties (inferred from soil texture) as an indicator of infiltration rate and water-holding capacity, presence of root-restricting layers, along with micro-topography (e.g., Gilgai) (Table 9.1). Use is also made of "Special Soil Groups" (Murphy et al., 2000; Table 9.1) as a basis for problem-oriented investigations, because the existing general-purpose soil classification systems do not adequately describe or account for them. In South Australia, Soil Groups and Soil Classes were developed using the presence of calcium carbonate (e.g., calcareous soils or depth to calcrete) at the highest level of classification, because calcareousness is a major component of South Australian soils, and has particular impact on specific needs for assessment and management work (Wetherby and Oades, 1975; Table 9.2). In contrast, many readily recognized soils with related management implications in Western Australia are distinguished by the relative abundance of ironstone gravels present (Figure 9.2).

Soil Groups and Classes of South Australia

Despite the trend toward mapping and land description based on specific soil and landscape attributes, there remains a need for a form of soil classification, which distinguishes key soils at the regional level. This is considered necessary for communication, and indeed for some GIS-based modeling applications. The Great Soil Groups described by Stace et al. (1968), although still widely used locally for convenience, are inadequate due partly to loose definitions, but mainly to the reality that many South Australian soils simply do not fit into any of the Groups. The Australian Soil Classification (Isbell, 1996) was extremely useful in the development of a labeling system for soil

Table 9.1 Broad Soil Groups and Special Soil Groups for New South Wales

Broad Soil Groups for Western Region of NSW[a]
Soft red soils
Hard red soils
Solonized brown soils
Texture contrast soils
Skeletal soils
Heavy soils
Granite soils
Brown gibber soils
Desert loam soils
Special Soil Groups for NSW[b]
Sodic and Dispersible Soils
Hardsetting Surface Soils or Fragile, Light-textured Surface Soils
Self-mulching Soils
Acid Surface Soils
Acid Sulfate Soils
Aggregated or Subplastic Clays
Expansive Soils
Saline Soils
Desert loam Soils

[a] Adapted from Murphy and Murphy (2000)
[b] Adapted from Murphy et al. (2000)

Table 9.2 Soil Groups for South Australia

Soil Groups for South Australia[a]
Calcareous soils
Shallow soils on calcrete
Gradational soils with highly calcareous lower subsoils
Hard red-brown texture contrast soils with highly calcareous lower subsoils
Cracking clay soils
Deep loamy texture contrast soils with brown or dark subsoils
Sand over clay soils
Deep sands
Highly leached sands
Ironstone soils
Shallow moderately deep acidic soils on rock
Shallow soils on rock
Deep uniform to gradational soils
Wet soils
Volcanic ash soils

[a] Adapted from PIRSA Land Information (2001)

landscape mapping units, but it is too complex for communication with those lacking a pedological background (the vast majority of users of land resource assessment).

The system developed in South Australia (PIRSA Land Information, 2001) sought to define a series of soil classes, which reflected the morphological characteristics having the greatest impact on agricultural land use and management in South Australia. The initial aim was to define sufficient classes to allow for meaningful discrimination of different soils, but not so many as to overwhelm the user. The target was 50 classes. In the end, 61 classes were defined.

The 61 classes are aggregated into 15 groups (Table 9.2). Classes within a group may vary in terms of surface or subsoil texture, presence of pans, pH trend, subsoil structure and so on. For example, calcareous soils are widespread in South Australia, as are soils with calcrete pans at

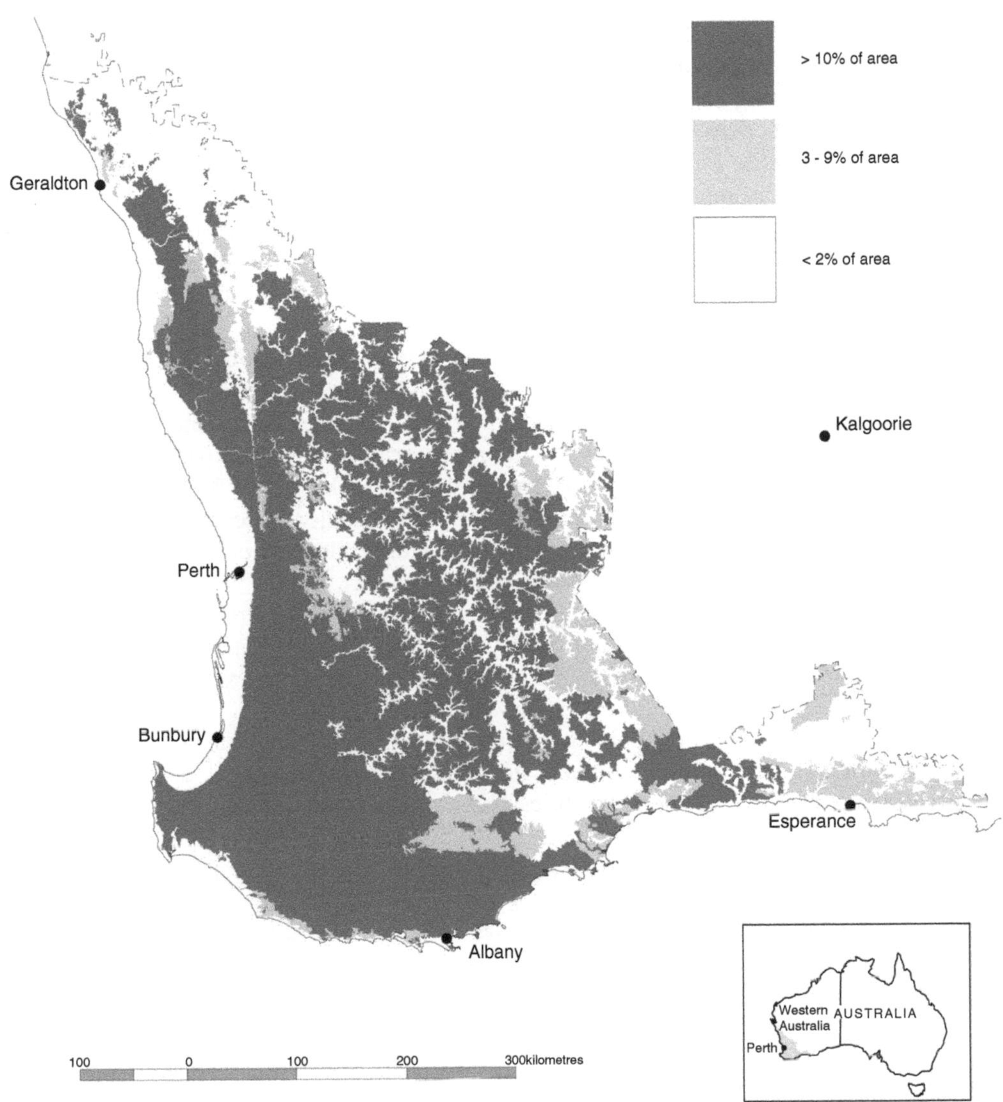

Figure 9.2 Distribution of ironstone gravelly soils in southwest Western Australia.

shallow (less than 50 cm) depth. While presence of a calcrete pan is used as a class definition at the Great Group or Subgroup level in the Australian Soil Classification, in the South Australian context this feature warrants separation at the highest level. Discriminators within this group of soils are the presence of fine earth carbonates above the calcrete, the presence of a more clayey subsoil, structure of subsoil, and surface texture. All soils in the group have the overriding limitation of restricted water holding capacity — the nine classes defined within the group separate other key properties related to fertility, drainage, and erosion potential.

The system is easy to use and understand, has real practical implications for users of agricultural land in South Australia, and complements the national system. The latter is essential for providing the higher resolution capacity needed to satisfy the classification demands of a much wider user-group over the whole continent.

Soil Supergroups and Groups of Western Australia

The Soil Groups of Western Australia (Schoknecht, 2001) were developed to satisfy a need for a simple, standardized and easy-to-understand way to recognize the most common soils in Western Australia—a state covering about 250 million hectares.

The three main objectives of Soil Groups are to provide common nomenclature to the main soils of Western Australia; provide a simple method of soil identification; and assist with the communication of soil information at a general level.

The Soil Groups aim to create common names that are more consistent than current terms, which are based on location, geology, and native vegetation. As a consequence, current common names describe very different soils in different geographic areas. Soil Group names have provided a standardized substitute for locally used common names.

Soil Groups can be allocated by non-technical people who usually have difficulty using formal soil classification systems such as the Australian Soil Classification (Isbell, 1996). It aims to categorize soils giving emphasis to soil characteristics that are relevant to land management and easy to identify by people with little soil description experience. For example, many soils in the southwest portion of the state are dominated by ironstone gravels (Figure 9.2). These gravels are used locally to identify these soils and are known to have a major effect on their management. Their presence is used as a primary division in the key for allocation to Soil Supergroups. The Soil Supergroups and their definitions are presented in Table 9.3.

Within the soil classification structure, two levels of generalization are identified: Soil Supergroups and Soil Groups. Thirteen Soil Supergroups for Western Australia are defined using three primary criteria: texture or profile permeability; coarse fragments (presence and nature); and water regime.

Soil Supergroups are further divided into 60 Soil Groups based on one or more of the following secondary and tertiary criteria: calcareous layer (presence of carbonates); color; depth of horizons/profile; pH; and structure. The definitions of the criteria for determining Soil Supergroups and Soil Groups are defined in Schoknecht (2001). An example of how one of the Soil Supergroups (Deep sands) is subdivided into Soil Groups using a simple key is presented in Table 9.4.

Table 9.3 The Soil Supergroups for Western Australia

Soil Supergroup[a]	Definition
Wet or waterlogged soils	Soils seasonally wet within 80 cm of surface for a major part of the year.
Rocky or stony soils	Soils, generally shallow, with > 50% of coarse fragments >20 mm in size (coarse gravels, cobbles, stones or boulders) throughout the profile. Includes areas of rock outcrop (all lithologies).
Ironstone gravelly soils	Soils that have an ironstone gravel layer (> 20% and > 20 cm thick) or duricrust/cemented gravels within the top 15 cm, and ironstone gravels a dominant feature of the profile.
Sandy duplexes	Soils with a sandy surface and a texture contrast or a permeability contrast (reticulite) at 3 to 80 cm.
Sandy earths	Soils with a sandy surface and grading to loam by 80 cm. May be clayey at depth.
Shallow sands	Sands ≤80 cm over rock, hardpan, or other cemented layer.
Deep sands	Sands >80 cm deep.
Loamy duplexes	Soils with a loamy surface and a texture contrast at 3 to 80 cm.
Loamy earths	Soils with loamy surface and either loamy throughout or grading to clay by 80 cm.
Shallow loams	Loams ≤ 80 cm over rock, hardpan, or other cemented layer.
Cracking clays	Soils with clayey surface at least 30 cm thick and crack strongly when dry.
Non-cracking clays	Soils with clayey surface at least 30 cm thick and do not crack strongly when dry.
Miscellaneous soils	Other soils.

[a] Adapted from Schoknecht (2001)

Table 9.4 Key to the Soil Groups[a] within the Deep Sands Soil Supergroup (Western Australia)

Calcareous within top 30 cm of surface, and usually throughout:	Calcareous deep sand
Yellow within top 30 cm, ironstone gravel common at depth:	Yellow deep sand
Brown within top 30 cm:	Brown deep sand
Red within top 30 cm:	Red deep sand
Gravelly below 15 cm and gravels a dominant feature of the profile, with a minimum gravel layer requirement of 30 cm thick and > 20% ironstone gravels starting within the top 80 cm. White, grey or Munsell value of 7 or greater (pale yellow) within top 30 cm. The sandy subsoil matrix may be colored:	Gravelly pale deep sand
White, grey or Munsell value of 7 or greater (pale yellow) within top 30 cm:	Pale deep sand (Figure 9.3)
Other deep sands:	(use DEEP SANDS Supergroup)

[a] Adapted from Schoknecht (2001)

The Soil Groups have proven to be a very useful and practical tool for describing and communicating soil information. The terms used are simple and easy to understand for most people. The system is not designed to replace the national classification system (Isbell, 1996), but is a user-friendly tool and complements it, as shown in Figure 9.3.

SPECIAL-PURPOSE SOIL CLASSIFICATION SYSTEMS

The following case studies have been selected to illustrate some of the developments in devising special-purpose soil classification systems in Australia, to solve a wide range of practical problems for end-users. The case studies discussed are selective, but various other classification systems have also been developed and summarized in Table 9.5. These technical or special-purpose soil classification systems rely mainly on soil attributes, but rarely also include other important environmental aspects such as geology, terrain, vegetation, or hydrology, which may be relevant to a particular end-user (Table 9.5). The combined use of soil and other information assists the understanding of how soils vary in landscapes so that strategies can be developed for managing both spatial and temporal changes within them.

Engineering Applications: Optical Fiber Cables

This case study summarizes the development of a suitable special-purpose soil classification system to minimize soil damage to the Australian telecommunication optical fiber cable network (Fitzpatrick et al., 1995; 2001). Some types of optical fiber cables can develop transmission faults by soil movements caused from soil shrink-swell properties, and corrosion from saline soil solutions. Such faults are very costly to repair and, if avoided, can save millions of dollars.

Field and laboratory investigations on a representative range of soils known to cause faults in optical fiber cables were undertaken (Fitzpatrick et al. 1995). Close liaison between soil scientists and engineers ensured that research investigations led to the development of a practical soil classification system comprising a user-friendly 1–10 rating of soil shrink-swell risk, which can be derived logically by using a series of questions and answers set out in a manual titled: "Soil Assessment Manual: A Practical Guide for Recognition of Soils and Climatic Features with Potential to Cause Faults in Optical Fibre Cables." The manual describes practical, surrogate methods to assist engineers in estimating soil shrink-swell indices by using either published soil maps in office assessments (Atlas of Australian Soils, Northcote, 1960–1968) or by undertaking simple visual observations and chemical measurements of soil properties in the field (Table 9.5). This information is incorporated in a planning operations and procedure manual for engineers.

Guided by the Manual, telecommunication engineers have learned how to use pedological, climatic, and soil chemical information to, first, avoid the shrink-swell and corrosive soils along

Pale deep sand

Sand >80 cm deep with white, grey or pale yellow topsoil

Characteristics

- White, grey or pale yellow (Munsell value 7 or higher) within top 30 cm
- Neutral to acid pH
- Ironstone gravel may be present, but not in large quantities
- Coffee rock, clay, or duricrust may occur at >80 cm
- A weak coffee rock layer may occur within 80 cm
- A colored sand may be present at 30-80 cm

Local names

Spillway sand, Gutless sand, Silver loam, Tincurrin/Harrismith sand, Christmas tree and Banksia sand, Deep mallee sand, Mungie sand

Typical Australian Soil Classification (ASC)

(dominant ASC in italics)

- *Bleached-Orthic Tenosol*
- Aeric Podosol

Main occurrences in Western Australia

- Common in the southwest agricultural area on the Swan Coastal Plain, the Scott River Plain and Cape Arid east of Esperance
- Also in broad valleys in lateritic terrain throughout the southwest, notably in the West Midlands north of Perth
- Scattered in other southwest areas

Soil attributes (dominant values)

Water repellence	**High**
Soil structure decline	Low
Subsurface acidification	Low to moderate
Surface condition	Soft to loose
Unrestricted rooting depth	Deep to very deep
Available water storage	Low

Subsurface compaction	**Low to moderate**
pH 0-10 cm	Neutral to acid
pH 50-80 cm	Neutral to acid
Soil permeability	Very rapid
Soil workability	Good
Wind erodibility	High

Land-use considerations

- Poor fertility and water-holding characteristics
- Nutrient leaching and groundwater recharge are significant issues
- Prone to wind erosion in exposed positions
- Prone to water repellence, especially after legume cropping

Figure 9.3 Example of a Soil Group description (pale deep sand).

optical fiber cable routes. Secondly, it enables them to use the appropriate type of cable for a particular soil, thereby using more expensive, heavy-duty cables in a cost-effective manner while ensuring the reliability of the optical link. Finally, they can then correct problems affecting cables previously installed in troublesome soil types, and rectify problems of reinstatement of fragile soils following cable burial.

Minesoils

A wide range of distinct types of minesoils forming on waste-rock and spoil dumps at open-cut mining operations in Australia has been recognized and described by Fitzpatrick and Hollingsworth

Table 9.5 Examples of Special-purpose Soil Classification Systems in Australia

Practical issue (Classification/Soil Classes)	Soil and Other Attributes Used	Reference (Section in Text)
Soil damage on the telecommunication optic fiber cable network. (10 soil shrink-swell risk classes/Soil assessment key to select cable type)	Soil shrink-swell and corrosion risk classes: rock type (geology), cracks, gilgai, soil color, structure (slickensides), texture, depth, dispersibility (sodicity), soil salinity. Soil assessment key: shrink-swell & corrosion risk, soil maps, vegetation, climate hydrology.	Fitzpatrick et al. (1995; 2001) (Special-Purpose Soil Classification Systems)
Minesoils on waste-rock and spoil dumps. (14 new subgroup classes of Spolic, Anthromorphic Anthroposols; Isbell, 1996)	Rockiness and stoniness, rock type, soil color and mottling, structure, texture, depth, dispersibility (sodicity), soil salinity, pH, acid sulfate soils, impermeable crusts, watertables.	Fitzpatrick and Hollingsworth (1995) (Special-Purpose Soil Classification Systems)
Saline soils across Australia. Linking hydrology and soil chemical hazards. (29 categories or classes of primary, secondary and transient salinity)	Halitic (sodium chloride dominant), gypsic (gypsum or calcium sulfate dominant), sulfidic (pyrite dominant), sulfuric (sulfuric acid dominant), and sodic (high exchangeable sodium on clay surfaces), hydrology (presence or absence of groundwater), water status (natural or primary as opposed to induced or secondary status).	Fitzpatrick et al. (2001; Table 9.6) (Special-Purpose Soil Classification Systems)
Saline and waterlogged soils in catchments. Linking to options for land use and remediation (8 soil classes).	Rockiness and stoniness, soil consistence (ease of excavation), color and mottling, structure, texture, depth, dispersibility (sodicity), soil salinity (EC), pH, sulfidic material, topography, watertables, vegetation type.	Fitzpatrick et al. (1997), Cox et al. (1999) (Special-Purpose Soil Classification Systems)
Viticultural soils of Australia. Identification of restrictive soil layers that limit effective root depth. (9 categories and 36 sub-categories)	Depth to characteristic changes in waterlogging (mottling), hard (non-rippable) or soft rock (rippable), rockiness and stoniness, soil consistence, color, texture and structure, calcareousness in different restrictive layers, cracking, texture change with depth: contrast (duplex character), uniform (little change) or gradational (gradual change).	Maschmedt et al. (2002) (Special-Purpose Soil Classification Systems)
Coastal ASS. Identify actual and potential acidification hazard. (2 soil classes: Actual ASS — AASS; Potential ASS — PASS).	Coastal acid sulfate soils (ASS): mangroves, reeds, rushes, estuaries, scalded, sulfurous smell. AASS: soil pH < 4, shells, yellow, jarositic horizons, water of pH < 5.5, iron stains, scalds. PASS: waterlogged, unripe muds, black to blue grey color, pH > 7, positive peroxide test, shells.	Ahern et al. (1998). (Special-Purpose Soil Classification Systems)
Site productivity for hardwood and softwood plantations in Tasmania. (4 site productivity classes).	Soil color, texture, depth of each soil layer to a minimum depth of 80 cm or to an impeding layer if shallower, native vegetation type and species, and rock type (geology), elevation, rainfall, soil drainage, tree-rooting conditions and nutrient availability	Laffan (1997)
Site productivity for Pinus radiata plantations.	Similar to Laffan (1997)	Turner et al. (1990)
Drainage and root growth. Irrigation/farming (6 carbonate classes).	Amount of calcrete fragments, fine carbonates, soil texture, and consistency.	Wetherby and Oades (1975)
Abrasive wear of cultivation equipment. Abrasive soils. (3 soil classes: highly and moderately abrasive; non-abrasive).	Highly abrasive soils: hardsetting, high bulk density, ploughpan, many rough-surfaced magnetic ironstone gravels, high silt and sand content. Moderately abrasive soils: few ironstone gravels; moderate organic matter, calcareous gravel, silt and sand contents. Non-abrasive soils: friable, no gravels; high clay, fine carbonate and organic matter contents.	Fitzpatrick et al. (1990); Fitzpatrick and Riley (1990)
Topdressing materials. Procedure for selection of suitable materials (3 classes: suitable, restricted and not suitable).	Soil structure, soil coherence, soil mottling, macrostructure, ped strength, soil texture, gravel and sand content, acidity, salt content, soil color, cutans, other toxic features (sulfides, metals, etc.).	Elliot and Reynolds (2000)

continued

Table 9.5 (continued) Examples of Special-purpose Soil Classification Systems in Australia

Practical issue (Classification/Soil Classes)	Soil and Other Attributes Used	Reference (Section in Text)
Urban planning. Capability, limitations. (5 primary classes, and several subclasses).	Soil properties: depth, permeability, shrink-swell potential, Gilgai, bearing strength, drainage properties, erodibility, salinity, and pH. Terrain properties: mass movement, watertables, subsidence and flooding.	Hicks and Hird (2002)
Urban planning.	Similar to above	Sheard and Bowman (1996)
Soil sampling medium for mineral exploration. (6 soil classes or materials).	Saline seepages, acid sulfate soils (ASS), iron- and aluminium-rich precipitates, sulfidic material, mottles in sulfuric horizons, salinity, pH, geochemical analyses.	Skwarnecki et al. (2002)

(1995). The formation of minesoils has proceeded in response to varying times of exposure (i.e., ranging from days to up to 20 years) of the materials to the surface weathering environment, and to varying rates of weathering depending on the nature of the rock types, in climates ranging from tropical to arid and Mediterranean. This information, together with proposed amendments to Soil Taxonomy (Smith and Sobek, 1978; Fanning and Fanning, 1989), has been used to develop preliminary proposals to modify both the Australian Soil Classification (Isbell, 1996) and Soil Taxonomy (Soil Survey Staff, 1999).

Minesoils generally have high contents of weathered rock fragments (ranging from 10% to >30%), a thin friable vesicular surface crust (5 to 10 mm), very weakly developed A and B horizons (100 mm thick and usually at depths of 50 to 500 mm) with silty-sandy-clay loam to silty light clay textures and saline to sodic properties. The following four broad categories of minesoils have been identified: lithosolic minesoils with minimal pedogenic development, stony/gravelly lags, vesicular crusts, and weak development of B horizons (e.g., Lithic and Typic Ustispolents/Xerispolents); polysequal soils indicative of erosion events (e.g., Fluventic and Aeric Ustispolents/Xerispolents); acid sulfate soils characteristic of waterlogged situations (e.g., Sulfic Spolaquents and Sulfic Ustispolents); and soils characteristic of seasonally waterlogged situations (e.g., Aquic Ustispolents). Major factors that influence the use of minesoils in rehabilitation strategies are stoniness, impermeable crust formation, and acid sulfate, saline, or sodic conditions.

Classification of the different types of minesoils using the proposed amendments to Soil Taxonomy has been undertaken (Fitzpatrick and Hollingsworth, 1995). As a result, suggestions for updating the classification of minesoils at subgroup level (Spolic, Anthromorphic Anthroposols) in the Australian Soil Classification (Isbell, 1996) have also been generated. This information can be used as the basis to construct a user-friendly technical soil classification for mine-site rehabilitation. The classification so proposed can include management-related properties.

Saline Soils

Several workers (e.g., Williams and Bullock, 1989) have attempted to categorize the wide range of dryland saline soils using hydrology (presence or absence of groundwater) and water status (natural or primary as opposed to induced or secondary status). Based on the Williams and Bullock (1989) system and newly acquired morphological, hydrological, and chemical information about saline soil-landscapes across Australia, a new classification has been developed to categorize saline soils using hydrological, soil water, and soil chemical status (sodicity or type of soluble salt) (Fitzpatrick et al., 2001; Tables 9.5 and 9.6). The important soil chemical features are halitic (sodium chloride dominant), gypsic (gypsum or calcium sulfate dominant), sulfidic (pyrite dominant), sulfuric (sulfuric acid dominant), and sodic (high exchangeable sodium on clay surfaces). Most of these terms are defined in Isbell (1996). Current general-purpose classification systems (e.g., Isbell,

Table 9.6 Categories of Saline Soils as Defined by Hydrology, Soil Water Status, and Soil Chemistry

Hydrology				
Groundwater/ Perched Water?	**Primary/ Secondary/ Transient?**	**Soil Water**	**Soil Chemistry**	**Saline Soil Category**
Groundwater present in root zone: salinity process driven by groundwater.	Primary salinity (natural)	Mangrove swamp	Sulfidic	Primary, mangrove swamp, sulfidic, saline soil
		Coastal marsh	Sulfidic	Primary, coastal marsh, sulfidic, saline soil
			Gypsic	Primary, coastal marsh, gypsic, saline soil
		Salt pan	Halitic	Primary, salt pan, halitic, saline soil
			Gypsic	Primary, salt pan, sulfidic, saline soil
		Seepage scald	Halitic	Primary, seepage scald, halitic, saline soil
			Sulfidic	Primary, seepage scald, sulfidic, saline soil
		Salt seepage	Halitic	Primary, salt seepage, halitic, saline soil
			Sulfidic	Primary, salt seepage, sulfidic, saline soil
	Secondary salinity (induced, dryland salinity)	Mangrove swamp	Sulfidic	Secondary, mangrove swamp, sulfidic, saline soil
		Drained mangrove swamp	Sulfuric	Secondary, drained mangrove swamp, sulfuric, saline soil
		Coastal marsh	Sulfidic	Secondary, coastal marsh, sulfidic, saline soil
			Gypsic	Secondary, drained coastal marsh, gypsic, saline soil
		Drained coastal marsh	Sulfuric	Secondary, drained coastal marsh, sulfuric, saline soil
			Gypsic	Secondary, drained coastal marsh, gypsic, saline soil
		Salt pan	Halitic	Secondary, salt pan, halitic, saline soil
			Gypsic	Secondary, salt pan, gypsic, saline soil
		Salt seepage	Halitic	Secondary, salt seepage, halitic, saline soil
			Sulfidic	Secondary, salt seepage, sulfidic, saline soil
			Sulfuric	Secondary, salt seepage, sulfuric, saline soil
		Eroded seepage scald	Halitic	Secondary, eroded seepage scald, halitic, saline soil
			Sulfuric	Secondary, eroded seepage scald, sulfuric, saline soil
Groundwater absent from root zone: salinity process driven by seasonal perched watertable in root zone.	Transient salinity (natural)	Topsoil "Magnesia" patch	Halitic	Transient, natural, topsoil, halitic, saline soil
			Sodic	Transient, natural, topsoil, sodic, saline soil
		Subsoil	Sodic	Transient, natural, subsoil, sodic, saline soil
	Transient salinity (induced)	Surface soil "Magnesia" patch	Halitic	Transient, induced, surface soil, halitic, saline soil
			Sodic	Transient, induced, surface soil, sodic, saline soil
		Subsoil	Sodic	Transient, induced, subsoil, sodic, saline soil
		Eroded	Halitic	Transient, induced, eroded, sodic, saline soil

Adapted from Fitzpatrick et al., 2001

1996; Soil Survey Staff, 1998) are not intended to distinguish between the causes of soil salinity. However, this classification conveys important information on hydrology, soil water status, and soil chemistry of saline soils to non-soil specialists, especially hydrologists dealing with rehabilitation of saline land.

The capacity to reverse established salinization in soils will depend strongly on the specific category or class of saline soil that is present (Fitzpatrick et al., 2001). The range of likely treatment options may also be very different between the salinity categories, as shown in Table 9.6. Management strategies for saline soils must be designed with processes of salt accumulation and mobilization in mind. Different management techniques are necessary for soils with different salt compositions and water regimes, as emphasized in this special-purpose classification.

Fitzpatrick et al. (2002) have recently summarized an approach and procedure for developing a user-friendly soil classification that is linked directly to land-use options and management in localized regions with specific types of salt-affected soils (Fitzpatrick et al., 1997; Cox et al., 1999; Table 9.5). These publications provide an account of how user-friendly soil classifications have been developed and applied to soils affected by dryland salinity and waterlogging in two landscapes in southern Australia (Mount Lofty Ranges, South Australia, and western Victoria). The descriptive soil information is pictorially integrated along toposequences (i.e., using colored cross-sectional diagrams and photographs of soils) and applied to also identify soil and hydrological features to overcome some of the perceived barriers to adoption of best management practices. In this way, complex scientific processes and terminology are more easily communicated to community groups.

Detailed field and laboratory investigations were used to develop the diagnostic soil attributes (Table 9.5) for use in the user-friendly soil classification system. Examples include the presence and amounts (or absence) of grey bleached and yellow colors in the form of distinct mottles (iron depletions), and black and red stains (iron concentrations), which are similar to the redoximorphic features described by Vepraskas (1992). These features have developed under particular conditions of water saturation, salinization, sodification, sulfidization, and water erosion, in surface and subsurface horizons, and are easily identified by farmers and land managers. The key enables classification of soils in relation to position along a toposequence and groundwater discharge areas. It was shown that information written in this format helped farmers and regional advisers to identify options for remediation of saline and waterlogged areas, and to improve planning at property and catchment scales. The approach is considered to have generic application to other regions in Australia (Fitzpatrick et al., 2002).

Viticultural Soils

All the Australian classification systems were found to lack user-friendly keys for identifying soil profiles in vineyards by people who are not experts in soil classification. It was clear from their lack of use in the viticulture industry in Australia that the existing systems were not suitable, because they were seen as too complex. Viticultural information in Australian and overseas literature, based on soils classified using these schemes, often could not be applied correctly to Australian conditions, because there was no means to link these identifiers to local understanding of the nature and properties of soils. Consequently, the Australian viticulture industry called for the development of a user-friendly soil key, which could be used by viticulturists to help select and match grapevine rootstocks to appropriate Australian soils (May, 1994). Using the data set provided by the survey of the rootstock trials and other resources (Cass et al., 2002), Maschmedt et al. (2002) developed an Australian Viticultural Soil Key. This key provides the means to describe Australian soils in terms of attributes meaningful to viticulture, and to correlate these attributes with local (Isbell, 1996; Stace et al., 1968) and international (Soil Classification Working Group, 1991; FAO, 1998; Soil Survey Staff, 1999) soil classification schemes. The key uses, as far as is possible, non-technical terms to categorize soils in terms of attributes that are important for vine growth. The soil features used in the key are easily recognized in the field by people with limited soil classification experience.

It focuses on the following viticulturally important and mostly visual diagnostic features: depth to certain characteristic changes in waterlogging, consistency, color, structure, calcareousness in different restrictive layers, cracking, and texture trends down profiles (e.g., texture contrast at A/B horizon boundary or duplex character) (Table 9.5). The key layout is bifurcating, based on the presence or absence of the particular keying property, which is usually a diagnostic property.

The key is seen as an important tool for correlating rootstock performance and soil properties, and as a vehicle for delivering soil-specific land development and soil management packages to grape growers.

Coastal Acid Sulfate Soils

Acid sulfate soils (ASS) are saline soils or sediments containing pyrites, which once drained (as part of land management or development measures) become acidic, and release large amounts of acidity and other contaminants to the environment with consequent adverse effects on plant growth, animal life, and urban infrastructures. These coastal ASS occur in tidal floodplains where sources of sulfates, iron, and other salts originate from seawater and estuarine sediments, which are less than 5 meters above sea level (AHD < 5 m). Nationally, there is an estimated 40,000 km^2 of coastal ASS, and potentially they contain over one billion tonnes of sulfuric acid. When undisturbed and saturated by water, ASS remain relatively benign.

The source of acid sulfate problems is pyrite, FeS_2, which when oxidized produces sulfuric acid that brings the pH below 4, sometimes even below 3, thereby causing minerals in soils to dissolve and liberate soluble and colloidal aluminium and iron, which may leak to drainage and floodwaters, killing vegetation and aquatic life; and steel and concrete infrastructures then corrode (White et al., 1995; Sammut et al., 1996). Drainage of acid sulfate soils also results in the substantial production of greenhouse gases, carbon dioxide, and N_2O (Hicks et al., 1999). Development and primary industries around Australia are facing a $10 billion legacy of acid sulfate soils (National Working Party on Acid Sulfate Soils, 2000). Public recognition of this serious problem has been reflected in government legislation in New South Wales, Queensland, and South Australia. There is much support from councils and industries to develop statutory requirements for rehabilitation.

Although the Australian Soil Classification (Isbell, 1996) does recognize ASS, there has been a need to develop a user-friendly and effective classification to meet the needs of local governments and coastal authorities in a policy and jurisdiction sense, as well as to guide potential users of these environments to information regarding processes of rehabilitation. The classification system needs to be used in the context of a product that enables clients to use technical information (Table 9.5). Although a plethora of complex terminology and standards have evolved in the literature in relation to acid sulfate soils, these have been simplified (Ahern et al., 1998; Department of Natural Resources, 2000) to support the planning process, and comprise the following:

- *Acid sulfate soils:* Soil or sediment containing highly acidic soil horizons or layers affected by the oxidation of iron sulfides (actual acid sulfate soils) and/or soil or sediment containing iron sulfides or other sulfidic material that has not been exposed to air and oxidized (potential acid sulfate soils). The term acid sulfate soil generally includes both actual and potential acid sulfate soils. Actual and potential acid sulfate soils are often found in the same soil profile, with actual acid sulfate soils generally overlying potential acid sulfate soil horizons.
- *Actual acid sulfate soils (AASS):* Soil or sediment containing highly acidic soil horizons or layers affected by the oxidation of soil materials that are rich in iron sulfides, primarily pyrite. This oxidation produces hydrogen ions in excess of the sediment's capacity to neutralize the acidity, resulting in soils of pH 4 or less. These soils can usually be identified by the presence of bright yellow or straw-colored mottles of jarosite.
- *Potential acid sulfate soils (PASS):* Soil or sediment containing iron sulfides or sulfidic material that have not been exposed to air and oxidized. The field pH of these soils in their undisturbed state is pH 4 or more, and may be neutral or slightly alkaline.

For management purposes, classification needs to account for the thickness and severity of acid sulfate soil horizons present, whether as actual or potential acid sulfate soil. These require specification of sulfide content, lime requirement, and an assessment of risk based on proposed volume of soil disturbance. This information is used to develop comprehensive, technically valid soil management plans.

Since the 1990s, this demand has led to the establishment of acid sulfate soils mapping programs, training programs on identification and management, and policy development to regulate use of these high-risk soils. These activities are most advanced in the states of New South Wales (Naylor et al., 1995) and Queensland, and include the establishment of advisory committees and acid sulfate soil-specific planning policies/regulations, technical management guidelines, and training programs. The mapping programs aided by research have led to a vast increase in knowledge about Australia's acid sulfate soils. This was not available when the Australian Soil Classification of Isbell (1996) was being constructed, but provides the basis for improvement in the future. Such new information (e.g., Ahern et al., 2000; Slavich, 2000) relates to better methods of field assessment, the hydrological and chemical changes and processes that occur following drainage or remediation, the variety of sedimentary facies possibly involved, improved methods of soil testing, and recognition of distinctions in various sulfur minerals and organic peats formed under contrasting pH and redox conditions. Acid sulfate soils with sapric material and intense reducing conditions (low redox potentials) have been identified in South Australia (Fitzpatrick et al., 1993). These conditions appear to originate from high input of detritus from seagrass and mangroves in the low energy environments and increased nutrient loads. This information was used to amend Soil Taxonomy (Fitzpatrick et al., 1993) and to develop preliminary proposals to modify the Australian Soil Classification (Isbell, 1996). In this case, soil properties and processes must be better understood if effective approaches to management are to be developed and changes to classification systems made.

FUTURE DIRECTIONS

In Australia, land managers, planners, and researchers often operate with poor spatial information on soil, and this limits their capacity to analyze and implement optimal systems of land use. Conventional survey programs and national soil classification schemes have not always provided appropriate information at relevant scales. New demands for information in Australia are exposing further weaknesses in the land resource survey coverage and the communication capacity of existing soil classification schemes.

Dryland salinity, soil acidification, declining stream water quality and a range of other emerging problems have created a strong environmental imperative for large-scale land-use change in Australia. Government, industry, and community groups wish to identify and analyze profitable production systems that alleviate serious land and water degradation. Likewise, a range of other natural resource management issues relating to international agreements (e.g., Kyoto Protocol, Montreal Convention), environmental regulation, and trading systems (e.g., for water, salt, carbon) are stimulating an unprecedented demand for analysis and prediction of soil and landscape function.

Methods of landscape analysis with a process-basis are required to provide spatial and temporal predictions of properties controlling the function of agro-ecological systems. There is a need to develop improved simulation models depicting spatial and temporal processes; measurement and prediction technologies to provide appropriate model inputs; and monitoring and experimentation to validate modeling predictions.

Both special-purpose and general-purpose soil classification systems are predicted to have a role in these developments, but in generally different directions. Like the examples presented in this chapter (Table 9.5), the special-purpose classification schemes will continue to be developed to meet the specific and practical needs of users of soil information. The general-purpose schemes, such as the Australian Soil Classification, will continue to evolve and have a place in the development of landscape models with spatial soil attributes.

Innovative methods for broad-scale mapping of land resources have been developed during the last decade using digital terrain analysis, geophysical remote sensing, direct field measurement and pedological knowledge (e.g., Cook et al., 1996; Fitzpatrick et al., 1999; McKenzie and Ryan, 1999; Davies et al., 2001; Merry et al., 2001). The new methods overcome many of the problems of conventional survey by providing predictions of individual soil and land properties (i.e., predict first, classify later). Substantial improvements to the methods are possible, and these will be necessary before they can be generalized to the broad range of landscapes occupying the productive zones of Australia. A critical requirement is the development of more process-based models of soil and regolith distribution. Current methods rely on statistical correlations between soil properties and various attributes computed from terrain models and related sources. While these are always assessed for their process significance, it would be much better to have a set of predictor variables and models that reflect processes active in landscape evolution at scales that can be applied across regions.

Powerful PC-based geographic information systems and web technology have created much better ways for communicating the results of landscape analysis for a range of purposes. Effective systems have been developed that provide simple access to these:

- Primary survey data
- Digital images of soil profiles and sites
- Summaries of laboratory data
- Spatial predictions of individual soil properties and land qualities
- High resolution digital surfaces of environmental variables
- Access to publications

A logical extension would be the integration of simulation modelling outputs. It is within this environment, driven by information technology, remote sensing, and conceptual models of landscape development, that field-based and practical soil classification schemes must operate if they are to be relevant.

SUMMARY

Soil classification systems in Australia are briefly reviewed. The recent Australian Soil Classification is a significant improvement over older systems, but it remains weak for soils from non-agricultural areas where few data exist, such as for acid sulfate, saline, and hydromorphic soils. Transfer of technology has been traditionally advocated and achieved by pedologists relying heavily on morphologically defined soil types—the morphological criteria have pedological significance, but their relevance to practical land management is less clear. The rationale is not necessarily true that general-purpose soil classification systems stratify the soil population in ways that are relevant to agronomic and engineering applications, and in sufficient detail to facilitate information transfer.

Transfer of soil information has become less pedocentric and is tailored more to suit client needs. This has involved improvements in the communication of general-purpose classification systems and the development of several special-purpose schemes tailored to particular problems (i.e., infocentric). Case studies are presented to demonstrate innovations incorporating these philosophies. They include improved communication of general-purpose classification systems through the development of a CD-based interactive key, and user-friendly soil classifications at state or regional levels in Western Australia (Soil Supergroup Groups), South Australia (Soil Groups and Classes), and NSW (Broad Soil Groups and Special Soil Groups) to assist with the communication of soils information and account for the occurrence of soils that impact specific needs for assessment and management work. Several user-friendly special-purpose classification systems, covering a variety of practical issues, were presented: engineering applications (optical fiber cable installations); minesoils; saline soils; soils used for viticulture; coastal acid sulfate soils (links to policy and jurisdiction); and others (e.g., forestry, abrasive soils, topdressing materials, urban planning, and mineral exploration). These special-purpose or technical classification systems all involve soil assessment criteria and recommendations for soil management practices to end-users.

DEDICATION

In 1996, The Australian Soil Classification was published by an eminent CSIRO pedologist, Ray Isbell. This achievement was remarkable, given the previous disarray of soil taxonomic work in Australia, and was due to extraordinary effort by Ray over more than a decade. Though he received support from many pedologists, it was Ray's dogged commitment, outstanding knowledge of Australian soils, and tremendous diplomacy that led to completion of the system against all odds. Sadly, Ray died at the end of 2001. This contribution is dedicated to Ray, who was a loyal colleague of us all.

ACKNOWLEDGMENTS

Richard Merry, Ken Lee, and Phil Slade from CSIRO Land and Water, Adelaide made useful corrections and suggestions to improve the manuscript.

REFERENCES

Ahern, C.R., Hey, K.M., Watling, K.M., and Eldershaw V.J., Eds. 2000. Acid Sulfate Soils: Environmental Issues, Assessment and Management, Technical Papers, Brisbane, 20–22 June, 2000. Department of Natural Resources, Indooroopilly, Queensland, Australia.

Ahern, C.R, Stone, Y., and Blunden, B. 1998. Acid Sulfate Soils Assessment Guidelines. Section 2: Acid Sulfate Soil Manual. Acid Sulfate Soils Management Advisory Committee (ASSMAC), Wollongbar, NSW, Australia.

Basher, L.R. 1997. Is pedology dead and buried? *Aust. J. Soil Res.* 35:979–994.

Beckett P.H.T., Webster R. 1971. Soil variability: a review. *Soils Fert.* 34:1–15.

Bouma, J. 1997. The role of quantitative approaches in soil science when interacting with stakeholders. *Geoderma.* 78:1–12.

Bridges, E.M., Batjes, N.H., and Nachtergaele, F.O. Eds. 1998. World Reference Base for Soil Resources: Atlas. ISRIC-FAO-ISSS. Acco: Leuven.

Burrough, P.A. 1993. Soil variability: A late 20th century view. *Soils Fert.* 56:529–562.

Butler, B.E. 1980. *Soil Classification for Soil Survey.* Oxford Univerity Press, Oxford, U.K.

Butler, B.E. and Hubble, G.D. 1977. Morphologic properties, in J.S. Russell and E.L. Greacen, Eds. *Soil Factors in Crop Production in a Semi-arid Environment.* Chapter 2. University of Queensland Press, St. Lucia, Queensland, 9-32.

Cass, A., Fitzpatrick, R., Maschmedt, D., Thomson, K., Dowley, A., and Van Goor, S. 2002. Soils of the Australian Rootstock Trials. The Australian and New Zealand Grapegrower & Winemaker, Annual Technical issue, 461A:40–49.

Chapman G.A. and Atkinson, G. 2000. Soil survey and mapping, in P.E.V. Charman and B.W. Murphy, Eds. *Soils: Their Properties and Management. Soil Conservation Commission of New South Wales.* Chapter 7. Oxford University Press, Australia, 106-132.

Cook, S.E., Corner, R.J., Grealish, G., Gessler, P.G., and Chartres, C.J. 1996. A rule based system to map soil properties. *J. Soil Soc. Am.* 60:1893–1900.

Cox, J.W., Fitzpatrick, R.W., Mintern, L., Bourne, J., and Whipp, G. 1999. Managing waterlogged and saline catchments in south-west Victoria: A soil-landscape and vegetation key with on-farm management options. Woorndoo Land Protection Group Area Case Study: Catchment Management Series No. 2. CRC for Soil & Land Management, CSIRO Publishing, Melbourne, Australia.

Dallwitz, M.J. 1980. A general system for coding taxonomic descriptions. *Taxon*. 29:41–6.

Dallwitz, M.J., Paine, T.A., and Zurcher, E.J. 1993 onward. User's Guide to the DELTA System: A General System for Processing Taxonomic Descriptions. 4th ed. Online at http://biodiversity.uno.edu/delta/.

Dallwitz, M.J., Paine, T.A., and Zurcher, E.J. 1995 onward. User's Guide to Intkey: A Program for Interactive Identification and Information Retrieval. 1st ed. Online at http://biodiversity.uno.edu/delta/.

Dallwitz, M.J., Paine, T.A., and Zurcher, E.J. 1998. Interactive keys, in P. Bridge, P. Jeffries, D.R. Morse, and P.R. Scott, Eds. *Information Technology, Plant Pathology and Biodiversity*. CAB International, Wallingford, 201-212.

Davies, P.J., Fitzpatrick, R.W., Bruce, D.A., Spouncer, L.R., and Merry, R.H. 2002. Land degradation assessment in the Mount Lofty Ranges: Upscaling from points to regions via toposequences, in T.R. McVicar, L. Rui, J. Walker, R.W. Fitzpatrick, and C. Liu, Eds. *Regional Water and Soil Assessment for Managing Sustainable Agriculture in China and Australia*. ACIAR Monog. 84. CSIRO Publishing, Melbourne, Australia, 291-303.

Deckers, J.A., Nachtergaele, F.O., and Spaargaren, O.C., Eds. 1998. World Reference Base for Soil Resources: Introduction. ISSS-ISRIC-FAO. Acco, Leuven.

Department of Natural Resources. 2000. State Planning Policy 1/00: Planning and management of coastal development involving acid sulfate soils. Queensland Government, Australia.

Dudal, R. 1987. The role of pedology in meeting the increasing demands on soils. *Soil Survey Land Evaluation*. 7:101–110.

Elliot, G.L. and Reynolds, K.C. 2000. Soils and extractive industries, in P.E.V. Charman and B.W. Murphy, Eds. *Soils: Their Properties and Management. Soil Conservation Commission of New South Wales*. Chapter 22. Oxford University Press, Australia, 390-397.

FAO. 1998. The world reference base for soil resources (WRB). World Soil Resources Report No. 84. Food and Agriculture Organisation for the United Nations (FAO)/ISSS/AISS/IBG/ISRIC, Rome, 1998.

Fanning, D.S. and Fanning, M.C. 1989. *Soil, Morphology, Genesis and Classification*. John Wiley & Sons, New York.

FitzPatrick, E.A. 1971. *Pedology: A Systematic Approach to Soil Science*. Oliver and Boyd, Edinburgh.

Fitzpatrick, R.W. 1996. Morphological indicators of soil health, in J. Walker and D.J. Reuter, Eds. *Indicators of Catchment Health: A Technical Perspective*. CSIRO Multidivisional Program on Dryland Farming Systems for Catchment Care. CSIRO Publishing, Melbourne, Australia, 75-88.

Fitzpatrick, R.W., Bruce, D.A., Davies, P.J., Spouncer, L.R., Merry, R.H., Fritsch, E., and Maschmedt, D. 1999. Soil Landscape Quality Assessment at Catchment and Regional Scale. Mount Lofty Ranges Pilot Project: National Land & Water Resources Audit. CSIRO Land & Water Technical Report. 28/99.

Fitzpatrick, R.W., Cox, J.W., and Bourne, J. 1997. Managing waterlogged and saline catchments in the Mt. Lofty Ranges, South Australia: A soil-landscape and vegetation key with on-farm management options. Catchment Management Series. CRC for Soil and Land Management. CSIRO Publishing, Melbourne, Australia.

Fitzpatrick, R.W., Cox, J.W., Munday, B., and Bourne, J. 2002. Development of soil-landscape and vegetation indicators for managing waterlogged and saline catchments. *Aust. J. Exp. Agric*. Special Issue featuring papers on "Application of Sustainability Indicators" (in press).

Fitzpatrick, R.W. and Hollingsworth, I.D. 1995. Towards a new classification of minesoils in Australia based on proposed amendments to Soil Taxonomy. Proceedings of the National cooperative soil survey conference. San Diego, California, July 10–14, 1995. U.S. Dept. of Agric., Natural Resources Conservation Service, Washington, DC., 87–97.

Fitzpatrick, R.W., Hudnall, W.H., Self, P.G., and Naidu, R. 1993. Origin and properties of inland and tidal saline acid sulfate soils in South Australia, in D.L. Dent and M.E.F. van Mensvoort, Eds. Selected papers of the Ho Chi Minh City Symposium on Acid Sulfate Soils. International Inst. Land Reclamation and Development Publication 53.

Fitzpatrick, R.W., McKenzie, N.J., and Maschmedt, D. 1999. Soil morphological indicators and their importance to soil fertility, in K. Peverell, L.A. Sparrow, and D.J. Reuter, Eds. *Soil Analysis: An Interpretation Manual*. CSIRO Publishing, Melbourne, Australia, 55-69.

Fitzpatrick, R.W., Rengasamy P., Merry R.H, and Cox, J.W. 2001. Is dryland soil salinisation reversible? National Dryland Salinity Program (NDSP). Online at http://www.ndsp.gov.au/10_NDSP_projects/05_project_descriptions/35_environment_protection/project_25.htm.

Fitzpatrick, R.W., Riley, T.W., Wright, M.J., Fielke, J.M., Butterworth, P.J., Richards, B.G., Peter, P., McThompson, J., Slattery, M.G., Chin, D., McClure, S.G., and Lundy, J. 1990. Highly abrasive soils and ground engaging tool performance. End of Grant Report (1988–1990) for the Australian Wheat Research Council (Project No. SAIT 1W), Canberra, Australia.

Fitzpatrick, R.W. and Riley, T.W. 1990. Abrasive soils in Australia: Mineralogical properties and classification. Transactions of the 14th International Soil Science Society Conference, Kyoto, Japan, August 1990. VII:404–405.

Fitzpatrick, R.W., Slade, P.M., and Hazelton, P. 2001. Soil-related engineering problems: identification and remedial measures, in V.A. Gostin, Ed. *Gondwana to Greenhouse: Australian Environmental Geoscience.* Chapter 3. Geological Society of Australia Special Publication 21. GSA, Australia.

Fitzpatrick, R.W., Wright, M.J., Slade, P.G., Hollingsworth, I.D., and Peter, P. 1995. Soil Assessment Manual: A Practical Guide for Recognition of Soils and Climatic Features with Potential to Cause Faults in Optical Fibre Cables. Confidential Report to Telstra. CSIRO Land and Water, Australia.

Gibbons, F.R. 1961. Some misconceptions about what soil surveys can do. *J. Soil Sci.* 12:96–100.

Gunn, R.H., Beatie, J.A., Reid, R.E., and van de Graaff, R.H.M., Eds. 1988. *Australian Soil and Land Survey Handbook: Guidelines for Conducting Surveys*, Inkata Press, Melbourne.

Hicks, R.W. and Hird, C. 2000. Soils and urban land use, in P.E.V. Charman and B.W. Murphy, Eds. *Soils: Their Properties and Management. Soil Conservation Commission of New South Wales.* Chapter 21. Oxford University Press, Australia, 378-389.

Hicks, W.S., Bowman, G.M., and Fitzpatrick, R.W. 1999. East Trinity Acid Sulfate Soils 1. Environmental hazards. CSIRO Land and Water Technical Report No. 14/99.

Isbell, R.F. 1992. A brief history of national soil classification in Australia since the 1920's. *Aust. J. Soil Sci.* 30:825–842.

Isbell, R.F. 1996. T*he Australian Soil Classification.* CSIRO Publishing, Melbourne, Australia.

Isbell, R.F., McDonald, W.S., and Ashton, L.J. 1997. Concepts and rationale of the Australian soil classification. CSIRO Land and Water, Australia.

Jacquier, D.W., McKenzie, N.J., Brown, K.L., Isbell, R.F., and Paine, T.A. 2001. *The Australian Soil Classification. An Interactive Key.* CSIRO Publishing, Melbourne.

Laffan, M. 1997. Site selection for hardwood and softwood plantations in Tasmania: A methodology for assessing site productivity and suitability for plantations using land resource information. Soils Technical Report 3, 2nd ed. Forestry Tasmania and the Forest Practices Board, Hobart.

Leeper, G.W. 1943. The classification and nomenclature of soils. *Aust. J. Soil Sci.* 6:48–51.

Leeper, G.W. 1956. The classification of soils. *J. Soil Sci.* 7:59–64.

MacArthur, W.M., Wheeler, J.L., and Goodall, D.W. 1966. The relative unimportance of certain soil properties as determinants of growth of forage oats. *Aust. J. Exp. Agric. Animal Husbandry.* 6:402–408.

Maschmedt, D.J., Fitzpatrick, R.W., and Cass, A. 2002. Key for identifying categories of vineyard soils in Australia. CSIRO Land and Water, Technical report No. 30/02.

May, P. 1994. Using grapevine rootstocks: The Australian perspective. Winetitles, Adelaide, South Australia.

McBratney A.B. 1994. Allocation of new individuals to continuous soil classes. *Aust. J. Soil Res.* 32:623–633.

McDonald, R.C., Isbell, R.F., Speight, J.G., Walker, J., and Hopkins, M.S. 1990. *Australian Soil and Land Survey Field Handbook,* 2nd ed. Inkata Press, Melbourne.

McKenzie, N.J. and Austin, M.P. 1989. Utility of the Factual Key and Soil Taxonomy in the Lower Macquarie Valley, N.S.W. *Aust. J. Soil Res.* 27:289–311.

McKenzie, N.J., Jacquier, D.W., Ashton, L.J., and Cresswell, H.P. 2000. Estimation of soil properties using the Atlas of Australian Soils. CSIRO Land and Water Technical Report 11/00, February 2000.

McKenzie, N.J. and McLeod, D.A. 1989. Relationships between soil morphology and soil properties relevant to irrigated and dryland agriculture. *Aust. J. Soil Res.* 27:235–258.

McKenzie, N.J. and Ryan, P.J. 1999. Spatial prediction of soil properties using environmental correlation. *Geoderma.* 89:67–94.

Merry, R.H., Spouncer, L.R., Fitzpatrick, R.W., Davies, P.J., and Bruce, D.A. 2002. Regional prediction of soil profile acidity and alkalinity, in T.R. McVicar, L. Rui, J. Walker, R.W. Fitzpatrick, and C.L. Changming, Eds. Regional Water and Soil Assessment for Managing Sustainable Agriculture in China and Australia. ACIAR Monograph 84. CSIRO Publishing, Melbourne, Australia, 155-164.

Moore A.W., Isbell, R.F., and Northcote, K.H. 1983. Classification of Australian soils, in *Soil—An Australian Viewpoint*. CSIRO Division of Soils, CSIRO, Melbourne/Academic Press, London, 253-266.

Murphy, B.W., Chapman, G.A., Eldridge, D.J., Atkinson, G., and McKane, D.J. 2000. Soils of New South Wales, in P.E.V. Charman and B.W. Murphy, Eds. *Soils: Their Properties and Management. Soil Conservation Commission of New South Wales*. Chapter 8. Oxford University Press, Australia, 133-149.

Murphy, B.W. and Murphy, C.L. 2000. Systems of soils classification, in P.E.V. Charman and B.W. Murphy, Eds. *Soils: Their Properties and Management. Soil Conservation Commission of New South Wales*. Chapter 6. Oxford University Press, Australia, 83-105.

National Working Party on Acid Sulfate Soils. 2000. National Strategy for the Management of Acid Sulfate Soils. NSW Agriculture, Wollongbar, New South Wales, Australia.

Naylor, S.D., Chapman, G.A., Atkinson, G., Murphy, C.L., Tulau, M.J., Flewin, T.C., Milford, H.B., and Morand, D.T. 1995. Guidelines for the use of acid sulfate soils risk maps. NSW Soil Conservation Service, New South Wales Department of Land and Water Conservation, Sydney.

NLWRA. 2001. Australian agricultural assessment 2001. National Land and Water Resources Audit, Canberra, Australia.

Northcote, K.H. 1979. *A Factual Key for the Recognition of Australian Soils*. 4th ed. Rellim, Adelaide.

Northcote K.H., Beckmann, G.G., Bettenay, E., Churchward, H.M., Van Dijk, D.C., Dimmock, G.M., Hubble, G.D., Isbell, R.F., MacArthur, W.M., Murtha, G.G., Nicholls, K.D., Paton, T.R., Thompson, C.H., Webb, A.A., and Wright, M.J. 1960–68. Atlas of Australian Soils. Sheets 1–10 with explanatory booklets. CSIRO Publishing, Melbourne, Australia. Soil maps may be purchased at AUSLIG Sales, PO Box 2, Belconnen, ACT 2616.

PIRSA Land Information. 2001. Soils of South Australia's Agricultural Lands. CD-ROM. Primary Industries and Resources South Australia.

Prescott, J.A. 1931. The soils of Australia in relation to vegetation and climate. CSIRO Bulletin 52.

Prescott, J.A. 1947. The soil map of Australia. Bulletin No. 177, CSIRO, Melbourne.

Sammut, J., Callinan, R.B., and Fraser, G.C. 1996. An overview of the ecological impacts of acid sulfate soils in Australia. In Proceedings of the 2nd National Conference of Acid Sulfate Soils, Coffs Harbour, 5–6 September, 1996. Robert J. Smith and Associates and Acid Sulfate Soils Management Advisory Committee, Australia.

Schoknecht, N.R., Ed. 2001. Soil Groups of Western Australia. Resource Management Technical Report 193. Agriculture Western Australia.

Sheard, M. and Bowman, G.B. 1996. Soils, stratigraphy and engineering geology of near surface materials of the Adelaide Plains. Department of Mines and Energy, South Australia. Report Book 94/9.

Skwarnecki, M., Fitzpatrick R.W., and Davies P.J. 2002. Geochemical dispersion at the Mount Torrens lead-zinc prospect, South Australia, with emphasis on acid sulfate soils. Cooperative Research Centre for Landscape Environments and Mineral Exploration (CRC LEME) Report 174. (plus 13 appendices), Perth, Australia.

Slavich, P.G., Ed. 2000. Remediation and assessment of broadacre acid sulfate soils. Acid Sulfate Soil Advisory Committee (ASSMAC), Wollongbar, New South Wales, Australia.

Smith, R.M. and Sobek, A.A. 1978. Physical and chemical properties of overburdons, spoils, wastes and new soils, in F.W. Schaller and P. Sutton, Eds. *Reclamation of Drastically Disturbed Lands*. American Society of Agronomy, Crop Science Society of America, and Soils Science Society of America, Madison, WI.

Soil Classification Working Group. 1991. Soil Classification: A Taxonomic System for South Africa. Memoirs on the Agricultural Natural Resources of South Africa No. 15.

Soil Survey Staff. 1999. Soil Taxonomy: A Basic System of Soil Classification for Making and Interpreting Soil Surveys, 2th ed. Agriculture Handbook No. 436. U.S. Dept. of Agric., Natural Resources Conservation Service, Washington, DC.

Stace, H.C.T., Hubble, G.D., Brewer, R., Northcote, K.H., Sleeman, K.H., Mulcahy, M.J., and Hallsworth, E.G. 1968. *A Handbook of Australian Soils*. Rellim, Glenside, South Australia.

Stephens, C.G. 1952, 1956, 1962. *A Manual of Australian Soils*. 1st, 2nd, and 3rd eds. CSIRO, Melbourne.

Turner, J., Thompson, C.H., Turvey, N.D., Hopman, S.P., and Ryan, P.J. 1990. A soil technical classification for *Pinus radiata* plantations 1. Development of the classification. *Aust. J. Soil Res*. 28:797–811.

Vepraskas, M.J. 1992. Redoximorphic features for identifying aquic conditions. North Carolina State University Technical Bulletin 301, NCSU, Raleigh, NC.

Webster, R. 1968. Fundamental objections to the 7th Approximation. *J. Soil Sci*. 19:354–366.

Webster, R. 1977. *Quantitative and Numerical Methods in Soil Classification and Survey*. Oxford University Press, Oxford, UK.

Webster, R. and Butler, B.E. 1976. Soil survey and classification studies at Ginninderra. *Aust. J. Soil Res*. 14:1–24.

Wetherby, K.G. and Oades, J.M. 1975. Classification of carbonate layers in highland soils of the Northern Murray Mallee, S.A., and their use in stratigraphic and land-use studies. *Aust. J. Soil Res*. 13:119–132.

White, I., Melville, M.D., Lin, C., Sammut, J., van Oploo, P., Wilson, B.P., and Yang, X. 1995. Fixing problems caused by acid sulfate estuarine soils, in C. Copeland, Ed. *Ecosystem Management: The Legacy of Science*. Halstead Press, Sydney.

White, R.E. 1993. The role of soil science in shaping policies for sustainable land management. *Soil News*. 93:1–4.

White, R.E. 1996. Soil science—raising the profile. Proceedings of the Australian and New Zealand Soils Conference 1996: Soil Science—Raising the Profile. Vol. 2, N. Uren, Ed. Plenary papers, Australian Society of Soil Science Inc., Parkville, 1-11.

Wilding, L.P. and Drees, L.R. 1983. Spatial variability and pedology, in L.P. Wilding, N.E. Smeck, and G.F. Hall, Eds. *Pedogenesis and Soil Taxonomy. I. Concepts and Interactions*. Developments in Soil Science 11A, Elsevier, Amsterdam.

Williams, B.G. and Bullock, P.R. 1989. The classification of salt-affected land in Australia. CSIRO Division of Water Resources, Technical Memorandum 89/8. Canberra, Australia.

Yaalon, D.H. 1996. Soil science in transition: Soil awareness and soil care research strategies. *Soil Sci*. 161:3–8.

Zinck, J.A. 1993. Introduction. *International Institute for Aerial Survey and Earth Sciences Journal*. 1993–1, 2–6.

CHAPTER 10

Development of Soil Classification in China

Zi-tong Gong, Gan-lin Zhang, and Zhi-cheng Chen

CONTENTS

0-8493-1339-2/03/$0.00+$1.50

INTRODUCTION

The development of soil classification in China has experienced a long progressive process. Soil characterization and attempts to group soils appeared in Chinese literature as early as two thousand years ago. Modern soil classification began around 1930, and was influenced by early American experiences. Since the mid-1980s, the Chinese Academy of Science, under the leadership of the Soil Science Institute at Nanjing, commenced working toward a Chinese Soil Taxonomic Classification system (CSTC or CST). This decision was made for several reasons. First, by then there were several systematic soil surveys of the country and many detailed soil investigations. Second, western and international classifications, which had been developed and refined since about 1950, had established new concepts and approaches in soil classification. In preparation for this effort, several international workshops and meetings were held in China, and Chinese scientists participated in many similar meetings in other countries.

A first step in the process was to establish the need for a classification system. With a land mass of about 9.6 million km^2 and a wide range of climatic and topographic conditions, the variety of soils that exist in the country is immense. International soil classification systems, such as the United States Soil Taxonomy (ST), could provide a name for almost all the soils. However, one fundamental difference between the soil systems in China and in many parts of the world, specifically the United States, is the influence of humans on land. CST felt that almost all other classification systems did not emphasize this. We felt strongly that the characteristics of our cultivated soils were largely determined by the management regimes in previous eons, and that a classification system that failed to express this has inherent weakness. For these reasons, it was decided that although we should incorporate many of the innovations of international systems, the identity of CST must be maintained with an appropriate emphasis on the specific conditions in China and the demands of our agricultural and environmental systems.

The First Proposal was distributed in 1991, and this was followed in 1995 by a second Revised Proposal. The 3rd Edition of the Chinese Soil Taxonomic Classification was published in 2001 as a set of Keys for field testing. A more detailed monograph describing the rationale and concepts was distributed earlier in 1999. The English edition of Chinese Soil Taxonomy appeared in 2001. With the acceptance of CST by the Chinese scientific community, it is now being used in universities and research institutions. However, it is recognized that the process of developing a classification system is dynamic, and a permanent committee has the responsibility to continue testing, modify and refine when necessary, and keep the system as current as possible.

The purpose of this paper is to describe the salient features of CST. For details of the system, the reader is invited to consult the larger monograph (CSTS, 2001).

BACKGROUND OF NATURAL ENVIRONMENT OF CHINA

Environmental Factors

China has an area of about 9,600,000 square kilometers, extending from about 22 to 55°N and 78 to 135°E. Within the vast territory, there are various soil types developed under different bioclimatic conditions and derived from various parent materials in diversified topographical environments. The relief of China can be divided into three steps or levels. The highest level is the Qinghai-Xizang Plateau with an altitude of more than 4000 m. The second level is from the plateau going northward and eastward to the line of the Da Hinggan Ling, the Taihang Mountains, the Wushan Mountains, east slope of the Yunnan-Guizhou Plateau, and most of the basins in the regions are between 1000 to 2000 m above sea level. The third level is composed of rolling hills and coastal plains where the hills rarely exceed 500 meters in height (Figure 10.1). This highly contrasting topography influences orographic changes and the penetration of the humid tropical Pacific air mass into the western inland. The range of climatic conditions superimposed on the geologic and topographic situations results in a wide variety of soils and soil conditions.

In the Tertiary and Quaternary Periods, geologic activities affected the soil formation predominantly, and consequently the formation of the main soil types is strongly influenced by their parent rocks. Thus in the mountainous regions of northeastern China, Inner Mongolia, and eastern and southern China, the soil-forming materials are mainly granite, rhyolite, andesite, and basalt. In the coastal and delta areas, they are mainly alluvial deposits, and in the northwestern plateau and the North China Plain, they are loessial deposits. Red earths of southern China are mainly derived from Quaternary red clay and Tertiary red sandstone, while the glacial deposits often appear on the Qinghai-Xizang Plateau.

Redox processes govern the geochemistry of soil formation and, as a result, the weathering products are divided into oxidizing and reducing crusts. The oxidizing crust is further divided into five types, i.e., fragmental crust resulting from mechanical disintegration of rocks; gypsum and salt crust in the inland basins and salt lakes of the desert areas in Xinjiang Basins and Qinghai-Xizang Plateau; carbonate crust in the Northwest Plateau, the North China Plain, and the central and southern parts of the Qinghai-Xizang Plateau; siallitic crust containing laminated clays of 2:1 type as dominant clay minerals associated with vermiculite and transitional clay minerals; and allitic crust in South China (Gong et al., 1997).

China has a very distinctive monsoon climate. In summer, high temperature prevails in both North and South China, but in winter, the climate is very cold in the north and rather warm and mild in the south. According to the temperature variations, five climate zones can be classified, i.e., the cool temperate zone, temperate zone, warm temperate zone, subtropical zone, and tropical zone. The regions are differentiated using an index called "accumulated temperature" which is the product of the number of days when the temperature is >10°C and the temperature of each of these days. Northern Heilongjiang Province, where the frostless season is about three months with an accumulated temperature of ≥10°C being less than 1,700°C belongs to the cool temperate zone.

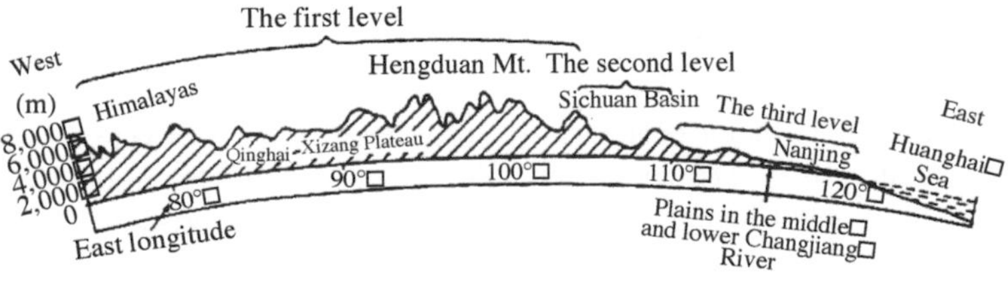

Figure 10.1 Sketch of topographic section of China. (From Xiong et al., 1987. With permission.)

Northeast China and Inner Mongolia having a frostless season of 4 to 7 months, with an accumulated temperature of ≥10°C being about 1,700 to 3,500°C, belong to the temperate zone. North China having a frostless season of 5 to 8 months, with an accumulated temperature of 10°C being 3,500–4,500°C, belongs to the warm temperate zone. The regions south of the Changjiang River are subtropical zones. The regions north of the Changjiang River and south of the Huaihe River having a frostless season of 8 months, with an accumulated temperature of ≥ 10°C being 4,500 to 5,300°C, are recognized as the northern subtropical zone. The vast areas between the Nanling Range and the Changjiang River having a frostless season of 9 to 10 months and an accumulated temperature of ≥10°C being 5,300 to 6,500°C belong to the middle subtropical zone. The regions south of the Nanling Range and Southern Yunnan Province, with no winter and no snow, with an accumulated temperature of ≥10°C being 6,500 to 8,000°C, belong to the southern subtropical zone. Hainan Province, the southern extremity of Guangdong Province Taiwan Province, and Yunnan Province, generally have no temperature below 0°C and no frost all year round; with an accumulated temperature of ≥10°C being 8,000 to 8,500°C or more than 8,500°C, these regions belong to the tropical zone.

In China, the coastal areas have a much higher annual rainfall, which decreases northwestwards, while the evaporation rate is changing in the reverse direction. According to annual precipitation and aridity, China is divided into four climatic regions, i.e., the humid region, semihumid region, semiarid region, and arid region. Southeast China and the northern part of Northeast China, where the annual precipitation is over 800 mm and the degree of aridity is less than 1.0, belong to the humid region, covered by good forests and with rare incidence of drought. The Northeast China and North China Plains and southern part of the Qinghai-Xizang Plateau where the annual precipitation is about 400 mm, with an aridity degree of about 1.5, belong to the semihumid region; in this region, the natural vegetations are forests and steppe (or meadow), the soils are mostly calcareous, and drought happens rather frequently in spring. The semiarid region covers Inner Mongolia, the Loess Plateau, and Qinghai-Xizang Plateau, where the annual precipitation is usually less than 400 mm, with an aridity degree of about 2.0. In this region, the soils are calcareous and are usually salinized, and drought is rather serious in spring. The natural vegetation is mainly steppe, and the soils are used for agriculture with appropriate irrigation. The arid desert region includes Xin-jiang, western Inner Mongolia Plateau, and northwestern Qinghai-Xizang Plateau, where the annual precipitation is less than 200 mm with an aridity degree of 2.0 to 8.0.

The natural vegetation in China is divided into three belts, i.e., the forest belt, the grassland belt, and the desert belt. The forest belt covers the conifer-broadleaf mixed forest areas in the northern part of Northeast China, the deciduous broadleaf forest areas in the hilly lands of Liaoning and the Shandong Peninsulas, the North China Plain and the Loess Plateau in warm temperate zone, the subtropical ever-green broadleaf forest areas south of the Qinling Mountains, and the Huaihe River and tropical rain forest areas of the southernmost China. The grassland belt is located in the northwest of the forest belt. It covers the Song-Liao Plain, the Hulunbeier Plateau, the Inner Mongolia Plateau, the central part of the Loess Plateau, and the central and southern parts of the Qinghai-Xizang Plateau. The desert belt is located in the inland basins of the Qinghai-Xizang Plateau and the Xinjiang Plateau. It is covered by desert bushes and steppe.

Major Ecological Regions of China

China is divided into six major ecological regions, based upon a broad climatic and physiographic zonation (Figure 10.2). The cool, subhumid Dongbei Plain (1a) in the Northeast and the humid temperate Huabei and Changjiang Plains (1b) in the East comprise the Chinese lowlands.

1. The cool, subhumid Da Hinggan, Xiao Hinggan, and Changbai Mountain ranges in the Northeast.
2. The semiarid to arid Nei Mongol Plateau with the adjacent, partly dissected, and subhumid Loess Plateau in the central northern part.

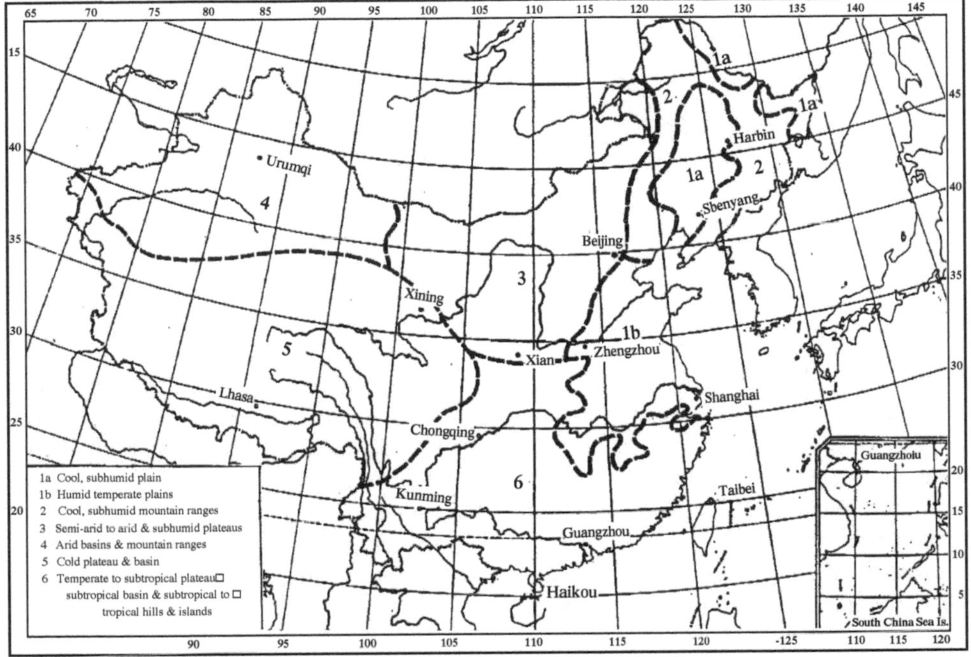

Figure 10.2 Ecological zones of China. (From Zhao, 1986. With permission.)

3. The arid Junggar and Tarim Basins and surrounding mountain ranges in the Northwest.
4. The cold Qinghai-Xizang Plateau (including the Qaidam basin) bordered by the Himalayan mountain range in the Southwest.
5. The temperate to subtropical Yunnan-Guizhou Plateau, the subtropical Sichuan Basin, and the subtropical to tropical Dongnan Hills and Hainan Island in the South and Southeast.

SOIL IDENTIFICATION IN ANCIENT CHINA

Long History of Agriculture

The relics of the Yangshao Culture and the Hemudu Culture in the New Stone Age were discovered in the Banpo village, Xi'an, located in the reaches of the Yellow River, and in many places in the delta of the Yangtze River, respectively, both of which are the Chinese culture cradles. The carbonized grains in those relics date back to 6000 to 7000 years ago. Much of the pottery of the New Stone Age, unearthed in the middle reaches of the Yangtze River in 1988, may be traced back to 8200 to 9100 years ago (Gong and Liu, 1994).

These and other archeological studies demonstrate that as early as the New Stone Age, our ancestors had already cultivated cereal crops. Afterwards, the farming activities gradually spread nationwide, with the propagation of Chinese culture.

Agriculture in Ancient China

During the Spring and Autumn Period and the Warring States Period (770 to 221 BC), some famous water projects were built, such as the Zhengguo Channel, Ximen Bao Channel, Dujiang Weir and Ling Channel. Many of these still play an important role in agricultural production today. During the Han Dynasty (206 BC to 8 AD), the technology of breeding and transplanting rice seedlings and planting vegetables in a greenhouse was introduced. Since then, the cultivation in

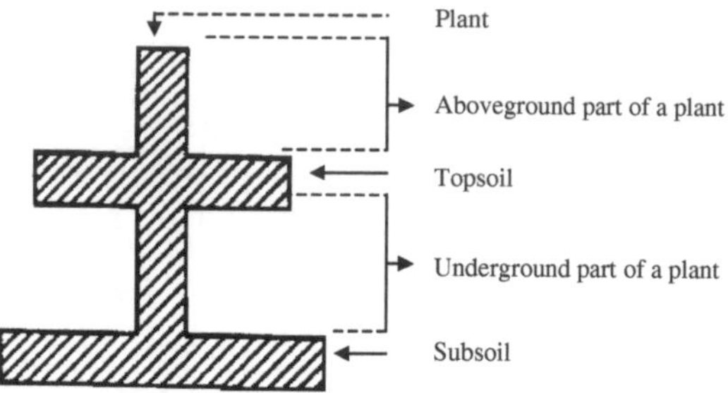

Figure 10.3 Interpretation of Chinese character ± (i.e., soil). (From Wang, 1980. With permission.)

the Yellow River Valley has developed gradually toward traditional agriculture with plowing, harrowing, and hoeing as the main cultivation techniques, applying huge amounts of farmyard manure, conserving soil and water, planting green-manure crops, carrying out the rotation system, and combining land use with land maintenance. Ancient people also created terraces to take full advantage of mountainous lands. In the aridic region, our ancestors built the Kanerjing wells for the development of irrigation agriculture. Ancient people also introduced the planting and use of green manure, improving cultivation measures and renewing soil fertility, which greatly increased agricultural productivity (Gong et al., 2000).

The Earliest Records of Soil Knowledge

Xushen (58–147 AD) pointed out that "the soil is that part of the land bearing living things." In the Chinese character ± (i.e., soil), the upper horizontal stroke refers to the topsoil, the lower horizontal stroke to the subsoil, and vertical stroke to both aboveground and underground parts of plants (Figure 10.3). It illustrates that the soil nurtures and supports plant growth, revealing the close relationship between the soil and plant, and shares the same meaning with the soil definition in modern pedology. This is probably regarded as the earliest scientific interpretation of soil and its integral function.

The Earliest Soil Classification

In Yugong, a book appeared about 2500 years ago in which soil fertility, soil color, soil texture, soil moisture regime and salinity were used as the criteria of soil classification. Based on these criteria, the soil distribution of soils in ancient China was described and is reproduced in Figure 10.4 and Table 10.1 (Lin, 1996).

The Earliest Soil Exhibition

Chinese emperors in many dynasties established "the land and grains altar" in order to show their reverence to the Land God and Grain God, for a good harvest and the security of the country. The altar was first established in the Zhou Dynasty (1000–771 BC). The five-color "sacrificial altar" established in 1421 of the Ming Dynasty has been preserved at Zhongshan Park, Beijing. Its highest layer is 15.8 m^2 in area, paved with five types of soils with different colors: soils in the east are blue, south—red, west—white, north—black, and center—yellow (Figure 10.5). It is consistent with the general soil distribution of the country: in the east of the country, most soils are blue because of gleization; in the south, the dominant soils, Ferrosols, are red; in the northwest, Aridosols and saline soils are white; and in the center, Cambosols in the Loess Plateau are yellow. Therefore, this is the preliminary, but scientific, knowledge of the soil classification and distribution in the country, and is also the earliest public portrayal of the nation's soil resources.

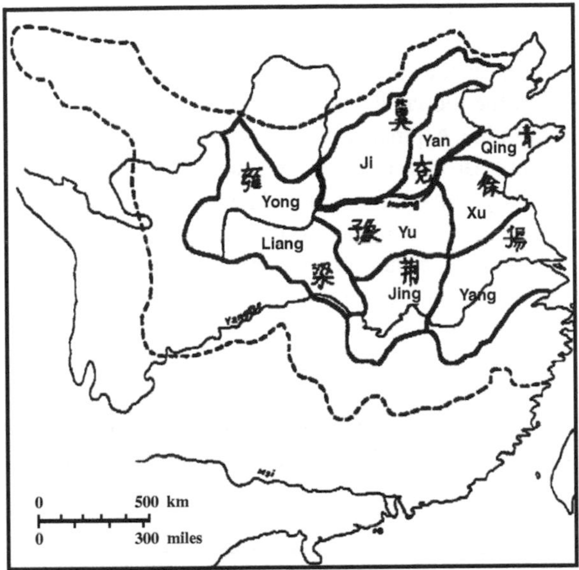

Figure 10.4 Sketch map of the nine provinces of China (Yugong, about 2500 B.P.) (Adapted from Lin, 1996. With permission.)

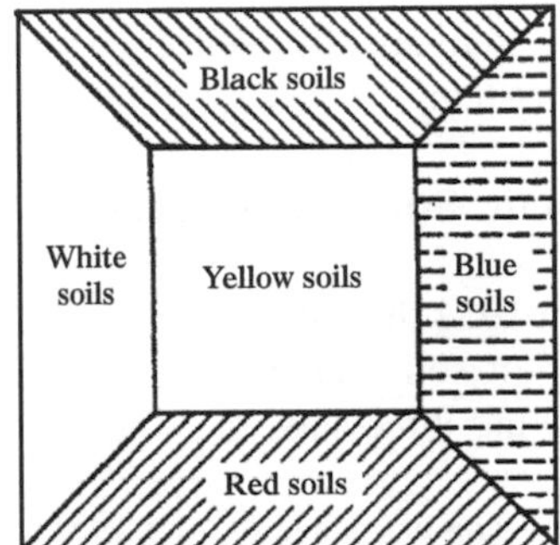

Figure 10.5 The sketch showing a five-color "sacrificial altar." (From Gong, 1999. With permis-

Table 10.1 Distribution of Soils in the Nine Provinces of Ancient China

Province of Ancient China	Description from Yugong (Lin,1996)			Corresponding Soils in Chinese Soil Taxonomy
	Soil	Color	Texture	
Ji	Bai Rang	White	Loam	Halosols, Aquic Cambosols
Jing	Tu Ni	Grey	Loamy clay (clay)	Gleyosols with Ferrosols
Liang	Qing Li	Dark	Loamy clay	Aquic Cambosols and Gleyosols
Qing	Bai Fen, Hai Bin Guan Chi	White	Clay loam	Argosols, Halosols
Xu	Chi Zhi Fen	Red	Loamy clay	Udic Argosols on limestone
Yan	Hei Fen	Black	Clay loam	Vertosols, Argosols
Yang	Tu Ni	Grey	Loamy clay (clay)	Gleyosols with Ferrosols
Yong	Huang Rang	Yellow	Silty loam	Ustic Cambosols on loess
Yu	Rang, Fen Lu	Yellow (black)	Loam	Ustic Argosols, Aquic Cambosols and Vertosols

From Lin, 1996. With permission.

DEVELOPMENT OF MODERN SOIL CLASSIFICATION IN CHINA

Early Marbut System

China has a longstanding history and rich experience in soil classification, but the study of modern soil classification only began in the 1930s. By combining the experience of Marbut's system developed in the United States (Marbut, 1927) with that of the Chinese, the concept of Great Groups was adopted, and more than 2,000 soil series were established. A comprehensive monograph—"Soil Geography of China"—was published to present a brief introduction to soils and their formation. Because of poor working conditions, simple equipment, and only a few people involved in the field of soil science before 1949, most of the work related to soil survey and soil mapping was done in eastern China. Soil classification made much progress after 1949, when many soil resource investigations, watershed planning, and general detailed soil surveys on a large scale were carried out.

At the beginning of the 1950s, Chinese classification still followed the Marbut concept according to Song Daquan in the paper, "Discussion on Chinese Soil Classification Criterion." A soil group was a basic soil unit, and soil series were classes of the lowest category. At the soil order level, there were Zonal, Azonal, and Intrazonal soils. At the second level (suborders) there were Pedocals, Pedalfers, Hydromorphic, Halomorphic, Calcimorphic, Alpine, and Young soils. These were divided into 18 soil groups, such as Chernozems, Chestnut, Brown, Red, and Yellow Soils. Soil subgroups and soil series were further subdivisions of the soil groups. The unique Shangdong brown, Shajiang black, and Paddy soils are soil groups of long standing in China, and the soil series established in this system are still well known by many soil scientists, although they are not directly used in the new system.

Genetic Soil Classification

In 1954 the Chinese economy was in the restoration and development period, and the working group of soil science was expanded. The resulting genetic classification can be divided into the following three stages:

1. The first stage is from 1954 to 1958. The pattern of the system followed that of the Soviet Union. A soil classification system was designed in the National Soil Congress (1954), which was based on the geographical genesis of soils and soil-forming conditions. The soil type was the basic unit, and the system had five levels: type, subtype, genus, species, and variety. After additional study, the following new soil types were proposed: Meadow, Drab, Yellow brown, Brown taiga, Black, Albic bleached, Heilu, Gray brown desert, Takyric, Latosols, Lateritic red, and Mountain meadow soils. According to zonality, soil distribution was identified, and soil maps at small and medium scales were compiled.
2. The second stage is from 1958 to 1978. Throughout the first General Detailed Soil Survey of the entire country (1958–1961) and the practice of soil improvement, in which more attention was paid to cultivated soils, it was deemed important to distinguish natural soils and cultivated soils at different categorical levels with their actual situation. Chao soils (cultivated fluviogenetic soils), Irrigated soils, Oasis soils, and Mien soils (loessial soils) were proposed, in addition to Paddy soils. In the monograph, "Soils of China" (1978), these soils were described in separate chapters. After the Cultural Revolution, the Chinese Soil Science Society held its first soil classification meeting in 1978 and made up a national soil classification system, "The Provisional Chinese Soil Classification," in which natural and cultivated soils were combined, and the Alpine and Phospho-calcic soils were established. Because of wide and open discussion, this system was accepted broadly and became the official system used in the Second General Detailed Soil Survey in whole country.
3. The third stage is from 1978 to 1984. With increasing international exchanges, Soil Taxonomy and the FAO Legend of Soil Map of the World were introduced into China. Although the soil classification system was still based on genetic concepts, it was influenced to a certain extent and also absorbed some ideas and names. In "Chinese Soil Classification System" (1984), which was edited by the Office of Second General Detailed Soil Survey, there were soil orders of Humosols and Primarisols, whose names are similar to other systems. Meanwhile, the Chinese forest soil scientists submitted evidence for 11 soil orders, including Frostic, Organic, Primitic, and Anthric Soils, and 13 diagnostic horizons were presented.

Undoubtedly, soil genetic concepts played vital roles in the research and application of soil science. Besides comprehensive monographs, "Soils of China" and "Agricultural Soils of China," a series of regional monographs, focusing on Northeast China, Xinjiang, Tibet, Inner Mongolia, North China, South China, Central China, East China, and Islands of South Sea, etc., were also published. Some soil groups were intensively studied—for example, Paddy, Black, Albic bleached,

Luo, Alpine meadow, Podzols, Purplish, and Phospho-calcic soils. For the national soil inventory, many soil maps at scales from 1:50,000 to 1:250,000 were compiled, in addition to national ones at scales of 1:4 million and 1:10 million. All were based on the existing genetic soil classification system. Thus the school of genetic classification has influenced Chinese thinking and will continue to do so.

Genetic classification is generally based on the soil-forming hypotheses, and the relationships are usually only qualitative. Consequently, people may classify the same soil into different soil groups with different understanding of genesis. In 1959, Albic bleached soils were classified as Podzols because of confusion about the spodic and bleaching processes. Genetic classification tended to emphasize bio-climatic conditions but ignored the time factor, and consequently there was sometimes confusion about which process was to be used for classification of a particular group of soils. For example, this permitted the Purplish soils to be mistakenly classified as Yellow soils. Most genetic classifications have described well the central concept of a soil group, but often there are not well-defined boundaries between soil groups. It is often difficult to properly place a soil, and the classification is ambiguous. The lack of quantitative criteria made it difficult to establish an information system or to develop an automated system of classification.

Taxonomic Soil Classification

In 1960, American soil scientists, in cooperation with others, created a classification system later to be known as Soil Taxonomy. It took 10 years and was revised seven times, based on the research results around the world. During that time, research on soil classification in China was at a standstill for 10 years. When the Chinese soil delegation took part in the 12th International Soil Science Society World Soil Congress, they were astonished that it was so difficult to exchange ideas with other soil scientists because of misunderstandings about Soil Taxonomy and the World Soil Map Legend, which were being widely used throughout the world. Consequently, Chinese soil scientists began to introduce ideas from abroad about progress in soil classification. To facilitate our understanding, important publications such as, *Keys to Soil Taxonomy*, *Rationale for Concepts in Soil Taxonomy*, and *Comments on International Soil Classification* were translated and published in Chinese.

In order to fill the gap, since 1984 the Institute of Soil Science, Chinese Academy of Sciences, has led a research group on Chinese Soil Taxonomic Classification. This collaborative effort comprises a number of researchers from more than 30 universities or institutions in the country. In 1985, China began research on a soil taxonomic classification. As mentioned above, Chinese Soil Taxonomic Classification (First proposal), the Chinese Soil Taxonomic Classification (Revised proposal), and Keys to Chinese Soil Taxonomic Classification (3rd Edition) were successively published in 1991, 1995, and 2001. Meanwhile, a comprehensive monograph of Chinese Soil Taxonomic Classification—Theory/Method/Practice—was published in 1999. But all of these publications are in Chinese only. In 1994, an English draft version of Chinese soil taxonomic classification was edited and distributed in some international meetings (Gong, 1994) and was used for soil study (ISSAS and ISRIC, 1995). The Chinese Soil Taxonomic Classification (Revised proposal) was fully translated and published in Japanese. Chinese Soil Taxonomy was published in English in 2001. Meanwhile this classification was introduced in detail by *Handbook of Soil Science* (Sumner, 2000) and *Soil Terminology and Correlation* (1999) in Russian.

The classification of Anthrosols in Chinese Soil Taxonomic Classification has been adopted as the classification of Anthrosols in World Reference Base for Soil Resources. In the last 10 years there have been more than 400 papers published in different journals and proceedings, not only in Chinese but also in English, Japanese, French, and Russian. Thus the classification became more acceptable both in China and abroad.

SPECIFIC CHARACTERISTICS OF CSTC

Based on the Diagnostic Horizons and Diagnostic Characteristics

Diagnostic horizons are the horizons used to identify the soil types and that have a series of quantified properties. They are the building blocks of the taxonomy.

There are several taxonomic soil classifications, developed by individuals, institutions, or through international cooperation. They all have features in common, but differ in some other major aspects. Most national systems tend to highlight soils that are of largest extent, form their major agricultural base, or even that have major constraining properties. In highlighting some, there is a concomitant effect to de-emphasize others, and this results in differences in the systems. Over the last few years, however, there has been increasing similarity in the approaches to identification and definition of diagnostic horizons. Consequently, a common language for international exchange and understanding on taxonomic soil classifications is now becoming available. In view of the practical needs of Chinese Soil Taxonomy, we have established a set of 33 diagnostic horizons and 25 diagnostic characteristics. These were established based on the following considerations:

- Some are adopted directly from other systems, if they are applicable to the Chinese situation; in such cases, we have used the original terms and definitions. Examples include mollic, umbric, and ochric epipedon; albic, glossic, spodic, hypersalic, gypsic, hypergypsic, calcic, hypercalcic horizons, lithic contact, paralithic contact, permafrost layer, *n* value, andic property, and alic property. Only the petrocalcic horizon was renamed as calcipan, according to the traditional use in China.
- In some cases, terms and definitions of other systems were modified or expanded to serve our local needs. These are histic epipedon, cambic horizon, ferralic horizon, agric horizon, agric horizon, alkalic horizon, salic horizon, sulfuric horizon, organic soil materials, vertic features, soil moisture regimes, soil temperature regimes, gleyic features, humic property, sodic property, calcaric property, and sulfidic materials.
- A number of diagnostic horizons and diagnostic characteristics, totaling 12 and 9, respectively, are proposed based on the special soil resources, especially anthropogenic soils in old agricultural areas, apline soils in Tibet plateau, extreme arid-desert soils in Xinjiang, etc., and their available information in China, such as mattic, irragric, cumulic, fimic, anthrostagnic, aridic epipedon, salic crust, LAC-ferric horizon, hydragric horizon, claypan, salipan, phosphipan, lithologic characteristics, anthro-silting materials, anthroturbic layer, frost thawic features, redoxic features, isohumic property, allitic property, phosphic property, and base saturation.

Using Soil Genetic Theories as the Guide

Following the Russian approach in soil genesis, soils were divided into clastic soils, halogenic soils, carbonate-genetic soils, siallitic soils, and ferrallitic soils, according to the developing stages of soil geochemistry. These divisions may be stable, but difficult to be identified in fields. Soil taxonomy paid great attention to the morphogenesis, which was not very stable. We emphasize the historical genesis in the First Proposal, and combine it with morphogenesis in our system, shown in the table below:

Historical Geno-sequence	Clastic Soil	Halogenic Soils	Carbonate-genetic Soils	Siallitic Soils	Ferrallitic Soil
CSTC system:	Primosols	Halosols Aridosols	Aridosols Isohumosols Cambosols	Cambosols Argosols Spodosols	Ferrosols Ferralosols

Linkages with International Soil Classification

Both plant classification and animal classification are world-uniform classifications. Thus soil classification should develop similarly. In addition to using the traditional diagnostic horizons and diagnostic characteristics, we had to create new ones, according to the same principles and methods, to meet our specific needs. Despite these additional horizons, higher categories in our system can still be compared with those of Soil Taxonomy, FAO/UNESCO map units, and WRB. We have tried to retain the international nomenclature as much as possible and retained the following terms: Histosols, Anthrosols, Spodosols, Andosols, Ferralosols, Vertosols, Aridosols, Halosols, Gleyosols, Isohumosols, Ferrosols, Argosols, Cambosols, and Primosols.

Unique Features of the Soils of China

Anthrosols. China is an ancient agricultural country in which the influences of human activities upon soil development is much greater in depth and intensity than that of any other country in the world, especially for the paddy soils that occupy one-fourth of the total area of the world paddy soils (Gong, 1986). Based on anthropogenic horizons as an anthrostagnic epipedon and hydragric horizon, a fimic epipedon and phos-agric horizon, and a irragric epipedon or cumulic epipedon, Anthrosols are grouped into Stagnic Anthrosols, Fimic Anthrosols, Cumulic Anthrosols, and Irragric Anthrosols.

Aridosols. The soils are characterized by a long dry period, high aridity degree, and possession of various types of special characteristics. Aridosols are identified by the aridic epipedon (Figure 10.6) and one or more of the following diagnostic horizons: salic, hypersalic, salipan, gypsic, hypergypsic, calcic, hypercalcic, calcicpan, argic, or cambic horizon, whose upper boundaries are within 100 cm of the mineral soil surface (Eswaran and Gong, 1991).

Subtropical soils. The soils possess the characteristics of strong leaching but not very strong weathering under the influence of monsoon climate. In tropical and subtropical China, there are Ferrosols based on low activity clay (LAC) ferric horizon (diagnostic subsurface horizon), which has low activity clays and is rich in free iron oxides (CEC_7 = 24 cmol (+) kg;$^{-1}$ free iron/total iron > 40%). Such soils occupy a total area of 2 million km^2.

Alpine soils. The soils resemble, but are not the same as, the arctic soils. Tibet area is called the third polar, where the climate is cold and dry. The soils are classified as Cryic Aridosols that have a cryic soil temperature regime, and Gelic Cambosols that have both a cryic or colder soil temperature regime and frost-thawic features (Gong and Gu, 1989).

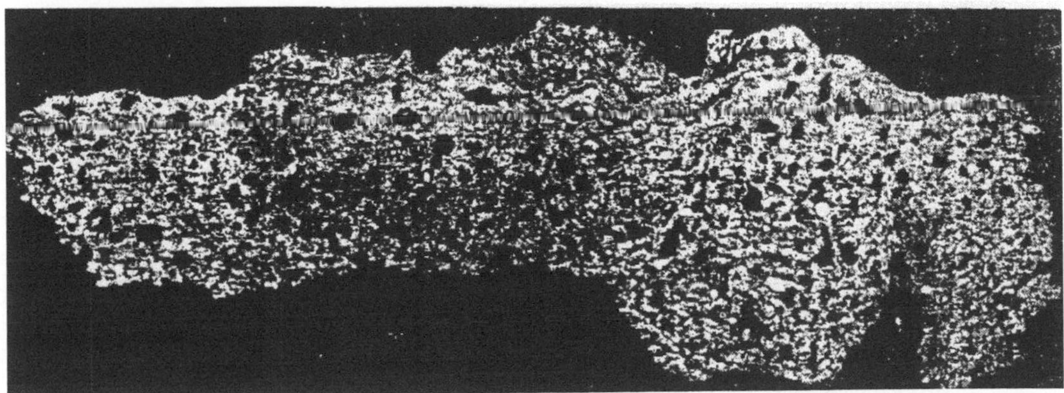

Figure 10.6 Thin section photo (_ 2.1) of vesicular crust (black parts being vesicles) of the aridic epipedon, 0–1.5(2) cm (Miquan, Xinjiang). (From Cao and Lei, 1995. With permission.)

SOIL ORDERS IN CSTC

Keys to Soil Orders

A. Soils that:

- Do not have andic soil properties in 60% or more of the thickness between the soil surface and either a depth of 60 cm, or a lithic or paralithic contact if shallower
- Have organic soil materials that meet one or more of the following:
  - Overlie cindery, fragmental, or pumiceous materials and/or fill their interstices, and directly below these materials either a lithic or paralithic contact
  - When added with underlying cindery, fragmental, or pumiceous materials, total 40 cm or more between the soil surface and a depth of 50 cm
  - Constitute two thirds or more of the total thickness of the soil to a lithic or paralithic contact, and minerals soils which, if present, have a total thickness of 10 cm or less
  - Are saturated with water for 6 months or more per year in most years (or artificially drained), have an upper boundary within 40 cm of the soil surface, and have a total thickness of either
    - 60 cm or more if three fourths or more of their volume consists of moss, or if their bulk density, moist, is less than 0.1 Mg/m^3
    - 40 cm or more if they consist either of sapric or hemic materials, or of fibric materials with less than three fourths (by volume) moss fibers and a bulk density, moist, of 0.1–0.4 Mg/m^3

Histosols

B. Other soils that have the following combination of horizons:

- Have an anthrostagnic epipedon and hydragric horizon
- Have a fimic epipedon and phos-agric horizon
- Have a siltigic epipedon or cumulic epipedon

Anthrosols

C. Other soils that do not have anthropic epipedon and do not have argic horizon above a spodic horizon, but have:

- A spodic horizon that has its upper boundary within 100 cm of the minerals soil surface
- No andic property in 60% or more of the thickness either
  - Within 60 cm either of the minerals soil surface, or of the top of an organic layer with andic property, whichever is shallower, if there is no lithic or paralithic contact within that depth
  - Between the minerals soil surface, or the top of an organic layer with andic properties, whichever is shallower, and a lithic or paralithic contact that is present within 60 cm either of the mineral soil surface or of the top of an organic layer with andic property, whichever is shallower

Spodosols

D. Other soils that have andic soil property in 60% or more of the thickness either

- Within 60 cm either of the mineral soil surface, or of the top of an organic layer with andic property, whichever is shallower, if there is no lithic or paralithic contact, within that depth
- Between the minerals soil surface, or the top of an organic layer with andic properties, whichever is shallower, and a lithic or paralithic contact that is present within 60 cm either of the mineral soil surface or of the top of an organic layer with andic property, whichever is shallower

Andosols

E. Other soils that have a ferralic horizon with its upper boundary within 150 cm of the mineral soil surface.

Ferralosols

F. Other soils that have:

- Vertic features within 100 cm of the mineral soil surface
- No lithic or paralithic contact within 50 cm of the mineral soils surface

Vertosols

G. Other soils that have:

- An aridic epipedon
- One or more of the following diagnostic horizons: salic, hypersalic, salipan, gypsic, hypergypsic, calcic, hypercalcic, calcipan, argic, or cambic horizon, whose upper boundaries are within 100 cm of the mineral soil surface

Aridosols

H. Other soils that have:

- A salic horizon whose upper boundary is within 30 cm of the mineral soil surface
- An alkalic horizon whose upper boundary is within 75 cm of the mineral soil surface

Halosols

I. Other soils that have gleyic features at least in some horizon 10 cm or more thick within 50 cm of the mineral soil surface

Gleyosols

J. Other soils that have:

- A mollic epipedon
- An isohumic property
- A base saturation of 50% or more (by NH_4OAc) in all horizons, either between the mineral soil surface and a depth of 180 cm, or between the upper boundary of any argic horizon and a depth of 125 cm below that boundary, or between the mineral soil surface and a lithic or paralithic contact, whichever depth is shallowest

Isohumosols

K. Other soils that have a LAC-ferric horizon with its upper boundary within 125 cm of the mineral soil surface

Ferrosols

L. Other soils that have either:

- An argic horizon which has its upper boundary within 100 cm from the mineral soil surface
- A claypan which has illuviation argillan 0.5 mm or more thick in some parts, and which has its upper boundary within 100 cm of the mineral soil surface

Argosols

M. Other soils that have:

- A cambic horizon
- Within 100 cm of the mineral soil surface, any one of the following: an albic, calcic, hypercalcic, gypsic, hypergypsic horizon, or a calcipan
- At least in one horizon (10 cm or more thick) between 20 cm and 50 cm below the mineral soil surface, either a *n* value of less than 0.7, or less than 80 g/kg clay in the fine-earth fraction; and one of the following: a histic, a mollic, or an umbric epipedon
- A permafrost layer and stagnic soil moisture regime within 50 cm of the mineral soil surface for at least 1 month per year in 6 or more out of 10 years

Cambosols

N. Other soils.

Primosols

Characteristics and Distribution of Soil Orders

In CSTC, 14 dominant soil orders are set up, based on diagnostic horizons and diagnostic characteristics.

Histosols. Histosols are soils composed mainly of organic soil materials that are accumulated from plant residuals, and they have a soil organic carbon content of at least 12–18%, depending on soil clay content. They occur dominantly in northeastern China as well as the eastern and northern margin areas of the Qinghai-Xizang Plateau, in cool temperate and temperate zones. Their total area is 14,000 km^2.

Anthrosols. Anthrosols are soils with an anthraquic epipedon and hydragric horizon, a fimic epipedon and phos-agric horizon, an irragric epipedon, or cumulic epipedon. Covering an area of about 313,000 km^2, these soils occur nearly throughout China. However, they mainly occur in the eastern and southern parts of China, especially in the deltas, which have intensive human activities and a long history of agriculture and dense population, such as the Yangtze Delta and Zhujiang Delta.

Spodosols. Spodosols are soils that have a spodic horizon within 100 cm of the mineral soil surface. Spodosols occur mostly in the cool and humid northern Da Hinggan mountains and in the Changbai Mountains and southern margins of Qinghai-Tibet Plateau. They are in very small areas, commonly covered by coniferous trees.

Andosols. Andosols are soils that developed from tephra or lava and that have andic soil properties. They occur in volcanic areas of northeastern and southern China, with a total area of only about 2,000 km^2.

Ferralosols. Ferralosols are soils that have a ferralic horizon within 150 cm of the mineral soil surface. This ferralic horizon is at least 30 cm thick. It has a texture of sandy loam or finer, and has a clay content of 80 g/kg or more. It has a CEC_7 of lower than 16 mol (+)/kg clay and an ECEC of lower than 12 mol (+)/kg clay, and it has a weatherable minerals content of <10% in fractions of 50–200 mm, or a total potassium content of <8 g/kg (K_2O <10 g/kg) in the fine-earth fraction. It has <5% by volume that has rock structure, or it has sesquans on the weatherable mineral fragments. These soils occur in low hills and terraces in hot and humid tropical and southern subtropical regions. Their area totals 103,000 km^2.

Vertosols. Vertosols are soils that have vertic features within 100 cm of the mineral soil surface. They are mainly derived from river and lacustrine deposit, and occur dominantly on the Huaibei Plain and in the Nanyang Basin and Xiangfan Basin, in the eastern and central parts of China. They cover an area of 52,000 km^2.

Aridosols. Aridosols are soils that have an aridic epipedon and one or more diagnostic subsurface horizons, such as salic horizon, salipan, gypsic horizon, calcic horizon, argic horizon, or cambic horizon, within 100 cm of the mineral soil surface. The aridic epipedon is a surface horizon with special morphological differentiation formed under an aridic moisture regime, and has one of the following land surface features:

- desert pavements consisting of gravel or stones with varnish or wind carving striae
- sand layer–sand gravel layer or small dune
- polygonal cracks and organic crusts with dark color
- bare land with polygonal cracks filled with sands and/or silts grains; very thin clay capping on the top surface of polygons; does not have a salic or natric puffed crust or the mixed layer of soils and salts from the surface; has a puffed crust with a thickness of 0.5 cm or more from the surface, or a sheet layer in squama structure

Aridosols occur mainly in arid regions of northwest China and in the Qinghai-Xizang Plateau. They cover a large area of 2,020,000 km^2.

Halosols. Halosols are soils that have a salic horizon within 30 cm of the mineral soil surface or an alkalic horizon within 75 cm of the mineral soil surface. The salic horizon is at least 15 cm thick, has a salt content of 20 g/kg or more in arid regions or 10 g/kg or more in other regions, and has a product of salt content multiplied by the thickness, equaling 600 or more. Halosols occur mainly in arid, semiarid, and desert regions of northern China. Their area totals about 312,000 km^2.

Gleyosols. Gleyosols are soils that have a horizon (>10 cm thick) with gleyic features (violently reduced features due to long-term water-logging) within 50 cm of the mineral soil surface. These soils are distributed in depressions, low land, flood land, and valleys. Their total area in China is 94,400 km^2.

Isohumosols. Isohumosols are soils that have a mollic epipedon and isohumic property and base saturation of 50% or more. They occur mostly in grassland and meadowland in semiarid and temperate regions. They cover an area of 624,000 km^2.

Ferrosols. Ferrosols are soils that have a LAC-ferric horizon within 125 cm of the mineral soil surface. This LAC-ferric horizon is a horizon formed in the preceding moderate ferralization, and which contains low-activity clay and free iron oxides, and is at least 30 cm thick. It has a texture of sand loam or finer, and has a clay content of 80 g/kg or more. It has a hue of 5YR or redder, or a DCB extractable iron content of 14 g/kg or more in fine-earth clay, and has CEC of < 24 cmol (+)/kg clay in subhorizons (10 cm or thicker). It does not meet the requirements of a ferralic horizon. These soils occur mainly in southern China, in central and southern subtropical and tropical regions. Their area totals about 350,000 km^2.

Argosols. Argosols are soils that have an argic horizon or claypan with illuviation cutans within 125 cm of the mineral soil surface. Argosols are distributed in the central and northern parts of eastern China, in areas influenced by monsoonal climate, mainly in low mountains and hills. Their area totals 558,000 km^2.

Cambosols. Cambosols are soils that have a cambic horizon, which has no illuviation of materials and no argillification, and has a color of brown or reddish-brown or red, yellow or purple, etc., and has a weakly developed B horizon, or has one diagnostic subsurface horizon (e.g., albic horizon, calcic horizon, etc.) within 100 cm of the mineral soil surface. These soils occur throughout the country, mainly in the east and south of the Qinghai-Xizang Plateau and in the Huang-Hui-Hai Plain. They cover the largest area of the 14 dominant soil orders, 2,320,000 km^2.

Primosols. Primosols are soils that have one surface layer and no diagnostic subsurface horizon. Primosols occur extensively in China, with an area totaling 608,000 km^2, commonly in the semiarid and arid regions of northern China and the Sichuan Basin. The Sketch Soil Map of China (compiled by CSTC Group, ISSAS) shows the broad distribution of the soil orders as mentioned above.

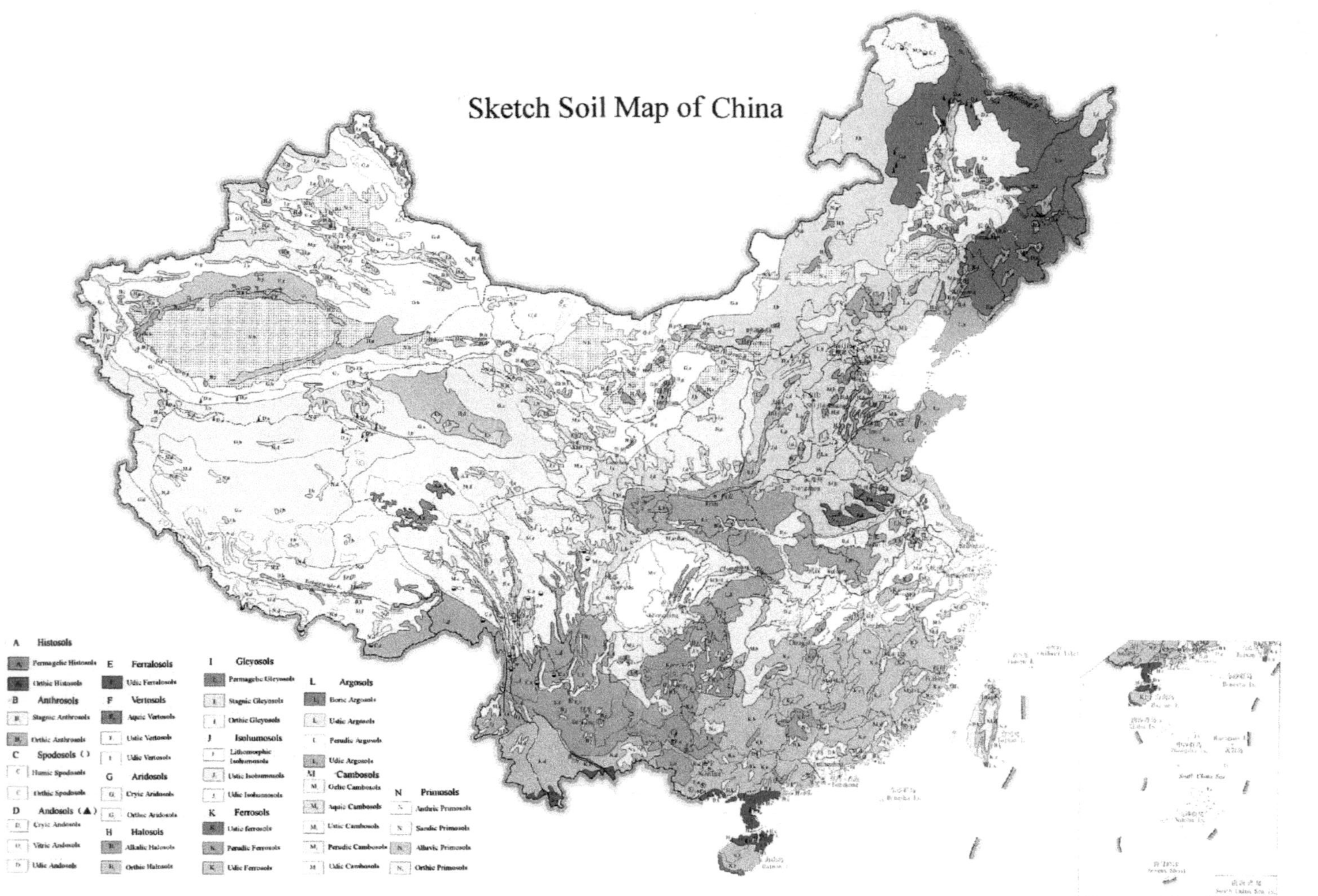

Figure 10.7 Sketch soil map of China. (Map compiled by the Chinese Soil Taxonomic Classification Research Group of the Institute of Soil Science, Academia Sinica.)

Correlation to Other Systems

Chinese Soil Taxonomic classification has become gradually known to the world after being suggested to its members by the Soil Science Society of China in 1996, translated into Japanese in Japan in 1997, and adopted by WRB, especially for the Anthrosols, in 1998. The correlation among CSTC, ST, and WRB systems is shown in Table 10.2.

Table 10.2 Correlation between CSTC, ST, and WRB Systems

Chinese Soil Taxonomy, 2001	ST, 1999	WRB.1998
Histosols	Histosols**	Histosols
Arithrosols	—	Anthrosols**
Spodosols	Spodosols	Podosols
Andosols	Andosols	Andosols**, Cryosols*
Ferralosols	Oxisols	Ferralosols**, Plinthosols*, Acrisols*, Lixisols*
Vertosols	Vertisols	Vertisols
Aridosols	Aridisols	Calcisols, Gypsisols
Halosos	Aridisols*, Alfisols*, Inceptisols*	Solonchaks, Solonetz
Gleyosols	Inceptisols, Gelisols, Entisols	Gleysols**, Cryosols*
Isohumosols	Mollisols	Chernozems, Kastanozems, Phaeozems
Ferrosols	Ultisols**, Alfisols*, Inceptisols*	Acrisols**, Lixisols*, Plinthosols*, Nitisols*
Argosols	Alfisols**, Ultisols*, Mollisols*	Luvisols**, Alisols*
Cambosols	Inceptsols**, Mollisols*, Gelisols*	Cainbisols**
Primosols	Entisols**, Gelisols*	Fluvisols, Leptisols, Arenosols, Regosols, Cryosols

** mostly corresponding
 * partly corresponding

ANTHROSOLS IN CSTC

Human Influence in Modern Times

The impact of human influences on soils is progressively increasing with major changes not only to the intrinsic properties of the soil but also its functions on the landscape. This impact is affected by farming systems, fertilization, irrigation, and tillage. To evaluate the extent of thc impact, an index has been proposed by Gong and Luo (1998).

- Pi/Max Pi
- Human Influence Index (HII)
- Pi: (population/farmland area)×(production output/farmland area)
- Max Pi: maximal HII within the period of statistics

From Figure 10.8, we can see that the HII in China is progressively increasing with time. The HII before 1700 AD was under 0.1, and went up exponentially afterwards, especially during the last several hundred years after several wars.

The national population of China has doubled in less than one hundred years, while chemical fertilizer input came up from less than 10×10^6 T(ton) to 41.46×10^6 T in 50 years. Organic fertilizer input and irrigation area increased 4 and 16 times during the same period. Production output increased from 1.0 T/ha 50 years ago to 4.5 T/ha today. With these changing trends, the anthropogenic soils have been formed. Simultaneously there has been some degree of land degradation because of increasing social pressures in land use.

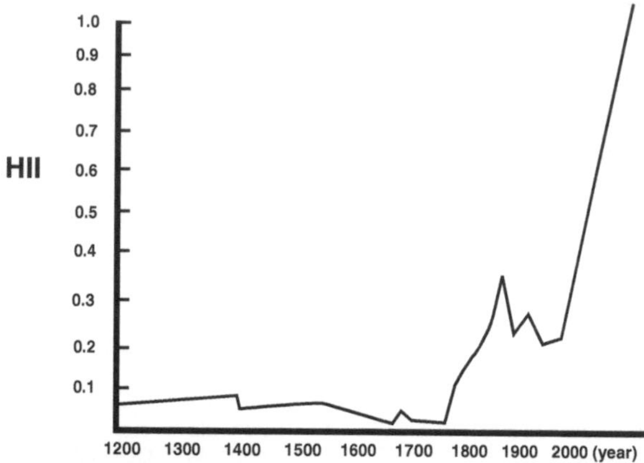

Figure 10.8 Historical change of Human Influence Index (HII). (From Gong et al., 1999. With permission.)

Diagnostic Horizons of Anthrosols

Under the condition of submerged cultivation, with the alternation of redox and fertilization, anthraquic horizon sequence has been formed, which includes anthraquic epipedon and hydragric horizon. Under the upland farming systems with irrigating siltigation or cumulating cultivation or planting vegetable cultivation, irragric epipedon, cumulic epipedon, or fimic epipedon has been formed (Gong, 1984; Gong and Zhang, 1998; Gong et al., 1999).

Anthraquic Epipedon

The anthraquic epipedon is an anthropic surface horizon that includes the cultivated horizon and the plow-pan formed under the condition of submerging cultivation (Figure 10.9). An anthraquic epipedon has all of the following characteristics:

- a thickness of 18 cm or more
- in most years, when the soil temperature is more than 5°C, an anthrostagnic moisture regime for at least 3 months per year
- in most years, when soil temperature is more than 5°C, in the upper part of the epipedon (the cultivated horizon), puddling for at least half a month per year due to soil mixing by submerging cultivation
- under a submerged condition, a color value (moist) of 4 or less and a chroma (moist) of 2 or less and a hue of 7.5 YR or yellower showing GY, B, or BG
- more rusty spots and rusty streaks after drainage
- in a dry condition after drainage, a ratio of the soil bulk density in the lower subhorizon (plowpan) to that in the upper subhorizon (cultivated horizon) being 1.10 or more

Hydragric Horizon

The hydragric horizon is a subsurface horizon formed by the reductive leaching of iron and manganese from surface layer or reductive upward movement of iron and manganese from underlying gleyic layer, and the successive oxidation of reduced materials. It meets the following requirements:

- a thickness of 20 cm or more with the upper limit located in the bottom of the surface layer
- having one or more of the following redox features:

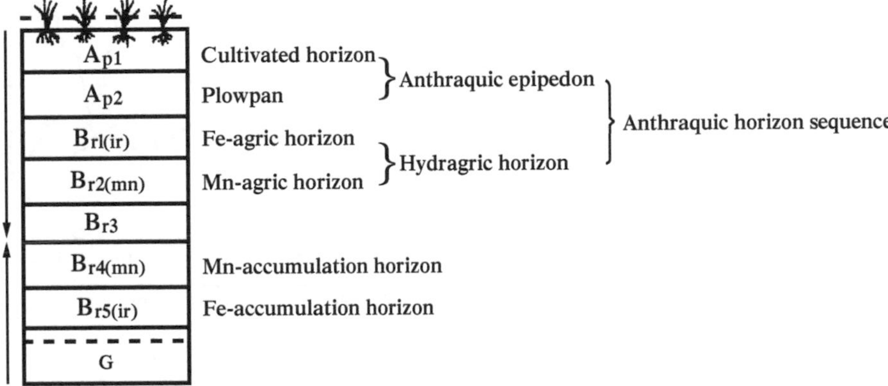

Figure 10.9 Sketch of the profile of typical Hydragric Anthrosols. (From Gong et al., 1999. With permission.)

  - nonobvious iron-manganese deposition, mainly in the form of mottles
  - an iron-manganese deposition differentiation formed by surface water, with the upper part containing mainly iron oxides concretion (mottles, concretions, and nodules), and the lower part black manganese oxides deposition (black spots, mottles, concretions, and nodules), besides new formation of iron oxides
  - iron-manganese oxides, newly formed by both surface water and underground water, and from top-to-bottom appearance of iron-manganese-manganese-iron deposition subhorizon
- a greyish iron-depleted subhorizon adjacently underlying anthraquic epipedon, which, however, does not meet the requirements of albic horizon, and the iron-depleted matrix with a hue of 10YR–7.5Y, a color value (moist) of 5–6, a chroma of 2 or less, and a few mottles
- except iron-depleted subhorizon, free iron content in the other layers being at least 1.5 times more than that of plough horizon
- grey humus-silt-clay cutans 0.5 mm or more in thickness on ped faces and voids
- an obvious prismatic and/or angular structure

Irragric Epipedon

The irragric epipedon (siltigic epipedon) is an anthropic surface horizon formed by gradual deposition of suspended particles during long-term irrigation and the mixture of cultivation. An irragric epipedon meets the following requirements:

- a thickness of 50 cm or more
- a uniform color, texture, structure, consistency, calcium carbonate content, etc., in the whole epipedon, and the textures of the adjacent subhorizons located in the adjacent positions of triangular textural diagram stipulated by the United States Department of Agriculture (USDA) (Figure 10.10)
- the weighted average of organic carbon being 4.5 g kg^{-1} or more within 50 cm of the soil surface, and gradually decreasing with increasing depth, but at least 3 g kg^{-1} at the bottom of the horizon
- after soaking for an hour and screening through a 0.2-mm wet sieve, showing sub-rounded and dense soil plates, within which the silted micro-beddings may be seen using hand lens; or micromorphologically, having the agripedoturbation features showing sub-rounded or rounded plasmic aggregates, inside which, sometimes, residual silted laminae may be found
- having coal cinders, charcoals, brick, or tile fragments and other artificial intrusions throughout the horizon

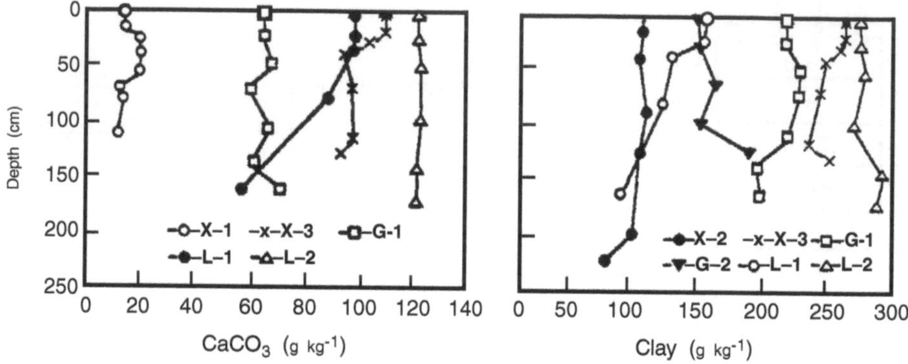

Figure 10.10 Distribution of $CaCO_3$ and clay contents in Irragric Anthrosols. X-l, X-2, X-3, G-l, G-2, L-l, and L-2 are Irragric Anthrosols sampled from Xinjiang, Gansu, and Ningxia, which were described in detail by Shi and Gong (1995). (From Gong et al., 1999. With permission.)

Cumulic Epipedon

The cumulic epipedon is an anthropic surface horizon formed by long-term intensive cultivation and frequent irrigation, application of large amounts of manure, earthy compost, pond muds, etc. (Figure 10.11). A cumulic epipedon meets the following requirements:

- a thickness of 50 cm or more
- a fairly uniform color, texture, structure, consistency, etc. throughout the whole epipedon, and the textures of the adjacent subhorizons located in the same or adjacent positions of the triangular texture table stipulated by USDA
- the weighted average of organic carbon being 4.5 g kg^{-1} or more within 50 cm of the soil surface
- besides the soil particle composition similar to that in the original soil, which is adjacent, influenced by the sources of cumulic materials, having one of the following characteristics: the characteristics of hydromorphic soils and half-hydromorphic soils: residual or recently formed rusty spots, rusty streaks, gley spots, or additional aquatic animal residuals such as shells of conches and shellfish (mud-cumulic features); or some diagnostic horizons and characteristics similar to the adjacent automorphic soils (earth-cumulic features)
- having coal cinders, charcoals, brick or tile fragments, ceramic pieces, and other artificial intrusions

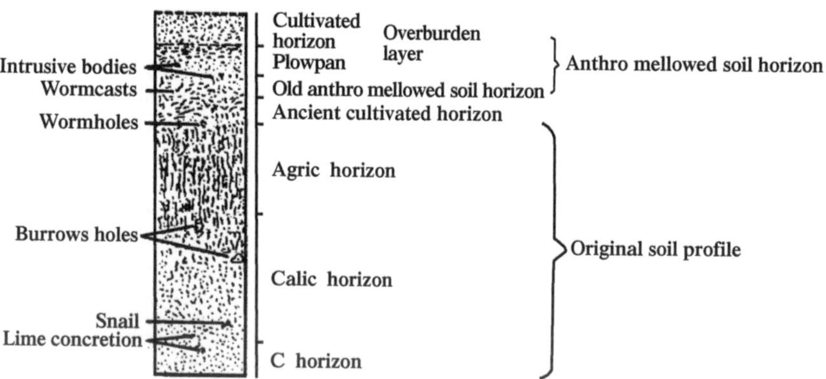

Figure 10.11 Sketch of the profile of typical Earth-cumuli-orthic Anthrosols. (From Gong et al., 1999. With permission.)

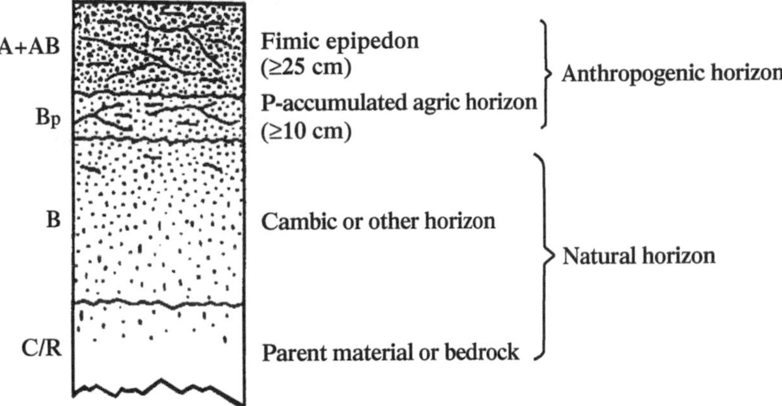

Figure 10.12 Sketch of the profile of Typic Fimic Anthrosols. (From Gong et al., 1999. With permission.)

Fimic Epipedon

The fimic epipedon is an anthropic surface horizon with high mellowness formed by long-term plantation of vegetables, application of large amounts of night soil, manure, organic trash, etc., and intensive cultivation and frequent irrigation (Figure 10.12).

A fimic epipedon meets the following requirements:

- a thickness of 25 cm or more, in which the upper part is highly fimic subhorizon and the lower part is transitional fimic subhorizon
- the weighted average of organic carbon being 6 g kg^{-1} or more
- the weighted average of extractable P contents (by 0.5 mol L^{-1} $NaHCO_3$) being 35 mg kg^{-1} or more (80 mg kg^{-1} or more P_2O_5)
- having large amounts of worm casts, and wormholes with intervals of less than 10 cm, which occupy half or more of the total
- having coal cinders, charcoals, brick or tile fragments, ceramic pieces, and other artificial intrusions

Agric Horizon

An illuviation horizon formed by cultivation in upland soils, the agric horizon directly underlies the plough horizon, and is generally derived from one or another diagnostic subsurface horizon. It meets the following requirements:

- a thickness of 10 cm or more
- one of the following:
  - macromorphologically, having humus-clay or humus-silt-clay coatings on voids and ped faces with both of color value and chroma lower than those of matrix, which are darker and as thick as 0.5 mm or more, and account for 5% or more by volume; or micromorphologically, with coatings which account for 1% or more in thin section, or higher color value and lower chroma in chromic soils than those of underlying layers, and the hue does not change or becomes a little yellower
  - a higher or much higher pH and base saturation in the acid soil in this horizon than in underlying horizon, which is not influenced by illuviation due to cultivation
  - a phosphorus content (extracted by 0.5 mol L^{-1} $NaHCO_3$) of 18 mg kg^{-1} or more in fimic soils, which is much higher than that in the underlying horizon

THE CLASSIFICATION OF ANTHROSOLS

Key to Suborders

Anthrosols that have an anthrostagnic moisture regime, and that have an anthraquic epipedon and a hydragric horizon.

Stagnic Anthrosols

Other Anthrosols.

Orthic Anthrosols

Key to Groups of Stagnic Anthrosols

Stagnic anthrosols that have gleyic features in some layers (10 cm or more thick) within 60 cm of the soil surface.

Gleyi-Stagnic Anthrosols

Other Stagnic Anthrosols that have an iron-percolated grayish subhorizon immediately beneath the anthrostagnic epipedon.

Fe-leachi-Stagnic Anthrosols

Other Stagnic Anthrosols in which hydragric horizon has at least 1.5 times as much DCB extractable Fe as the the anthrostagnic epipedon.

Fe-accumuli-Stagnic Anthrosols

Other Stagnic Anthrosols

Hapli-Stagnic Anthrosols

Key to Groups of Orthic Anthrosols

Orthic Anthrosols that have a fimic epipedon and a phos-agric horizon.

Fimi-Orthic Anthrosols

Other Orthic Anthrosols that have an irragric epipedon.

Siltigi-Orthic Anthrosols

Other Orthic Anthrosols that have a cumulic epipedon with hydromorphic characteristics, and have both of the following: rusty spots and rusty streaks with a high chroma value in reduced materials with a low chroma; and relicts of aquatic animals, such as snails and shells.

Mud-cumuli-Orthic Anthrosols

Other Orthic Anthrosols.

Earth-cumuli-Orthic Anthrosols

The classification of Anthrosols in CSTC (Chinese Soil Taxonomy Research Group, 1995) has been adopted as the classification of Anthrosols in WRB (ISSS/ISRIC/FAO, 1998). But there are no Anthrosols as an independent soil order in Soil Taxonomy (ST) (Soil Survey Staff, 1999).

The Relationship of Anthrosols with Other Soil Orders

The genetic relationship between Anthrosols and other soil orders is given in Figure 10.13. On the natural soils, by the irrigation-accumulation and plow, the Irragri-Orthic Anthrosols were

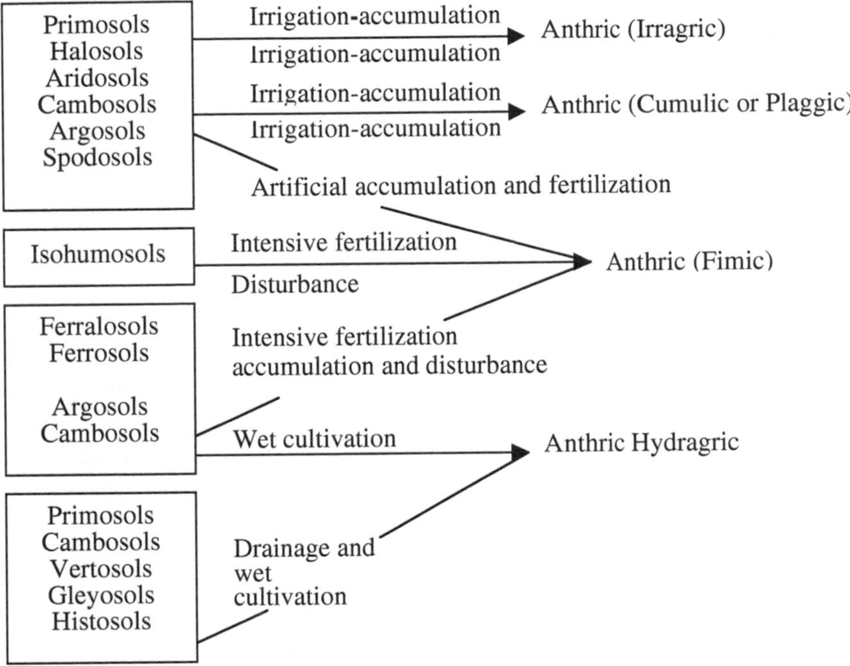

Figure 10.13 Linkage of Anthrosols with other soil orders as related to Anthropogenesis. (From Gong et al., With permission.)

formed. By the transport accumulation and intensive fertilization, the Cumuli-Orthic and Fimi-Orthic anthrosols were formed, and by the wet cultivation, the Stagnic Anthrosols were formed (Gong, 1994). The spatial relationship between Anthrosols and other soil orders is given in Figure 10.14. By the anthropogenic activity, various Anthrosols have been formed, although the original soils are quite different.

ACKNOWLEDGMENT

Chinese Soil Taxonomic classification is a key project in the soil science field of China, which is continuously supported by both the National Natural Science Foundation of China and the Chinese Academy of Sciences for almost 15 years. More than 100 soil scientists from 35 institutes and universities are involved. Although it is organized by the Institute of Soil Science (the Chinese Academy of Sciences), it is a joint achievement belonging to all the soil scientists in China, as well as many international colleagues.

We would like to express our thanks to both the National Natural Science Foundation of China and the Chinese Academy of Sciences for their financial support, and we should remember nationally famous soil scientists Li Qingkui, Li Lianjie, Sun Honglie, Zhao Qiguo, and Tang Yaoxian for their guidance and concern for the project. We extend our sincere thanks to international colleagues R.W. Arnold, H. Eswaran, J.M. Kimble, R. Ahrens (USDA), L.P. Wilding (Texas A&M University), and staff from ISRIC and WRB for their valuable assistance.

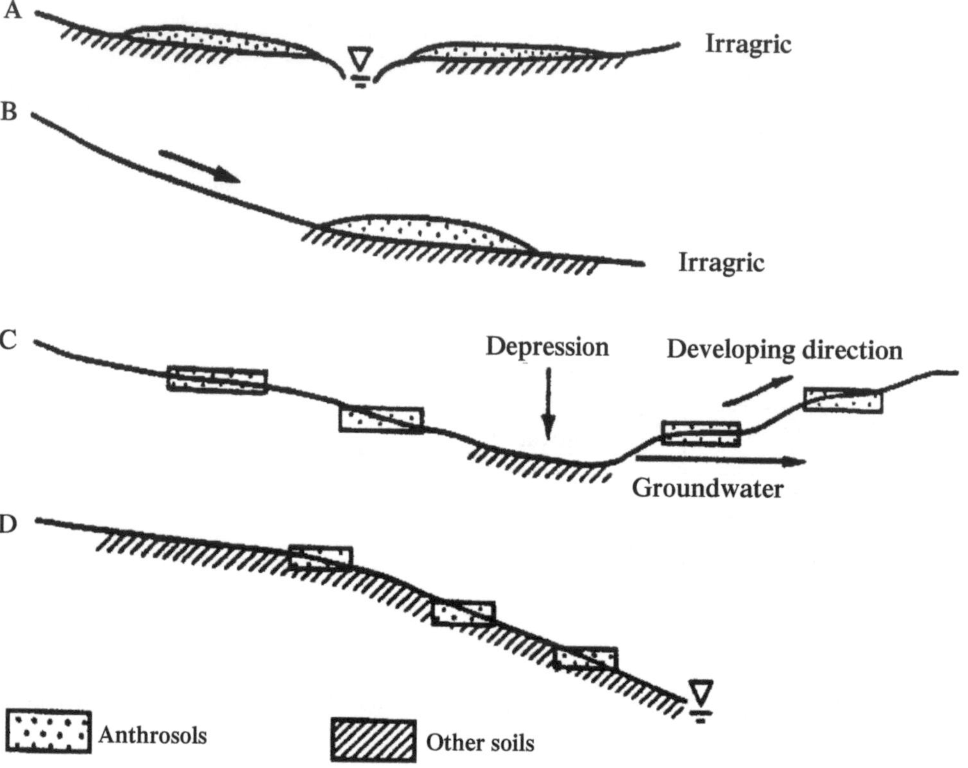

Figure 10.14 Spatial linkage and limits of Anthrosols with other soil orders: Primosols, Aridosol, and Cambosols (A); Halosols, Primosols, Cambosols (B); gleysols, Cambosols and Primosols (C); and Ferrosols, Argosols, and Cambosols (D).

REFERENCES

Cao-Shenggeng and Wenjing, L. 1995. Characteristics of aridic epipedon and its significance in soil taxonomic classification (with English summary), *Acta Pedologica Sinica.* 32(l):34-39.

Chinese Soil Taxonomy Research Group. 1995. *Chinese Soil Taxonomy* (revised version), China Agricultural Science and Technology Press, Beijing.

CSTC (Cooperative Research Group on Chinese Soil Taxonomy). 2001. *Chinese Soil Taxonomy,* Science Press, Beijing, New York.

Eswaran, H. and Zitong, G. 1991. Properties, classification and distribution of soils with gypsum. Occurrence, Characteristics, and Genesis of Carbonate, Gypsum and Silica Accumulation in Soil. SSSA, Special Publication No. 26:89-119.

Gong, Z. 1986. Origin, evalution and classification of paddy soils in China. *Ad. Soil Sci.* 5:179 – 200.

Gong, Z. and Guoan, G. 1989. Altocryic Aridisols in China. Proceedings of Sixth International Soil Correlation Meeting (VI ISCOM). Characterization, Classification and Utilization of Cold Aridisols and Vertisols SMSS. Washington, DC. 54-60.

Gong, Z. 1994. Formation and Classification of Anthrosols: China's Perspective. Transaction of World Congress of Soil Science. Acapulco, Mexico. 6a:120-128.

Gong, Z. and Liangwu, L. 1994. Soil and archaeological research. Transaction of World Congress of Soil Science. Acapulco, Mexico. 6a:369-370.

Gong, Z., Ed. 1994. Chinese Soil Taxonomy (First Proposal). Institute of Soil Science, Academia Sinica, Nanjing, China.

Gong, Z., Hongzhao, C., and Helin, W. 1997. The distribution regularity of higher categories in Chinese Soil Taxonomy. Chinese Geographical Science, 7(1):1–10.

Gong, Z., Xiaopu, Z., Guobao, L., Han, S., and Spaargaren, O. 1997. Extractable phosphorus in soils with fimic epipedon. *Geoderma.* 75:289-296

Gong, Z. and Ganlin, Z. 1998. New horizon formed by human activities, in Proceedings CD-ROM of 16th World Soil Congress, Symposium 16, No. 1430.

Gong, Z. and Guobao, L. 1998. The human-induced soil degradation in China. International Symposium on Soil, Human and Environment Interaction. China Science & Technology Press, 152-160.

Gong, Z., Ganlin, Z. and Guobao, L. 1999. Diversity of Anthrosols in China. *Pedosphere.* 9(3):193–204.

Gong, Z., Putian, L., Jie, C., and Xuefeng, H. 2000. Classical farming systems of China. *J. Crop Production.* 3:11–21.

ISSAS & ISRIC. 1995. Reference Soil Profiles of the People's Republic of China, Field and Analytical Data. Country Report 2 (compiled by Gong Zitong, Luo Guobao and Zhang Ganlin (ISSAS) and O.C. Spaargarent and J. H. Kauffman (ISRIC), Wageningen, Netherlands.

ISSS/ISRIC/FAO. 1998. World Reference Base for Soil Resource. World Soil Resources Reports 84, FAO, Rome.

Lin, P. 1996. *Soil Classification and Land Utilization in Ancient China* (with English Summary). Science Press, Beijing, 1–183.

Marbut, C.F. 1927. A scheme for soil classification. *Proc. First Int. Cong. Soil Sci.* 4:1–31.

Soil Survey Staff, 1999. Soil Taxonomy, 2nd ed. Agriculture Handbook 436. U.S. Government Printing Office, Washington, DC.

Sumner, M.E., Ed. 2000. *Handbook of Soil Science.* CRC Press, Boca Raton, FL, 161.

Wang, Y. 1980. *Soil Science in Ancient China* (in Chinese), Science Press, Beijing.

Xiong, Y. et al., 1987. *Soils of China.* Science Press. Beijing.

Zhao, S, 1986. *Physical Geography of China.* Science Press. Beijing.

CHAPTER 11

The Brazilian Soil Classification System

Francesco Palmieri, Humberto G. dos Santos, Idarê A. Gomes, José F. Lumbreras, and Mario L.D. Aglio

CONTENTS

ABSTRACT

Systematic surveys of Brazilian soils began in the decade of 1950, and since the beginning they have been influenced by both U.S. system and FAO-UNESCO legend. The main objective was to make an inventory of Brazilian soils. This was accomplished with the publication of the Brazilian Soil Map at scale 1:5,000,000 in 1981. During the course of subsequent soil surveys, criteria and attributes for characterization of soil classes under varied environments were refined. This process also provided the information and data for developing the next version of the Brazilian Soil Classification and the Soil Map of Brazil. Phases of soil classes were also devised to highlight limitations for land use, constraints for specific uses, and the ecological relationship concerning potential land use. The current Brazilian system, published in 2000, presents adaptations of criteria and concepts established in Soil Taxonomy and in the World Reference Base for Soil Classification. However, the structure of the system and the emphasis given to various criteria differ from these international systems. These differences reflect the experience of Brazilian soil scientists, specifically in the context of use and management of the soils. It is structured in six categorical levels, arranged in a hierarchy of increasing information content: Order, Suborder, Great Group, Subgroup, Family, and Series.

0-8493-1339-2/03/$0.00+$1.50

INTRODUCTION

Brazil, with a land area of about 8,511,940 km^2 (IBGE, 1987), has a wide variety of soils. The soil moisture regime (Soil Survey Staff, 1999) ranges from perudic to aridic, and the soil temperature regime ranges from isohyperthermic to thermic. The geologic history indicates rocks ranging in age from Pre-Cambrian to recent volcanic. Geomorphologically, a significant part of the country is occupied by extensive peneplains. Much of the country is also part of the catchment of the Amazon River, which has helped to shape the landscape, particularly in the north part of the country. The volcanic Andes mountains aligned north-south on the western part of the continent have active volcanoes, and pyroclastics have been added to many of the soils of the country. All of these factors provide the complexity of conditions that has resulted in the array of soils.

The extensive planation surfaces are the sites of highly weathered soils, and as Brazil has the largest contiguous extent of such soils in the world, international soil classification systems have relied on Brazilian contributions toward the evaluation and classification of such soils. The soils are called Latosols, a term introduced in the United States in 1949 (Cline, 1975). With the advent of modern classification systems, they are now referred to as Oxisols in Soil Taxonomy, and Ferralsols in the legend of the FAO-UNESCO (1974) Soil Map of the World. In the recent World Reference Base for Soil Classification (Nachtergaele et al., 2000), they are still referred to as Ferralsols. Transitions to and from these weathered soils are also extensive, and thus the country presents a natural laboratory for the study of such soils, particularly soils in their pristine environment.

Systematic surveys of Brazilian soils began in the decade of 1950, and with them the use of soil classification criteria and principles developed in the United States by Baldwin et al. (1938), Thorp and Smith (1949), and Cline (1949). Bennema and Camargo (1964) reviewed the existing Brazilian soil survey data, and proposed a partial classification of Brazilian soils titled "Second Partial Draft of the Brazilian Soil Classification." Twelve soil classes were recognized at the highest categorical level. The classes were defined based on the presence or absence of diagnostic horizons, most of them equivalent to those of the 7th Approximation (Soil Survey Staff, 1960). An attempt was made to develop the lower categories of the well-drained Latosol class on the basis of clay activity, base saturation, Al extracted by KCl, type of A horizon, color, texture, amount of total Fe_2O_3, silica/alumina ratio, and alumina/total Fe_2O_3 ratio. The nonhydromorphic soils with textural B horizons were split into two classes at the highest level. One class was made up of soils with clay activity of less than 24 me/100 g of clay and either aluminum saturation >50% or base saturation <35% (approximately equivalent to Ultisols). The other classes were differentiated by having a clay activity of >24 me/100 g of clay and either aluminum saturation <50% or base saturation >35% (approximately equivalent to Alfisols). In this scheme, the diagnostic clay activity criteria (cation exchange capacity, CEC, by NH_4OAC/pH 7) does not include the CEC of the organic matter.

The Brazilian soil survey program was accelerated in 1965 in order to make an inventory of Brazilian soils. A number of soil surveys at several levels have been completed and interpreted for agricultural purposes. During this soil survey program, fundamental concepts and criteria of other systems (Soil Survey Staff, 1975; FAO-UNESCO, 1974) were incorporated and adapted for developing the Brazilian soil classification approaches, along with several soil field special investigations and technical meetings. This goal was reached with the publication of the Brazilian Soil Map at scale 1:5,000,000 in 1981 (Serviço Nacional de Levantamento e Conservação de Solos, 1981).

During the course of subsequent soil surveys, criteria and attributes for characterization of soil classes were refined to cope with soil characteristics under varied environments. Phases of soil classes were also devised to highlight limitations for land use, constraints for specific uses, and ecological relationship concerning potential land use.

On a field trip for special soil investigation and correlation, driving on a newly open gravel road of about 800 km, from Porto Velho to Manaus in the Amazon region, a predominant occurrence of soils with plinthite was observed. For this reason, it received a nickname of Plinthic Route by the soil scientists. Because of the great geographic extent of soils with plinthite, it was recognized as a unique soil class in the new Brazilian Soil Classification System (EMBRAPA, 1982).

AN OVERVIEW OF THE CURRENT SYSTEM

The current Brazilian Soil Classification System, developed and adapted by soil survey staff of EMBRAPA and its partners, is believed to cover all known soils of the Brazilian Territory. It was published in 1999 (EMBRAPA, 1999) and presents adaptations of criteria and concepts established in Soil Taxonomy (Soil Survey Staff, 1975; 1999) as well as in the World Reference Base for Soil Classification (FAO-UNESCO, 1988; FAO/ISRIC/ISSS, 1998). However, the structure of the system and the emphasis given to criteria differ from these international systems. These differences reflect the experience of Brazilian soil scientists, specifically in the context of use and management of the soils.

The system defines a number of diagnostic characteristics that form the building blocks of the classification. As the characteristics have quantitative limits, class definitions are simplified. Class-defining characteristics are incorporated into the key of the hierarchy, and in this manner, the presence or absence of diagnostic characteristics defines a class. Position in the hierarchy also ensures that mutually exclusive classes are structured. Subordinate expression of the characteristics is expressed through the use of intergrades and extragrades, particularly at the subgroup level. By quantification and through the use of keys, ambiguity is reduced, enabling all users to arrive at the same classification for a soil.

All classifications attempt to partition nature, and there are no perfect systems. The ultimate test of a system is when multiple users arrive at the same classification and the same kinds of interpretations, with respect to use and management. Research and discussions toward a more perfect system will be focused on better ways to characterize the soils, the rationale for prioritizing the classes, and the ultimate challenge of meeting the developmental needs of the nation. Differences in emphasis and variations in the structure between the Brazilian system and other national and international systems will become evident in this paper. The differences stem from perceptions, experiences, and the desire to serve the objectives of the classification system. The intent in this paper is to elaborate on some of these.

THE BRAZILIAN SOIL CLASSIFICATION SYSTEM

Presently it is structured in six categorical levels, arranged in a hierarchy of increasing information content. It comprises 14 classes at Order level, separated by presence or absence of diagnostic horizons and attributes that reflect predominant soil-forming processes or mechanisms. Characteristics or properties that represent important variations within the Order level differentiate the 44 classes at Suborder level. These characteristics include color, accumulation of organic matter, complexed organic matter and sesquioxide, degree of salinity, presence of sulfuric horizon, etc. The 150 classes at Great Group are defined on the basis of kind and arrangement of horizons with depth, exchangeable bases and aluminum saturation, clay activity, presence of sodium or soluble salts, presence of iron or aluminum oxides, carbonates, etc. Finally, the 580 classes at Subgroup level comprise the central concept (the typic) and all kinds of intergrades and extragrades soil classes. Table 11.1 presents the categories and the classes recognized within each category.

Family and Soil Series levels have not yet been developed. They will be defined on the basis of physical, chemical, and mineralogical characteristics that reflect pedon-environment conditions that are very important for interpreting soil classes for agriculture and non-agricultural purposes.

Every soil class definition is based on diagnostic soil attributes and properties, according to a classification key that reaches to the subgroup level.

Table 11.1 Outline of the Four Highest Categories of the Brazilian Soil Classification System

Order	Suborder	Great Group	Subgroup
Organossolos	Thiomorphic	Fibric	Saline; Solodic; Terric; Typic
		Hemic	Saline; Solodic; Terric; Typic
		Sapric	Saline; Solodic; Terric; Typic
	Folic	Fibric	Lithic; Typic
	Mesic	Hemic	Salic; Saline; Sodic; Solodic; Carbonatic; Terric; Typic
		Sapric	Salic; Saline; Sodic; Solodic; Carbonatic; Terric; Typic
	Haplic	Fibric	Solodic; Terric; Typic
		Hemic	Salic; Saline; Solodic; Carbonatic; Terric; Typic
		Sapric	Salic; Saline; Solodic; Terric; Typic
Neossolos	Litholic	Histic	Typic
		Humic	Spodic; Typic
		Carbonatic	Typic
		Psamitic	Typic
		Eutrophic	Chernosolic; Typic
		Dystrophic	Typic
	Fluvic	Salic	Solodic;Typic
		Sodic	Vertic; Saline; Typic
		Carbonatic	Typic;
		Psamitic	Typic
		Tb[1] Dystrophic	Gleyic;Typic
		Tb[1] Eutrophic	Gleyic; Solodic; Typic
		Ta[2] Eutrophic	Gleyic; Vertic-solodic; Vertic; Solodic; Solodic-saline; Calcaric; Typic
	Regolithic	Psamitic	Fragic-leptic; Fragic; Leptic; Gleyic-solodic; Solodic; Typic
		Dystrophic	Fragic-leptic; Fragic; Humic-leptic; Humic; Typic
		Eutrophic	Fragic-leptic; Fragic; Solodic-leptic; Leptic; Solodic; Typic
	Quartzarenic	Hydromorphic	Histic; Spodic; Plinthic; Typic
		Orthic	Humic; Fragic; Solodic; Eutric; Leptic; Spodic; Plinthic; Gleyic; Latosolic; Argisolic; Typic
Vertissolos	Hydromorphic	Sodic	Saline; Carbonatic; Typic
		Salic	Solodic; Typic
		Carbonatic	Solodic; Typic
		Orthic	Chernosolic; Solodic; Typic
	Ebanic	Sodic	Saline; Typic
		Carbonatic	Chernosolic; Typic
		Orthic	Solodic; Chernosolic; Typic
	Chromic	Salic	Lithic; Gleyic-solodic; Gleyic; Solodic; Typic
		Sodic	Lithic; Gleyic; Saline; Typic
		Carbonatic	Lithic; Chernosolic; Gleyic-solodic; Gleyic; Solodic; Typic
		Orthic	Lithic; Chernosolic-solodic; Chernosolic-gleyic; Chernosolic; Saline; Solodic; Gleyic; Typic
Espodossolos	Carbic	Hydromorphic	Histic; Duric; Arenic; Typic
		Hypertick	Typic
		Orthic	Duric; Duripanic; Fragic; Espesarenic-solodic; Espesarenic; Arenic; Typic

Table 11.1 (continued) Outline of the Four Highest Categories of the Brazilian Soil Classification System

Order	Suborder	Great Group	Subgroup
	Ferrocarbic	Hydromorphic	Histic; Duric; Arenic; Typic
		Hypertick	Typic
		Orthic	Duric; Duripanic; Fragic; Carbonatic; Solodic; Eutric; Espesarenic; Arenic-solodic; Argiluvic; Typic
Alissolos	Chromic	Humic	Cambic; Typic
		Argiluvic	Epiaquic; Abruptic; Typic
		Orthic	Nitosolic; Typic
	Hypochromic	Argiluvic	Epiaquic; Abruptic; Typic
		Orthic	Nitosolic; Typic
Planossolos	Natric	Carbonatic	Vertic; Typic
		Salic	Arenic; Duripanic; Fluvic; Typic
		Orthic	Espesarenic; Arenic; Vertic; Plinthic; Duripanic; Typic
	Hydromorphic	Salic	Arenic; Solodic; Plinthic; Typic
		Eutrophic	Espesarenic; Arenic; Solodic; Plinthic; Typic
		Dystrophic	Arenic; Solodic; Plinthic; Typic
	Haplic	Carbonatic	Solodic; Vertic; Typic
		Salic	Arenic; Solodic; Vertic; Typic
		Eutrophic	Espesarenic; Arenic; Solodic; Chernosolic-vertic; Vertic; Typic
		Dystrophic	Espesarenic; Arenic; Plinthic; Solodic; Aluminic; Typic
Gleissolos	Thiomorphic	Histic	Sodic; Solodic; Typic
		Humic	Sodic; Salic-solodic; Solodic; Typic
		Orthic	Sodic; Salic-solodic; Solodic; Anthropogenic; Typic
	Salic	Sodic	Thionic; Argisolic; Typic
		Orthic	Vertic; Solodic; Typic
	Melanic	Aluminic	Histic; Inseptic; Typic
		Dystrophic	Histic; Inseptic; Plinthic; Typic
		Carbonatic	Vertic; Inseptic; Typic
		Eutrophic	Leptic; Vertic; Inseptic-vertic; Inseptic; Typic
	Haplic	Ta[2] Aluminic	Typic
		Ta[2] Dystrophic	Solodic; Typic
		Tb[1] Dystrophic	Solodic; Plinthic; Argisolic-fragic; Argisolic; Inseptic; Typic
		Ta[2] Carbonatic	Typic
		Ta[2] Eutrophic	Vertic; Luvisolic; Plinthic; Solodic; Typic
		Tb[1] Eutrophic	Argisolic; Inseptic; Thionic; Plinthic; Typic
Nitossolos	Red	Dystroferric	Latosolic; Typic
		Dystrophic	Argisolic; Latosolic; Typic
		Eutroferric	Chernosolic; Plinthic; Latosolic; Typic
		Eutrophic	Leptic; Latosolic; Typic
	Haplic	Aluminic	Latosolic; Typic
		Dystrophic	Humic; Argisolic; Latosolic; Typic
		Eutrophic	Ferric-chernosolic; Chernosolic; Leptic; Typic
Latossolos	Brown	Acric	Humic; Typic
		Aluminic	Humic-cambic; Humic; Cambic; Typic

Table 11.1 (continued) Outline of the Four Highest Categories of the Brazilian Soil Classification System

Order	Suborder	Great Group	Subgroup
		Dystrophic	Humic-cambic; Humic; Cambic; Typic
	Yellow	Cohesive	Anthropic; Humic; Cambic; Argisolic; Petroplinthic; Plinthic; Lithoplinthic; Typic
		Acriferric	Humic; Argisolic; Typic
		Acric	Humic; Argisolic; Petroplinthic; Plinthic; Typic
		Dystroferric	Humic; Typic
		Dystrophic	Humic; Cambic; Psamitics; Argisolic; Petroplinthic; Typic
		Eutrophic	Cambic; Argisolic; Typic
	Red	Perferric	Humic; Cambic; Typic
		Alumiferric	Humic; Cambic; Typic
		Acriferric	Humic; Cambic; Typic
		Dystroferric	Humic; Cambic; Nitosolic; Plinthic; Typic
		Eutroferric	Cambic; Chernozemic; Nitosolic; Typic
		Acric	Humic; Cambic; Argisolic; Typic
		Dystrophic	Humic; Cambic; Psamitic; Argisolic; Typic
		Eutrophic	Cambic; Psamitic; Argisolic; Chernozemic; Typic
	Red yellow	Acriferric	Cambic; Argisolic; Typic
		Acric	Humic; Cambic; Argisolic; Typic
		Dystroferric	Cambic; Argisolic; Typic
		Dystrophic	Humic; Psamitic; Cambic; Plinthic; Nitosolic; Argisolic; Typic
		Eutrophic	Psamitics; Cambic; Argisolic; Typic
Chernossolos	Rendzic	Lithic	Typic
		Saprolithic	Typic
	Ebanic	Carbonatic	Vertic; Typic
		Orthic	Vertic; Typic
	Argiluvic	Ferric	Saprolithic; Typic
		Carbonatic	Vertic; Abruptic; Saprolithic; Typic
		Orthic	Leptic; Abruptic-saprolithic; Saprolithic; Abruptic-vertic; Vertic; Abruptic-solodic; Solodic; Epiaquic; Typic
	Haplic	Ferric	Typic
		Carbonatic	Vertic; Leptic; Nitosolic; Saprolithic; Typic
		Orthic	Vertic; Leptic; Nitosolic; Typic
Cambissolos	Histic	Aluminic	Leptic; Spodic; Typic
		Dystrophic	Leptic; Typic
	Humic	Aluminiferric	Leptic; Latosolic; Spodic; Typic
		Aluminic	Leptic; Gleyic; Sombric, Latosolic; Typic
		Dystroferric	Leptic; Latosolic; Typic
		Dystrophic	Leptic; Gleyic; Latosolic; Typic
	Haplic	Aluminic	Gleyic; Argisolic; Typic
		Carbonatic	Vertic-saprolitic; Vertic; Leptic; Typic
		Salic	Gleyic-solodic; Gleyic; Solodic; Leptic; Typic
		Sodic	Gleyic-saline; Gleyic; Saline; Vertic; Leptic; Typic
		Dystroferric	Leptic; Typic

Table 11.1 (continued) Outline of the Four Highest Categories of the Brazilian Soil Classification System

Order	Suborder	Great Group	Subgroup
		Eutroferric	Leptic; Latosolic; Typic
		Perferric	Latosolic; Typic
		Ta[2] Eutrophic	Lithic; Leptic-calcaric; Leptic; Vertic; Gleyic; Solodic; Typic
		Ta[2] Dystrophic	Lithic; Leptic; Gleyic; Typic
		Tb[1] Eutrophic	Leptic; Gleyic; Latosolic-petroplinthic; Latosolic; Argisolic; Typic
		Tb Dystrophic	Leptic; Gleyic; Latosolic; Argisolic; Plinthic; Typic
Plintossolos	Petric	Lithoplinthic	Arenic; Typic
		Concretionary Dystrophic	Lithic; Leptic; Typic
		Concretionary Eutrophic	Lithic; Leptic; Typic
	Argiluvic	Aluminic	Abruptic; Typic
		Dystrophic	Espesarenic; Arenic; Abruptic-solodic; Abruptic; Solodic Typic
		Eutrophic	Espesarenic; Arenic; Abruptic-solodic; Abruptic; Solodic; Typic
	Haplic	Dystrophic	Lithic; Solodic; Typic
		Eutrophic	Lithic; Solodic; Typic
Luvissolos	Chromic	Carbonatic	Vertic; Planosolic; Typic
		Paleic	Planosolic; Arenic; Abruptic-plinthic; Abruptic; Cambic; Saprolithic; Typic
		Orthic	Lithic; Planosolic-vertic; Planosolic-solodic; Planosolic; Vertic-solodic; Vertic; Solodic-saline; Solodic; Typic
	Hypochromic	Carbonatic	Typic
		Orthic	Planosolic; Aluminic; Typic
Argissolos	Gray	Dystrophic	Espesarenic; Arenic-fragic; Arenic; Abruptic-fragic; Abruptic; Fragic; Latosolic; Plinthic; Typic
		Eutrophic	Abruptic; Duripanic; Typic
	Yellow	Dystrophic	Arenic-fragic; Arenic; Planosolic-fragic; Planosolic; Abruptic-fragic-spodic; Abruptic-fragic; Abruptic-spodic; Abruptic-petroplinthic; Abruptic-plinthic; Abruptic-solodic; Abruptic; Fragic-spodic; Fragic-plinthic; Fragic; Epiaquic; Spodic; Plinthic; Latosolic; Cambic; Typic
		Eutrophic	Abruptic-leptic; Abruptic-chernosolic; Abruptic-plinthic; Abruptic; Plinthic; Leptic; Typic
	Red yellow	Aluminic	Alisolic; Typic
		Dystrophic	Espesarenic; Arenic-fragic; Arenic; Planosolic; Abruptic-fragic; Abruptic; Fragic; Plinthic; Latosolic; Typic
		Eutrophic	Planosolic-fragic; Planosolic; Abruptic-fragic; Abruptic-leptic; Abruptic; Fragic; Latosolic; Typic
	Red	Dystrophic	Arenic; Planosolic; Abruptic-plinthic; Abruptic; Latosolic; Plinthic; Typic
		Eutroferric	Abruptic-saprolithic; Abruptic; Chernosolic; Latosolic; Typic
		Eutrophic	Espesarenic; Arenic; Planosolic; Abruptic-chernosolic; Abruptic-plinthic-solodic; Abruptic-plinthic; Abruptic-solodic; Abruptic; Leptic; Latosolic; Chernosolic; Saprolithic; Cambic; Typic

[1]Low-activity clay—CEC less than 27 $cmol_c$/kg of clay.

[2]High-activity clay—CEC equal to or higher than 27 $cmol_c$/kg of clay.

ATTRIBUTES OF THE ORDER CLASSES

Organossolos—Grouping of soils formed by organic soil materials with variable decomposition stages under an environment saturated with water for long periods. They have at least 30 cm of soil organic material when overlying a lithic contact; 40 cm or more thickness either from the surface or taken cumulatively within the upper 80 cm of the surface; 60 cm or more if the organic material consists mainly of sphagnum or moss fibers or has a bulk density of less than 0.15 g cm^{-3}.

This class includes mainly those soils that have been previously identified as Solos Orgânicos, Semi-Orgânico, Solos Tiomórficos Turfosos, and Solos Litólicos with histic horizon. They are not representative in small scale map.

Neossolos—Grouping of soils showing absence of diagnostic B horizons and weakly developed, because of either reduced performance of pedological processes or predominant characteristics inherited from the parent material. Insufficient manifestation of diagnostic attributes characteristic of soil-forming processes. Having little horizon differentiation, other than an A horizon followed by C horizon or R.

This class includes mainly those soils that have been previously identified as Litossolos, Solos Litólicos, Regossolos, Solos Aluviais, and Areias Quartzosas (Distróficas, Marinhas, and Hidromórficas).

The spatial distribution of this soil class secluded from the 1:5,000,000 scale of the Brazilian soil map is shown in Figure 11.1. It comprises 1,251,255 km^2 corresponding to 14.7% of the country.

Vertissolos—Soils showing a diagnostic vertic horizon with and without gilgai. Limited horizon development because of the shrinking and swelling nature of clay. They have 30% or more clay in all horizons, show vertical cracks during the dry season that are at least 1 cm wide to a depth of 50 cm, except for shallow soils in which the minimum depth is 30 cm. Further, they have slinckensides or parallelepipedic structural aggregates and COLE index equal to or more than 0.06 and/or linear expansibility of 6 cm or more. They also do not have any diagnostic B horizon above the vertic horizon.

This class includes mainly those soils that have been previously identified as Vertissolos, including the hydromorphic ones.

The geographical distribution of this soil class secluded from the 1:5,000,000 scale of the Brazilian soil map is shown in Figure 11.2. It comprises 8,512 km^2 corresponding to 0.1% of the country.

Figure 11.1 A map of Brazil showing the main distribution of Neossolos.

Figure 11.2 A map of Brazil showing the main distribution of Vertissolos.

Espodossolos—Grouping of soils with spodic horizon below an eluvial E horizon or surface H horizon less than 40 cm thick or subjacent to any other kind of A horizon. The soil is characterized by a predominance of podzolization process showing eluviation, followed by accumulation of aluminum compounds along with humic acids, with or without iron.

This class includes mainly those soils that have been previously identified as Podzol, including the hydromorphic ones.

The geographical distribution of this soil class secluded from the 1:5,000,000 scale of the Brazilian soil map is shown in Figure 11.3. It comprises 136,191 km^2 corresponding to 1.6% of the country.

Alissolos—Soils having a textural B or nitic horizon with CEC equal to or higher than 20 cmol$_c$/kg of clay and aluminum saturation equal to or more than 50% and extractable aluminum Al^{3+} equal to or more than 4.0 cmol$_c$/kg of soil. Plinthic and/or gley horizons, if present, they are below the

Figure 11.3 A map of Brazil showing the main distribution of Espodossolos.

Figure 11.4 A map of Brazil showing the main distribution of Alissolos.

upper boundary of the B horizon or 50 cm of the surface. They lack horizon diagnostic for Plintossolos and Gleissolos, respectively.

This class includes mainly those soils that have been previously identified as Rubrozem, Podzólico Bruno-Acinzentado Distrófico or Álico, Podzólico Vermelho-Amarelo Distrófico, or Álico Ta and a few Podzólicos Vermelho-Amarelos Distróficos or Álicos Tb (with clay activity equal to or more than 20 $cmol_c/kg$ of clay).

The geographical distribution of this soil class secluded from the 1:5,000,000 scale of the Brazilian soil map is shown in Figure 11.4. It comprises 374,525 km^2 corresponding to 4.4% of the country.

Planossolos—Soils with a planic B horizon immediately below an A or E horizon and abrupt textural change; soils having permeability restrictions and evidence of reduction processes with or without mottles of iron segregation. They lack a spodic B horizon.

This class includes mainly those soils that have been previously identified as Planossolos, Solonetz-Solodizado, and Hidromórficos Cinzentos with abrupt textural change.

The geographical distribution of this soil class secluded from the 1:5,000,000 scale of the Brazilian soil map is shown in Figure 11.5. It comprises 153,214 km^2 corresponding to 1.8% of the country.

Gleissolos—Grouping of soils with gleyzation characteristics within 50 cm of the surface and usually saturated permanently or seasonally by groundwater. Lacking diagnostic characteristics of Vertissolos, Espodossolos, Planossolos, Plintossolos, or Organossolos.

This class includes mainly those soils that have been previously identified as Glei Pouco Húmico, Glei Húmico, Hidromórfico Cinzento without abrupt textural change, Glei Tiomórfico, and Solonchak with gley horizon.

The geographical distribution of this soil class secluded from the 1:5,000,000 scale of the Brazilian soil map is shown in Figure 11.6. It comprises 323,453 km^2 corresponding to 3.8% of the country.

Nitossolos—Clayey soils showing an advanced development of pedological ferralitization process along with formation of a kaolinitic-oxidic colloid. They have a nitic B horizon with CEC less than 27$cmol_c/kg$ of clay at least within the upper 50 cm of the nitic B horizon, and show gradual to diffuse horizon boundaries between A and B horizons. They have moderate to strong

Figure 11.5 A map of Brazil showing the main distribution of Planossolos.

development of blocky and prismatic soil structure, and shiny ped faces that are either clay skins or pressure faces.

This class includes mainly those soils that have been previously identified as Terra Roxa Estruturada, Terra Roxa Estruturada Similar, Terra Bruna Estruturada, Terra Bruna Estruturada Similar, and a few of both Podzólicos Vermelho-Escuros Tb and Podzólico Vermelhos-Amarelos Tb.

The geographical distribution of this soil class secluded from the 1:5,000,000 scale of the Brazilian soil map is shown in Figure 11.7. It comprises 110,655 km^2 corresponding to 1.3% of the country.

Figure 11.6 A map of Brazil showing the main distribution of Gleissolos.

Figure 11.7 A map of Brazil showing the main distribution of Nitossolos.

Latossolos—Grouping of soils showing strongly weathered pedological development along with kaolinitic-oxidic soil genesis, resulting in intensive weathering of both less resistant primary and secondary minerals, except for hydroxy interlayer vermiculites. They have a latosolic B horizon, which is at least 50 cm thick immediately below any surface diagnostic horizon except histic H horizon. The CEC is less than 17 $cmol_c$/kg of clay; texture sandy loam or finer; has less than 4% of weatherable minerals and less than 6% of muscovite in the sand fraction. They do not have more than trace of smectite and have less than 5% by volume showing rock structure. The silt-clay ratio is less than 0.6 for soils having more than 35% clay, and 0.7 for soils with less than 35% clay. They also have gradual to diffuse boundaries between the subhorizons.

This class includes mainly those soils that have been previously identified as Latossolos, except a few varieties identified as Latossolos Plinticos.

The geographical distribution of this soil class secluded from the 1:5,000,000 scale of the Brazilian soil map is shown in Figure 11.8. It comprises 3,277,096 km^2 corresponding to 38.5% of the country.

Chernossolos—Grouping of soils showing rather moderate weathering resulting from bisialitization process, resulting in a highly saturated exchange complex with mainly calcium as basic exchangeable cation. They have a chernozemic A horizon overlying a cambic B, textural B, or a nitic B horizon. All B horizons have both CEC of the clay fraction equal to or higher than 27$cmol_c$/kg of clay, and base saturation of 50% or more, with or without either a calcic horizon or carbonatic properties. They lack diagnostic characteristics of Vertissolos, Planossolos, and Gleissolos.

This class includes mainly those soils that have been previously identified as Brunizem, Rendzina, Brunizem Avermelhado, Brunizem Hidromórfico.

The geographical distribution of this soil class secluded from the 1:5,000,000 scale of the Brazilian soil map is shown in Figure 11.9. It comprises 42,559 km^2 corresponding to 0.5% of the country.

Cambissolos—Grouping of weakly developed soils having a cambic B horizon. Little development with features representing genetic soil development without extreme weathering; showing stronger chroma and redder hue than the underlying horizon. If the cambic B horizon occurs under a chernozemic A horizon, it must have either a low-activity clay (CEC less than 27 $cmol_c$/kg of clay) or base saturation less than 50%. The cambic horizon may occur immediately below any kind

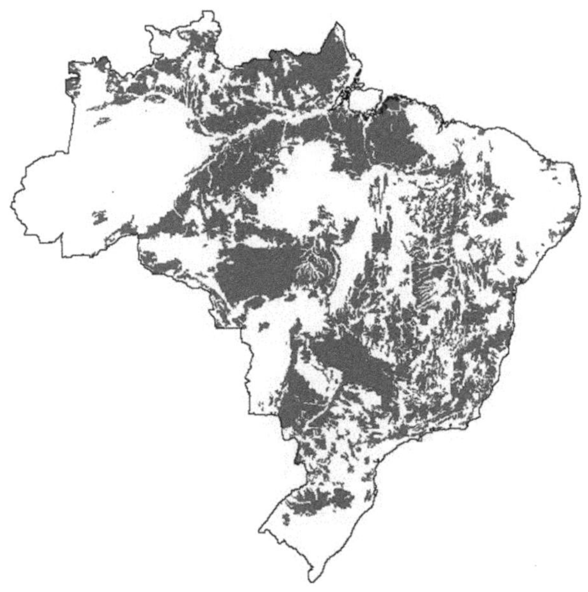

Figure 11.8 A map of Brazil showing the main distribution of Latossolos.

of A horizon; if a histic A horizon is present, it is less than 40 cm. They lack diagnostic characteristics of Vertissolos, Chernossolos, Plintossolos, or Gleissolos.

This class includes mainly those soils that have been previously identified as Cambissolos Eutróficos, Distróficos e Álicos all Ta and Tb, except those with chernozemic A horizon and eutrophic cambic B with high-activity clay.

The geographical distribution of this soil class secluded from the 1:5,000,000 scale of the Brazilian soil map is shown in Figure 11.10. It comprises 229,822 km^2 corresponding to 2.7% of the country.

Figure 11.9 A map of Brazil showing the main distribution of Chernossolos.

Figure 11.10 A map of Brazil showing the main distribution of Cambissolos.

Plintossolos—Grouping of soils with a dominance of plinthite and with or without petroplinthite. Soils having within 40 cm of the surface a lithoplinthic or a plinthic horizon showing 15% or more plinthite by volume or petroplinthic material that is at least 15 cm thick. The above diagnostic characteristics may occur within a depth of 200 cm when underlying horizons show evidence of reduction processes and/or gleyic conditions.

This class includes mainly those soils that have been previously identified as Lateritas Hidromórficas; part of Podzólicos Plinticos, part of Glei Húmico Plintico and Glei Pouco Húmico Plintico; and a few Latossolos Plinticos.

The geographical distribution of this soil class secluded from the 1:5,000,000 scale of the Brazilian soil map is shown in Figure 11.11. It comprises 510,716 km^2 corresponding to 6.0% of the country.

Luvissolos—Soils lacking a chernozemic A horizon but having a nitic B or textural B horizon with CEC of the clay fraction equal to or higher than 27 cmol$_c$/kg of clay and base saturation of 50% or more. Nitic or textural B horizon may occur immediately below an albic E horizon or a weak, a moderate, or a prominent A horizon. Plinthic and/or gley horizons, if present, occur below the upper boundary of the B horizon or 50 cm of the surface; therefore they are not diagnostic for Plintossolos and Gleissolos, respectively.

This class includes mainly those soils that have been previously identified as Bruno Não Cálcico, Podzólico Vermelho-Amarelo Eutrófico with high-activity clay, Podzólico Bruno-Acinzetado Eutrófico, and Podzólico Vermelho-Escuro Eutrófico with high-activity clay.

The geographical distribution of this soil class secluded from the 1:5,000,000 scale of the Brazilian soil map is shown in Figure 11.12. It comprises 178,610 km^2 corresponding to 2.7% of the country.

Argissolos—Grouping of soils showing a rather moderate development of pedological ferralitization process along with kaolinitic-oxidic soil genesis. Have a textural B horizon with CEC less than 27 cmol$_c$/kg of clay. Plinthic and/or gley horizons, if present, occur below the upper boundary of the B horizon or 50 cm of the surface, and are therefore not diagnostic for Plintossolos and Gleissolos, respectively. They lack diagnostic characteristics of Alissolos and Planossolos.

Figure 11.11 A map of Brazil showing the main distribution of Plintossolos.

This class includes mainly those soils that have been previously identified as Podzólico Vermelho-Amarelo; Podzólico Vermelho-Escuro; Podzólico Amarelo, all with low-activity clay; small part of Terra Roxa Estruturada and Terra Roxa Estruturada Similar, Terra Bruna Estruturada; and Terra Bruna Estruturada Similar, all of them with textural B horizon.

The geographical distribution of this soil class secluded from the 1:5,000,000 scale of the Brazilian soil map is shown in Figure 11.13. It comprises 1,711,040 km^2 corresponding to 19.8% of the country.

Figure 11.12 A map of Brazil showing the main distribution of Luvissolos.

Figure 11.13 A map of Brazil showing the main distribution of Argissolos.

APPROXIMATE EQUIVALENTS WITH OTHER SYSTEMS

Although the Brazilian system of soil classification is a national system, there has been a consistent desire to relate it to international systems to permit easy translations. As there are differences in definitions and the structure, direct equivalence may not always be possible. Table 11.2 attempts to provide broad equivalence with Soil Taxonomy and the units of the World Reference Base.

As evident in Table 11.2, there are direct equivalents for some classes, while others may have more than one. Vertissolos and Latossolos are examples where all systems have equivalent definitions. The exact classification in any system is best made for a soil based on its unique properties.

Table 11.2 Equivalents of the Order Category in the Brazilian System

Brazilian System	Soil Taxonomy	World Rreference Base
Organossolos	Histosols	Histosols
Neossolos	Entisols	Fluvisols, Leptosols, Regosols, Arenosols
Vertissolos	Vertisols	Vertisols
Espodossolos	Spodosols	Podzols
Alissolos	Ultisols	Acrisols, Alisols
Planossolos	Alfisols, Ultisols, Mollisols, Aridisols	Planosols
Gleissolos	Inceptisols, Ultisols, Mollisols, Alfisols, Entisols	Fluvisols, Gleysols
Nitossolos	Ultisols, Alfisols	Nitisols
Latossolos	Oxisols	Ferralsols
Chernossolos	Mollisols	Leptosols, Kastanozems, Greyzems, Chernozems, Phaeozems
Cambissolos	Inceptisols	Cambisols
Plintossolos	Oxisols, Ultisols, Inceptisols, Entisols, Alfisols	Plinthosols
Luvissolos	Alfisols	Luvisols
Argissolos	Ultisols, Alfisols	Acrisols, Lixisols

THE RATIONALE OF THE SYSTEM

The rationale of the Brazilian system follows the current understanding of soil genesis. The knowledge of the Brazilian soil scientist regarding soil genesis under equatorial, tropical, and subtropical conditions has grown, but even so, the soil scientists agree that using the effects of the soil genesis expressed by soil morphology is basic for land suitability appraisal for agricultural and non-agricultural purposes. Grouping soils in terms of morphogenetics attributes and properties is a good mechanism for building, fence-like, around a track of land in order to get somewhat homogenous and similar individual soil classes. The differences in relation to other international systems arise because of the emphasis placed on specific attributes from the point of view of use and management. The system is still highly biased toward agricultural use of soils.

There are some fundamental changes that have been introduced since our previous version, and also with respect to the international systems. Unlike other systems, we have preferred to evaluate soils by using the degree of expression and relative importance of diagnostic horizons that are readily identified or inferred in the field. This follows the mental process of the field soil scientist who looks for diagnostic horizons and, in general, an ease of recognition.

The orders have been set apart based on the presence or absence of diagnostic horizons, attributes, and properties believed to be the expression(s) of predominant(s) soil-forming process(es) or mechanism(s). The logic for distinguishing the fourteen orders will be sketched briefly.

The first order to be keyed out, the Organossolos, is not a mineral soil, but rather organic soils, distinguished from the others by having a high content of organic matter at several decomposition stages, and developed under environments saturated with water for long periods. Thereafter is followed by mineral soil classes. The second order, the Neossolos, consists of mineral soils with low horizonation, a minimal expression, or no diagnostic horizons. This is followed by an order in which the soil shows intense internal stresses because of its special mineralogical composition: the Vertissolos. The vertic properties are considered a special expression of the cambic horizon, though cambic horizons are expressions of an initial stage of soil formation. The next soils keyed out, the Espodossolos, are characterized by having a subsurface E horizon with a subsurface horizon of illuvial accumulation of organic matter and aluminum, with or without iron. They are followed by the Alissolos, with high content of aluminum, and Planossolos, with an abrupt textural change. The Gleissolos is the next order which is saturated with water for long periods, and shows evidence of reduction processes.

These are followed by groups of soils showing expressions of intensive weathering processes, Nitossolos and Latossolos, unique to the tropics and extensive in Brazil. The Nitossolos show the blocky structure of the B horizon with shiny ped faces that are either clay coatings or pressure faces. In some respects, these are with properties transitional to the Latossolos, but they are "better" soils than the Latossolos and, for this reason, we have made the definition of the Latossolos more restrictive than Soil Taxonomy. The latosolic B horizon must be more than 50 cm to be diagnostic for the Latossolos. At the same time, we tried to ensure that the charge is because of the mineral colloid, and not to organic matter. We have also included a factor (silt/clay ratio) that is an indicator of the weathering intensity. This ensures that the low activity is less influenced by the parent material and is essentially due to intense weathering. The Latossolos are weathered, generally with low fertility, excessively to moderately well drained, with deep solum generally thicker than 2 m. They are most common on flat to gentle slopes on old geomorphic surfaces, and very extensive in Brazil.

The Chernossolos keyed out next are characterized by thick organic-rich surface horizon, solum high in base saturation and B horizon showing high-activity clay with expressive amounts of calcium, with or without accumulation of calcium carbonate. The next group of soils, the Cambissolos, is characterized by a subsurface incipient B horizon showing higher chroma and redder hue than the underlying horizon. These are followed by Plintossolos, a group of soil having plinthite as predominant feature. The next group, the Luvissolos, comprises soils lacking chernozemic A

Table 11.3 Frequency of Occurrence of Soil Order Classes in Brazil

Soil Order	Area (km^2)	Percent	Soil Order	Area (km^2)	Percent
Alissolos	374,525	4.4	Luvissolos	178,610	2.7
Argissolos	1,711,040	19.8	Neossolos	1,251,255	14.7
Cambissolos	229,822	2.7	Nitossolos	110,655	1.3
Chernossolos	42,559	0.5	Organossolos[1]	—	Not Significant
Espodossolos	136,191	1.6	Planossolos	153,214	1.8
Gleissolos	323,453	3.8	Plintossolos	510,716	6.0
Latossolos	3,277,096	38.5	Vertissolos	8512	0.1

[1]Usually associated with Gleissolos.

horizon and having a textural B horizon with high-activity clay. Finally we have a group of soils with both a textural B horizon and low-activity clay.

This is a conventional approach and we believe that, with a small amount of training, most soil scientists and other users of the soil information can identify the soil order without too much difficulty. This is important for agronomists, soil fertility specialists, and land evaluation specialists, as much of the land use recommendations are still based on the highest level of the classification. At this moment, environmental applications of soil information is still limited. But as our knowledge base on environmental and other applications of soils grows, we will modify the system accordingly.

DISTRIBUTION OF SOILS IN BRAZIL

The Latossolos order is the most extensive soil class comprising about 38.5% of the Brazilian Territory. Considerable areas are on the central plateaus, on the lower portion of the Amazon River Basin, and significant areas on the upper part of the Paraná River Basin in the South region. This class is followed by the Argissolos, with about 19.8%, and they occur mainly among the latosolic plateaus on the Brazilian rock shield. Together they represent about 58% of Brazilian soil classes. The Neossolos Class is the third in area, representing 14.7% of the total area. The Plintossolos class is the fourth most extensive, performing 6% of Brazilian soils. This class commonly occurs on pediments and/or on the footslopes of plateaus. The other classes occur at lower percentages, as shown in Table 11.3. The Organossolos, in terms of Brazilian soils, occupy very small areas, and they always occur in association with the Gleissolos. The distribution of the soil classes at Order category is shown in Figure 11.14.

CONCLUSION

The elaboration of any soil classification is influenced by the knowledge of soil formation and genesis. The selection of differentiating criteria and attributes for setting apart soil classes in the higher categories was guided by grouping similar soils based upon ease of identification of morphogenetic features in the soil environment. These are believed to carry out expressive amounts of accessory characteristics that commonly have great importance for agricultural use and management requirements of each class. The current Brazilian system represents a major reorganization of the outline of the restricted distribution of the fourth approximation, which circulated in 1997 among the partners. The system comprises 14 orders, frameworked in six categorical levels in a hierarchy of increasing information content for the known Brazilian soils. It is an open system that allows inclusion of classes if it becomes necessary to introduce new one. The Brazilian system was published in 2000, and presents adaptations of criteria and concepts established in Soil Taxonomy and in the World Reference Base for Soil Classification. However, the structure of the system and emphasis given to criteria differ from these international systems. The system is not considered a

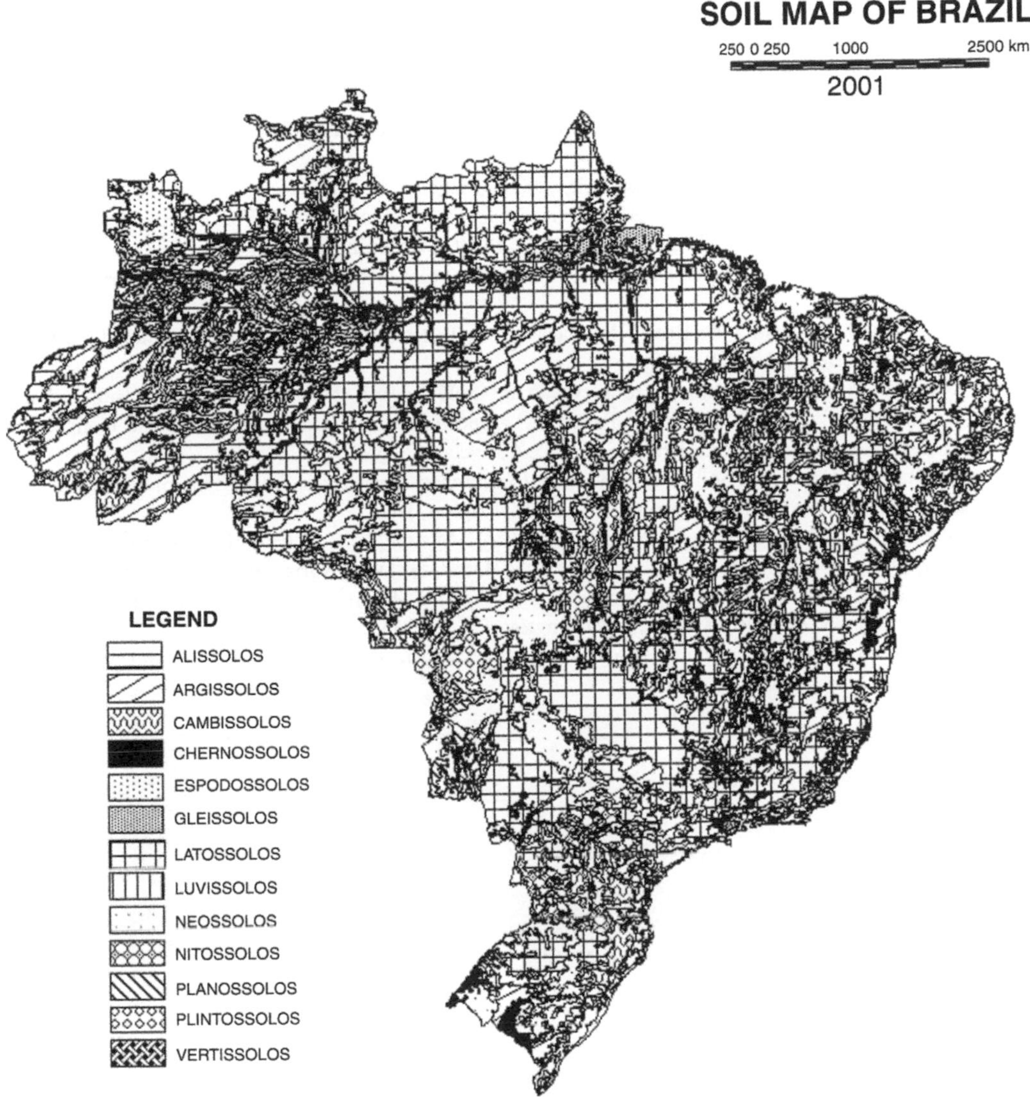

Figure 11.14 A map of Brazil showing the main distribution of soil order classes.

final one; it was schemed to permit changes as knowledge of Brazilian soils increases. During the July 2001 Brazilian Soil Science Society meetings, various proposals were made and questions were raised that are believed to improve the system. All of them are being examined by the soil classification committee.

REFERENCES

Baldwin, M., Kellog, C.E., and Thorp, J. 1938. Soil classification, in Soils and Men. Yearbook of Agriculture. USDA. U.S. Government Printing Office, Washington, DC.

Bennema, J. and Camargo, M.N. 1964. Segundo esbôço parcial de classificação de solos brasileiros. Divisão de Pedologia e Feritlilidade do Solo. Seção de Pedologia e Levantamento de Solos. Ministério da Agricultura. Departamento de Pesquisa e Experimentação Agropecuária. Rio de Janeiro, RJ., Brazil.

Cline, M.G. 1949. Basic principles of soil classification. *Soil Sci.* 67:81–91.

EMBRAPA. 1982. Conceituação sumária de algumas classes de solos recém reconhecidas nos levantamentos e estudos de correlação do SNLCS. EMBRAPA-SNLCS, Circular Técnica, 1. Rio de Janeiro, RJ., Brazil.

EMBRAPA. 1999. Sistema Brasileiro de Classificação de Solos. Embrapa: Produção de Informação. Brasilia, D.F. Rio de Janeiro: Embrapa Solos.

FAO-UNESCO. 1974. Soil Map of the World. 1:5,000,000 legend. Paris: Unesco.

FAO-UNESCO. 1988. Soil Map of the World. Revised Legend, with corrections. World Resources Report 60, FAO, Rome. Reprinted as Technical Paper, 20, ISRIC, Wageningen, 1994.

FAO/ISRIC/ISSS. 1998. World Reference Base for Soil Resources. World Soil Resources Reports, 84, FAO. Rome, Italy.

IBGE. 1987. Anuário estatístico do Brasil. Fundação Instituto Brasileiro de Geografia e Estatística. Rio de Janeiro, RJ, Brasil. Vol. 48, p.20.

Serviço Nacional de Levantamento e Conservação de Solos. 1981. Mapa de Solos do Brasil. Escala 1:5.000.000. Empresa Brasileira de Pesquisa Agropecuária-EMBRAPA. Sev. Nac. Lev. Cons. Solos-SNLCS. Rio de Janeiro, RJ, Brazil.

Soil Survey Staff, USDA. 1960. Soil classification, A Comprehensive System. 7th Approximation. U.S. Dept. of Agric. U.S. Government Printing Office, Washington, DC.

Soil Survey Staff, USDA. 1975. Soil Taxonomy. A Basic System of Soil Classification for making and Interpreting Soil Surveys. Agric. Handbook No. 436. U.S. Dept. Agric. Washington, DC.

Soil Survey Staff, USDA. 1999. Soil Taxonomy. A basic system of soil classification for making and interpreting soil surveys. USDA-NRCS, Washington, DC. AH-436.

Thorp, J. and Smith, G.D. 1949. Higher categories of soil classification: Order, suborder, and great groups. *Soil Sci.* 67:117–126.

CHAPTER 12

The Future of the FAO Legend and the FAO/UNESCO Soil Map of the World

Freddy O.F. Nachtergaele

CONTENTS

ABSTRACT

The development of the FAO legend is intimately linked with the development of the FAO/UNESCO Soil Map of the World. Therefore the historical development of the map is discussed first, including its present update under the global SOTER (Soil and Terrain Database) program. Then the evolution from the FAO Legend toward the World Reference Base (WRB), an internationally endorsed soil correlation system, is documented. Anticipated and desirable future developments for both the global SOTER and the WRB are highlighted.

DEVELOPMENT OF THE SOIL MAP OF THE WORLD AND THE GLOBAL SOTER DATABASE

At the global level, the 1:5 million scale FAO/UNESCO Soil Map of the World (FAO, 1971–1981) is still, 20 years after its finalization, the only worldwide, consistent, harmonized soil inventory that is readily available in digital format, and comes with a set of estimated soil properties for each mapping unit.

The project started in 1961 and was completed over a span of 20 years. The first draft of the Soil Map of the World was presented to the Ninth Congress of the ISSS in Adelaide, Australia, in 1968. The first sheets, those covering South America, were issued in 1971. The results of field correlation in different parts of the world and the various drafts of the legend were published as

0-8493-1339-2/03/$0.00+$1.50

issues of FAO World Soil Resources Reports (FAO, 1961–1971). The final Legend for the map was published in 1974. The last and final map sheets for Europe appeared in 1981.

With the rapidly advancing computer technology and the expansion of geographical information systems during the 1980s, the Soil Map of the World was first digitized by ESRI (1984) in vector format, and contained a number of different layers of land resource-related information (vegetation, geology), often incomplete and not fully elaborated upon. In 1984, a first rasterized version of the soil map was prepared by Zöbler, using the ESRI map as a base and using 1° × 1° grid cells. Only the dominant FAO soil unit in each cell was indicated. This digital product gained popularity because of its simplicity and ease of use, and is still considered a standard, particularly in the United States, in spite of its drawbacks.

In 1993, FAO and the International Soil Reference and Information Centre (ISRIC) jointly produced a raster map with a 30′ × 30′ cell size in the interest of the WISE (World Inventory of Soil Emissions) project (Batjes et al., 1995). This database contains the distribution of up to ten different soil units and their percentages in each cell.

In 1995, FAO produced its own raster version, which had the finest resolution with a 5′ × 5′ cell size (9 km × 9 km at the equator), and contained a full database corresponding with the information in the paper map in terms of composition of the soil units, topsoil texture, slope class, and soil phase in each of the more than 5000 mapping units. In addition to the vector and raster maps discussed above, the CD-ROM contains a large number of databases and digital maps of statistically derived (or expert guestimates of) soil properties (pH, OC, C/N, soil moisture storage capacity, soil depth, etc.). The CD-ROM also contains interpretation by country on the extent of specific problem soils, the fertility capability classification results by country, and corresponding maps. For more information, see *http://www.fao.org/WAICENT/FAOINFO/AGRICULT/AGL/lwdms.htm*

An overview of the publication stages of the paper Soil Map and its digitized version is given in Table 12.1.

The development of the SOTER (SO = SOil, TER = TERrain) program started in 1986 (The ISSS Congress in Hamburg) with the aim of providing the framework for an orderly arrangement of natural resource data, in such a way that these data can be readily accessed, combined, and analyzed. Fundamental in the SOTER approach is the mapping of areas with a distinctive, often repetitive pattern of landform, morphology, slope, parent material, and soils at 1:1 million scale (SOTER units). Each SOTER unit is linked through a geographic information system with a computerized database containing, ideally, all available attributes on topography, landform and terrain, soils, climate, vegetation, and land use. In this way, each type of information or each combination of attributes can be displayed spatially as a separate layer or overlay or in tabular form.

The SOTER methodology was developed by the International Soil Reference and Information Centre (ISRIC) in close cooperation with the Land Resources Research Centre of Canada, FAO,

Table 12.1 Important Dates in the Development of the Soil Map of the World

1960	ISSS recommends the preparation of the soil maps of continents
1961	FAO and UNESCO start the Soil Map of the World project
1971	Publication of the first sheet of the paper map (South America)
1974	Publication of the FAO Legend
1981	Publication of the last sheet of the paper map (Europe)
1984	ESRI digitizes the map and other information in vector format
1988	Publication of the Revised Legend
1989	Zöbler produces a 1° x 1° raster version
1991	FAO produces an Arc/Info vector map including country boundaries
1993	ISRIC produces a 30′x 30′ raster version under the WISE project
1995	FAO produces a CD ROM raster (5′ x 5′) and vector map with derived soil properties
1998	FAO-UNESCO reissues the digital version with derived soil properties, including corrections

and ISSS. After initial testing in three areas, involving Argentina, Brazil, Uruguay, the United States, and Canada, the methodology was endorsed by the ISSS Working Group on World Soils and Terrain Digital Database (DM). In 1993, the Procedures Manual for Global and National Soils and Terrain Digital Databases was jointly published by UNEP, ISSS, FAO, and ISRIC, thus obtaining international recognition. The Procedures Manual is available in English, French, and Spanish (UNEP/ISRIC/FAO/ISSS, 1995).

The SOTER concept was originally developed for application at country (national) scale and national SOTER maps have been prepared, with ISRIC's assistance, for Uruguay (1:1 M), Kenya (1:1 M), Hungary (1:500,000), and Jordan and Syria (1:500,000). More information is available from ISRIC's SOTER website at *http://www.isric.nl/SOTER.htm*; the Kenya data are downloadable at *http://www.isric.nl/SOTER/KenSOTER.zip.*

The original idea of SOTER was to develop this system worldwide at an equivalent scale of 1:1 M in order to replace the paper Soil Map of the World (Sombroek, 1984). It soon became obvious that the resources were lacking to tackle and complete this huge task in a reasonable timeframe. However, this still remains the long-term objective pursued on a country-by-country basis.

In the early 1990s, FAO recognized that a rapid update of the Soil Map of the World would be a feasible option if the original map scale of 1:5 M were retained, and started, together with UNEP, to fund national updates at 1:5 M scale of soil maps in Latin America and Northern Asia. At the same time, FAO tested the physiographic SOTER approach in Asia (Van Lynden, 1994), Africa (Eschweiler, 1993), Latin America (Wen, 1993), the CIS, the Baltic States, and Mongolia (Stolbovoy, 1996).

These parallel programs of ISRIC, UNEP, and FAO merged together in mid-1995, when at a meeting in Rome the three major partners agreed to join the resources devoted to these programs and work toward a common world SOTER approach, covering the globe at 1:5 M scale by the 17th IUSS Congress of 2002, to be held in Thailand. Since then, other international organizations have shown support and collaborated to develop SOTER databases for specific regions. This is, for instance, the case for Northern and Central Eurasia, where the International Institute for Applied System Analysis (IIASA) joined FAO and the national institutes that were involved, and for the European Soils Bureau (ESB) in the countries of the European Union. The ongoing and planned activities are summarized in Table 12.2 and Figure 12.1.

Table 12.2 Operational Plan for a World SOTER: 1995–2006

Region	Status	Main Agencies Involved	Published Date
Latin America and the Caribbean	Published	ISRIC, UNEP, FAO, CIAT, national soil institutes	1998
Northeastern Africa	Published	FAO-IGAD	1998
South and Central Africa	Ongoing	FAO-ISRIC-national inst.	2002
North and Central Eurasia	Published	IIASA, Dokuchaev Institute, Academia Sinica, FAO	1999
Eastern Europe	Published	FAO-ISRIC-Dutch Government-national inst.	2000
Western Europe	Ongoing	ESB-FAO-national inst.	2002
Mahreb	Ongoing	ESB-FAO-ISRIC-national inst.	2004
West Africa	Proposal submitted	Awaits funding (ISRIC, IITA)	
Southeast Asia	Proposal discussed	Awaits funding	
U.S. and Canada	Own Effort	NRCS	Own effort
Australia	Own Effort	CSIRO	Own effort

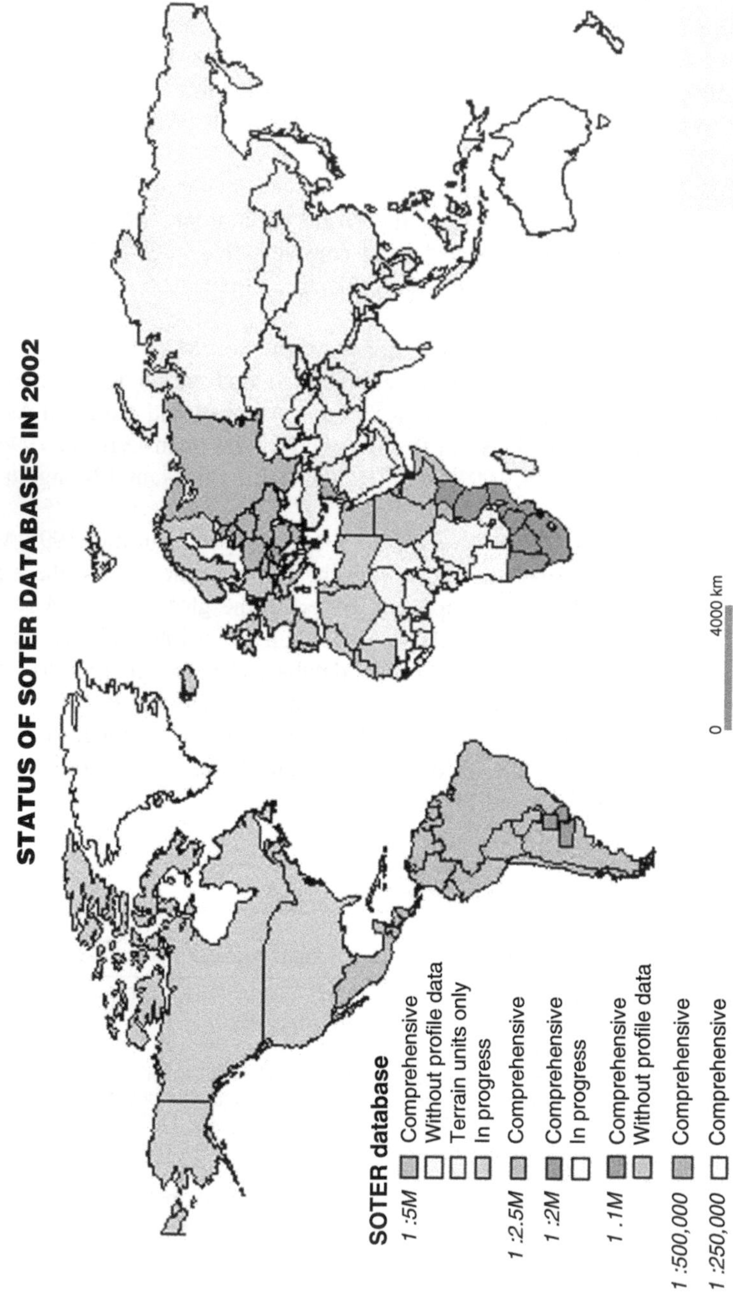

Figure 12.1 World Soter status (October, 2001).

It should be noted that although the information is collected according to the same SOTER methodology, the specific level of information in each region results in a variable scale of the end products presented. The soils and terrain database for northeastern Africa, for instance, contains information at equivalent scales between 1:1 million and 1:2 million, but the soil profile information is not fully georeferenced (reference is done by polygon or mapping unit, not by latitude and longitude). The same will be true for the soil profile information to be provided by ESB for the European Community, to be released as a SOTER database at 1:5 M. For north and central Eurasia, profile information contained in the CD-ROM is very limited. Fully comprehensive SOTER information is available for South and Central America and the Caribbean (1:5 M scale), and includes more than 1800 georeferenced soil profiles. Data are available from *http://www.isric.nl/SOTER/LACData.zip* and viewable using a viewer program at *http://www.isric.nl/SOTER/Viewer102b.exe)*. For Central and Eastern Europe (1:2.5 M scale), this SOTER database contains more than 600 georeferenced soil profiles and the results are also available on CD-ROM (URL: *http://www.fao.org/catalog/book_review/giii/x8322-e.htm)*.

Notable gaps in harmonized soil information occur in southern and eastern Asia and in western and northern Africa. Northern America and Australia are expected to join the program (Figure 12.1). The effort to harmonize these various formats in a global SOTER product should not be underestimated.

It should also be noted that the SOTER methodology and manual are in need of a thorough review, given the recent developments in remotely sensed technologies and the availability of high-resolution digital elevation models, which should permit refinements and automation of the physiographic terrain units to be introduced. It has also been recommended that WRB become the standard classification to be used for the soil profiles stored in SOTER, instead of the FAO revised legend.

Finally, attention should be drawn to the fact that SOTER is not a methodology to carry out soil surveys or to produce maps, but is in essence a relational database storing, in a systematic way, related physiographic and soil information. This reflects the modern practice of digital storage of existing soil information. On the other hand, when soil information is incomplete or inexistent, the difficulties of creating SOTER databases should not be underestimated.

DEVELOPMENT OF THE FAO LEGEND

The FAO Legend (FAO/UNESCO, 1974) for the Soil Map of the World had the following characteristics:

It was largely based on the diagnostic horizon approach developed under Soil Taxonomy (Soil Survey Staff, 1960) by the USDA during the 1950s and 1960s. Similar horizons were defined, and where definitions of the diagnostic horizons were slightly simplified, different names were used for comparable horizons such as the ferralic horizon equivalent to the oxic horizon, or the argic horizon equivalent to the argillic horizon in Soil Taxonomy.

As the Legend was the outcome of a vast international collaboration of soil scientists, it was by necessity a compromise. Certain historical soil names were retained to accommodate some national sensitivities. Examples of these at the highest level were Rendzina's, Solonetzes, Solonchaks, and Chernozems. Some of the names had a dubious scientific connotation (such as the Podzoluvisols in which no Podzolization takes place), while others were nearly identical as those developed in Soil Taxonomy, such as the Vertisols.

In contrast with Soil Taxonomy, climatic characteristics were not retained in the FAO Legend, although the Xerosols and Yermosols largely coincided with soils developed under an aridic moisture regime.

The FAO Legend of 1974 recognized 26 Great Soil Groups subdivided in 106 Soil Units, which were the lowest category recognized on the world soil map. In addition, twelve soil phases were recognized, three general texture classes (fine, medium, and coarse) and three general slope classes (with slopes less than 8%, 8 to 30%, and more than 30%). Most soil mapping units were in fact soil associations, the composition of which was indicated at the back of each paper map sheet. The dominant soil unit gave its name (and appropriate color) to the mapping unit, followed by a number unique to the associated soils and inclusions. Texture (1, 2, and 3) and slope symbols (a, b, and c) were included in the mapping unit symbol.

In 1984, during the initial Agro-ecological Zones study (FAO, 1978), the mapping composition was formalized, and the proportional distribution of soil units and their corresponding soil phases (if any), textures, and slopes in the mapping unit could be expressed as a percentage. The rules used to do this were logical (a function of the number of soil units associated and included), but obviously resulted only in a very rough estimate of the proportions of each unit present in the mapping unit.

Although initially developed as a Legend for a specific map, not a soil classification system, the FAO Legend found quick acceptance as an international soil correlation system, and was used, for instance, as a basis for national soil classifications (Bangladesh, Brammer et al., 1988; Kenya, Sombroek et al., 1982), and regional soil inventories as in the soil map of the European Union (CEC, 1985).

With the applications as a soil classification, numerous comments and suggestions were received to improve the coherence of the system. In fact, some combinations of diagnostic horizons could not be classified, while one soil unit identified in the key (gelic Planosols) did not occur in the map. The revision effort undertaken in the 1980s finally resulted in the publication of the Revised Legend of the FAO/UNESCO Soil Map of the World (FAO/UNESCO/ISRIC, 1988). This revised legend was applied to the World Soil Resources Map at 1:25,000,000 scale, accompanied by a report (FAO, 1993) and presented at the Kyoto ISSS Congress.

The revised legend retained many features of the original FAO legend. The number of Great Soil groupings increased from 26 to 28: the Rankers and Rendzinas were grouped with the Leptosols, the "aridic" Yermosols and Xerosols disappeared, and new Great Soil groups of Calcisols and Gypsisols were created. The Luvisols (Alfisols)–Acrisols (Ultisols) division was further divided according to the activity of the clay fraction resulting in four symmetric groups (Luvisols, high base saturation, high-activity clays; Acrisols, low base saturation, low-activity clays; Lixisols, high base saturation, low-activity clays; and Alisols with low base saturation and high-activity clays). The revised legend also created at the highest level the Anthrosols, grouping soils strongly influenced by human activities. The number of soil units increased from 106 to 152. Texture and slope classes remained unchanged, but were not represented on the map produced for obvious reasons of scale. A start was made with the development of a set of names at the third level. The latter were further elaborated in a comprehensive set of third level "qualifiers" by Nachtergaele et al. (1994) and presented at the 15th ISSS Congress in Acapulco.

As had happened with the Legend, the Revised Legend was used as a basis for the national map legends as in Botswana (Remmelzwaal and Verbeek, 1990).

In a parallel development, a working group of the ISSS had been active in the development of an internationally acceptable soil classification system (first meeting in Sofia, 1982). In 1992, at a meeting of the Working Group RB (Reference Base), the strong recommendation was made that rather than developing a fully new soil classification system, the Working Group should consider the FAO Revised Legend as a base and give it more scientific depth and coherence. This principle was accepted, and the first draft of the World Reference Base (WRB) appeared in 1994 (ISSS/ISRIC/FAO, 1994), still showing large similarities with the FAO revised legend.

WRB aims to satisfy at the same time two very different kinds of users of soil information: one, the occasional interested user, who should be able to differentiate the 30 reference soil groups, and second, the professional soil scientist who needs a universal nomenclature for soils in a simple system that enables him to communicate about the soils classified in his national soil classification.

The first version of WRB (Working Group RB, 1998a, b, c) was presented in 1998 at the 16th ISSS Congress in Montpelier and was endorsed, in a historical motion, as the official soil correlation system of the International Union of Soil Sciences. The detailed characteristics of the system (Nachtergaele et al., 2000; Deckers et al., 2001) are described by others and will not be treated here, but major changes with the first draft (and implicitly the FAO Revised Legend) include the following:

- The development of a two-tier system in which the first level describes and defines, in simple terms, 30 Soil Reference Groups and classifies them with a key. The second level is composed of a list of uniquely defined qualifiers (121 of them) that can be attached to the Soil Reference Groups. For international soil correlation purposes, a specific preferred priority ranking is given for each Soil Reference Group. This approach results in a very compact system.
- The definition of a number of prefixes that permit the precise description for where a soil phenomena occurs (epi, endo, bathy, etc.) or how strongly it is expressed (hypo, hyper, petri). This results in the ability to give a very precise characterization of the soil.
- The explicit emphasis put on morphological characterization of soils rather than analytical procedures. This aim is not always fully realized, but it is at least an attempt to permit soil classification to take place in the field, rather than in the laboratory.
- The parallel publication of a WRB topsoil characterization system (Purnell et al., 1994; URL: *ftp://ftp.fao.org/agl/agll/docs/topsoil.pdf*), with about 70 different topsoil combinations recognized, that permits classification in a much more elaborated way than possible with six epipedons, the most important portion of the soil for most of its uses.

The WRB system as such has met with considerable success since its official endorsement by the IUSS in 1998. In the 3 years since its appearance, 3000 WRB books have been sold or distributed, a web site was established (*http://www.fao.org/ag/AGL/agll/wrb/wrbhome.htm*), the system has been translated into 10 languages, it has been adopted in several regions (European Union, West Africa, Global SOTER) and countries (Italy, Lithuania) for national correlations or classification purposes, and has led to a revival of the interest in soil classification and soil nomenclature.

The developments described above and the association of FAO with the WRB effort indicate that the latter has effectively replaced the FAO legend with a FAO-backed and internationally endorsed World Soil Reference system.

FUTURE DEVELOPMENTS

The World Reference Base for Soil Resources will be re-edited as a one-volume student version that includes some minor changes and additions (Driessen et al., 2001), including rules for arriving at the preferred ranking of qualifiers. The book will be followed by a CD-ROM containing more than 1500 pictures of landscapes and soil profiles classified using WRB.

Future developments of WRB could include the following:

- A much more explicit emphasis on soil characteristics as the base for soil classification and correlation, rather than continuing to provide lip service to pedogenesis as the underlying central factor for combining soils in classes or for subdividing them.
- An explicit recognition that the subdivisions of the Soil Reference Groups are largely arbitrary and that the hierarchical ranking of these subdivisions makes little scientific sense. This can be achieved by using all relevant qualifiers as a single "flat" second level. The order in which the qualifiers are used would be left open and be a function of the purpose of the soil survey. For international soil correlation purposes, a reasoned preferred ranking order is proposed (Nachtergaele et al., 2001). The application would result in greater transparency, permit further simplifications, and, more importantly, would allow a full transfer of soil profile information, which is

often hampered in other classification systems by hierarchical rules (permitting only a very limited number of predefined names per hierarchical level).

- The adaptation of soil classification systems to soil map legends in smaller scale maps rather than the reverse. When soil map legends are based on soil classification criteria alone, the natural context of their occurrence in a landscape is sometimes lost, and the resulting maps have units with boundaries that have little relationship to the environment in which they occur (Soil map of Rwanda, AGCD, personal communication). The linkage between the scale of the soil map and the hierarchical level of the soil classification used puts a further burden on the transfer of information. A SOTER-like approach largely avoids this problem, and WRB would be the ideal tool to classify the soil profiles characterizing the SOTER units.
- A better definition and characterization of anthropogenic horizons to deal with Anthrosols that are growing at a faster rate than any other soil group because of population growth and continued intensification of agriculture. Concurrently, more marginal lands are brought under cultivation, changing forever the characteristics of the soils and hence their nomenclature and classification.
- Much greater attention paid to topsoil characterization and to subsoil physical soil characteristics than has been the case in the past, if soil information is to be taken seriously by agricultural users, environmental planners, and engineers.
- Incorporation of certain "family" level criteria of Soil Taxonomy. It is suggested that the texture and mineralogy classes, as presently defined, should be tested in a WRB context.

Future developments of the global SOTER database would include the following:

- A correlation of all soil profiles present in the database with the WRB soil correlation system.
- A review of the physiographic elements incorporating new data and techniques related to DEM.
- A completed harmonized World SOTER database to be presented at the 18th IUSS Congress in 2006.

None of the above developments will lead anywhere without a general revival of awareness among users that soil information does make a difference in economic terms for the projects in which they are involved.

CONCLUSION

The FAO Legend for the Soil Map of the World has evolved over time, and its ideas are presently incorporated into the World Reference Base for Soil Resources, a universal soil correlation system endorsed by the IUSS.

The application of WRB to the global SOTER database and the completion of this database is a task presently undertaken by FAO and ISRIC with the assistance of many international and national organizations. Progress has been adequate, but has slowed because of a lack of resources to finalize large areas in South and Southeast Asia and West and Central Africa.

In the foreseeable future, the Global SOTER database will replace the FAO/UNESCO Soil Map of the World.

REFERENCES

Batjes, N.H., Bridges, E.M., and Nachtergaele, F.O. 1995. World Inventory of Soil Emission Potentials: Development of a Global Soil Database of Process Controlling Factors, in *Climatic Change and Rice*. S. Peng et al., Eds. Springer Verlag, Heidelberg, 110–115.

Brammer H., Antoine, J., Kassam, A.H., and van Velthuizen, H.T. 1988. Land Resources Appraisal of Bangladesh for Agricultural Development. Report 3. Land Resources Data Base, Vol. II. Soil, Landform and Hydrological Database. UNDP/FAO, Rome.

CEC (Commission of the European Communities). 1985. Soil Map of the European Communities 1:1,000,000. Directorate General for Agriculture, Coordination of Agricultural Research, Luxemburg.

Deckers, J., Driessen, P., Nachtergaele, F., Spaargaren, O., and Berding, F. 2001. Anticipated Developments of the World Reference Base for Soil Resources (in this volume).

Driessen, P., Deckers, J., Spaargaren, O., and Nachtergaele, F. 2001. Lecture Notes on the Major Soils of the World. World Soil Resources Report #94. FAO, Rome (in preparation).

Eschweiler, H. 1993: Draft Physiographic Map of Africa.Working Paper. FAO/AGLS, Rome.

ESRI. 1984. Final Report. UNEP/FAO World and Africa GIS Database. Environmental Systems Research Institute, Redlands, CA.

FAO. 1978. Report on the Agro-ecological Zones project, Vol 1. Methodology and results forAfrica. World Soil Resources Reports #48. FAO, Rome.

FAO. 1993. World Soil Resources. An explanatory note on the FAO World Soil Resources map at 1:25,000,000 scale, 1991. Rev. 1993. World Soil Resources Reports # 66. FAO, Rome.

FAO. 1995. The Digitized Soil Map of the World Including Derived Soil Properties (version 3.5). FAO Land and Water Digital Media Series 1. FAO, Rome.

FAO/IGAD/Italian Cooperation. 1998. Soil and terrain database for northeastern Africa and crop production zones. Land and Water Digital Media Series 2. FAO, Rome.

FAO/IIASA/Dokuchaiev Institute/Academia Sinica. 1989. Soil and Terrain. Database for northand central Eurasia at 1:5 million scale. FAO Land and Water Digital Media series 7. FAO, Rome.

FAO/ISRIC. 2000. Soil and terrain database, soil degradation status, and soil vulnerability assessments for Central and Eastern Europe (scale 1:2.5 million; version 1.0). Land and Water Digital Media Series 10, FAO, Rome.

FAO/UNEP/ISRIC/CIATT. 1998. Soil and terrain digital database for Latin America and the Caribbean at 1:5 million scale. FAO Land and Water Digital Media series 5. FAO, Rome.

FAO/UNESCO. 1974. FAO-UNESCO Soil Map of the World. Volume 1 Legend. UNESCO, Paris.

FAO/UNESCO. 1971–1981. The FAO-UNESCO Soil Map of the World. Legend and 9 volumes. UNESCO, Paris.

FAO/UNESCO/ISRIC. 1988. FAO-UNESCO Soil Map of the World. Revised Legend. World Soil Resources Report #60. FAO, Rome.

ISSS/ISRIC/FAO. 1994. World Reference Base for Soil Resources. Draft. O. Spaargaren, Ed. Wageningen/Rome.

Nachtergaele, F.O., Berding, F.R., and Deckers, J. 2001. Pondering hierarchical soil classification systems. Proceedings “Soil Classification 2001.” University of Godollo, Hungary.

Nachtergaele, F.O., Remmelzwaal, A., Hof, J., van Wambeke, J., Souirji, A., and Brinkman, R. 1994. Guidelines for distinguishing soil subunits. In Transactions 15th World Congress for Soil Science. Volume 6a. Commission V: Symposia, Etchevers, BJD, Ed. Instituto Nacional de Estadistica, Geografia e informatica. Mexico, 818–833.

Nachtergaele, F.O., Spaargaren, O., Deckers, J.A., and Ahrens, R. 2000. New developments in soil classification. World Reference Base for Soil Resources. *Geoderma*. 96:345–357.

Purnell, M.F., Nachtergaele, F.O., Spaargaren, O.C., and Hebel, A. 1994. Practical topsoil classification–FAO proposal. Transactions 15th International Congress of Soil Science, Acapulco. 6b:360.

Remmelzwaal, A. and Verbeek, K., 1990: Revised general soil legend of Botswana. AG/BOT/85/011. Field Document 32. Ministry of Agriculture, Gaberone, Botswana.

Soil Survey Staff, 1960. Soil Classification, a Comprehensive System. 7th Approximation. SCS, USDA. U.S. Government Printing Office, Washington, DC.

Soil Survey Staff. 1999. Soil Taxonomy: A basic system of soil classification for making and interpreting soil surveys. 2nd ed. Agricultural Handbook 436. NCS, USDA. U.S. Government Printing Office, Washington, DC.

Sombroek, W.G. 1984. Towards a Global Soil Resources Inventory at Scale 1:1 Million. Discussion Paper. ISRIC, Wageningen, The Netherlands.

Sombroek, W.G., Braun, H.M.H., and van der Pouw, B.J.A. 1982. Exploratory soil map and agro-climatic zone map of Kenya, 1980. Scale 1:1,000,000.

Stolbovoy, V. 1996. Draft Physiographic Map of the Former Soviet Union and Mongolia at scale 1:5 million. Working Paper. FAO/AGLS, Rome.

UNEP/ISRIC/FAO/ISSS. 1995. Global and National Soils and Terrain Digital Databases (SOTER) Procedures Manual. World Soil Resources Report No. 74. FAO, Rome.

Van Lynden, G.W.J. 1994. Draft Physiographic Map of Asia. FAO/ISRIC Working Paper. FAO/AGLS. Rome.

Wen Ting-Tiang. 1993. Draft Physiographic Map of Latin America. FAO/ISRIC Working Paper. FAO/AGLS, Rome.

Working Group RB. 1998a. World Reference Base for Soil Resources. Introduction, in J. Deckers, F.O. Nachtergaele, and O. Spaargaren, Eds. Acco, Leuven.

Working Group RB. 1998b. World Reference for Soil Resources. Atlas, in N.H. Batjes, M. Bridges, and F. Nachtergaele, Eds. Acco, Leuven.

Working Group RB. 1998c. World Reference Base for Soil Resources (ISSS/ISRIC/FAO). World Soil Resources Report #84. FAO, Rome.

Zöbler, L. 1989. A World Soil Hydrology File for Global Climate Modeling, in Proceedings, International Geographic Information Systems (IGIS) Symposium: The Research Agenda.

CHAPTER 13

The Current French Approach to a Soilscapes Typology

Marcel Jamagne and Dominique King

CONTENTS

0-8493-1339-2/03/$0.00+$1.50

ABSTRACT

The Soil Science Department of the French National Institute for Agronomic Research is currently undertaking a test to better understand and formulate the arrangement of soils in landscapes, no longer done solely on soil definitions according to various taxonomies or referentials, but based on the relationships between the main components of the soil cover. Our approach stresses a three-dimensional and spatial view of the soil cover. We introduce the concepts of Soilscapes and Soil-systems and discuss their usefulness in landscapes where the functions of the soil cover can be detemined. The first part of the paper reviews the concepts of Soil Cover and Soilscape, followed by comments on recent developments in taxonomic and cartographic fields, stressing the introduction of modern techniques. We then summarize the most current ideas of French pedologists on the use of spatial analysis in pedology. This is based on the analysis of successive organizations of soil components in the continuum of the Soil Cover in the landscape, leading to the notions of "soil typological units," "soil mapping units," "soil functioning units," and "spatial organization models of soils." We also discuss the feasibility and advisability of classifying different soil sequences on the basis of the dominant pedogenetic factors involved in their differentiation (e.g., relief for toposequences, parent material for lithosequences, time for chronosequences). Finally, we detail an approach that progressively builds an actual typology of the main Soilscapes in the French territory. Priority was given, for perhaps the first time, to toposequences within catchments in which the functions of the soil cover can be at least partly determined, and which can be considered as Soil-systems. Three examples from representative areas of the French territory are described: a loessic region of the Paris Basin, a detritic zone of the Centre, and a mountainous granite area of the Vosges Massif. These examples represent a first application of a new typology of French Soilscapes, which could therefore be based, in the first instance, on geomorphological and lithological parameters. An attempt to extend the representativeness of the three presented Soil-systems has been carried out. The basis has first been a delimitation of watersheds with DEM, combined with dominant parent materials selected from the Geographical Soil Database of France. The present work remains to be completed and validated. However, this general approach is being applied to other important natural regions of the French territory. In conclusion, the definition of Soilscapes and Soil-systems makes it necessary to consider all of the parameters used in soil taxonomy. There is a strong linkage between the proposed approach and the different soil classification systems.

INTRODUCTION

The Soil Research Unit of the French National Institute for Agronomic Research (INRA) is currently attempting to better understand and formulate the arrangement of soils in landscapes. The approach is no longer based solely on soil definitions according to various taxonomies or reference systems, but on the relationships between the main components of the soil mantle; in other words, it is based on a three-dimensional and spatial view of this mantle.

Most soil classifications are based on the study of the pedon, or pedological profile, according to a selective and mainly point-specific view. These rarely take into account the three-dimensional soil arrangements in a space that is sufficiently large to demonstrate the origin of both the horizontal and vertical differentiations of soil horizons. Henceforth, we must evaluate soil distributions more dynamically, considering both their spatial and temporal variabilities.

As a starting point, the approach consists of studying the soils in the field with the notion of "spatial organization models" of soils, which leads to the analysis of the structural arrangement of the landscape's pedological volumes—essentially the horizons, the sola or pedons, and the "pedo-landscapes" or "soilscapes."

It is known that many mapping units correspond with organized systems for which the distribution of soil units, like pedons, is not random but follows the laws of pedological evolution. Consequently, different soil types are not independent entities. Instead, their relative geographical position frequently indicates the pedogenetic relationships in the arrangement of the soil mantle. These spatially linked structures influence present soil and landscape functions, such as the orientation and importance of the vertical or lateral flow of water and matter. The aim is to determine the existence of structures at appropriate levels that may range from the basic catchment area to the regional landscape unit. Thus we attempt to reveal the role played by this geographic arrangement in the general functioning of natural systems. This has led to the notion of "pedological systems" or "soil-systems" linked, within the landscapes, to geomorphologic units formed by toposequences and catchment areas of different size.

SHORT REVIEW OF THE CONCEPTS OF SOILSCAPE AND SOIL-SYSTEMS

From the early inception of the idea of soilscape in the 1970s, the importance of soil/landscape relationships was brought to the fore by a number of authors, relating them to the concepts of "Soil landscape systems" and "Soil-systems," and stressing the importance of modelling these arrangements (Hugget, 1975). The formalization of the concept of "Soil Cover" also dates from this time (Fridland, 1975; 1976), with different levels of "soil combinations."

A structured approach to the analysis of the relationships between landscapes and soilscapes, the latter being a geographic subdivision of the former, was proposed by Hole (1978). He indicated the advantage of this approach for structuring the whole of the existing cartographic data. Northcote (1984) elaborated on the relationships between Soilscapes or "Soil-landscapes," taxonomic units, and soil profiles.

However, the understanding of the arrangement of soils within their soilscapes raises numerous methodological problems (Wilding, 1989). The study of the morphogenetic processes of the landscapes themselves should enable the prediction of soil distribution. It seems evident that the landscapes should be described in three dimensions, including vertical and lateral changes in the composition of parent materials (Hall and Olson, 1991). The authors appear to consider the "catena" model (Milne, 1935; 1936) as the best model for this type of analysis.

This period sees the development of various models. Hewitt (1993) highlights in particular the existence of two types of complementary models for spatial analysis: one concerning soilscapes and the other taking into account soil properties. These modelling approaches are widely used by Slater et al. (1994) and McSweeney et al. (1994) to introduce different types of models. They reveal the necessity of combining pedogenetic models with models describing the overall landscape context. The "soil-landscape models" offer a genuine spatial context for the application of pedogenetic models. They propose a methodological and conceptual approach, consisting of several steps, for a three-dimensional modelling of the "soil-landscape continuum." The subsequent soil-landscape modelling enables the prediction of certain soil properties, particularly by the use of appropriate indexes (Gessler et al., 1995).

Fridland's ideas have been further elaborated by other Russian scientists working on the structure of the soil cover (Kozlovskiy and Goryachkin, 1994). In 1997, a comprehensive evaluation of the approaches was made by Sommer and Schlichting (1997), who synthesized the many works carried out on the "catena" concept. The authors propose here a preliminary approach to the structuring and typology of toposequences in relation to their functioning.

The Global Soil and Terrain Database (SOTER) programme (FAO-ISRIC, 1995), whose aim is to elaborate a worldwide geographic database, uses in part the ideas that have been mentioned. More recently, the European program "Georeferenced Soil Database for Europe" is based on the concepts of "Soil Body," "Soilscape," and "Soil Regions" (EC, 1998), stressing the necessity of describing with precision the spatial structure of the soils within mapping units.

French researchers have been particularly active in this field for some years. Erhart (1956) and later on Kilian (1974) introduced the importance of geomorphology in soil genesis, and that the assessment of morphogenesis/pedogenesis is at the basis of the functioning of soils on slopes.

A team from the Scientific Research Office for Overseas Territories (ORSTOM), led by Boulet (Boulet et al., 1982), revealed, essentially in tropical environments, the importance of differentiating between lateral transfers within a toposequence and those of the "transformation systems" at the level of the interfaces between horizons (Boulet et al., 1984; Fritsch et al., 1986). These studies were carried out through an approach titled "structural analysis," based mainly on the observation of morphological parameters. At this time, Boulaine (1986) described several "pedosystems" stemming from the lateral dispersion of certain constituents. Therefore a methodology was proposed that considers the need for knowledge of soil arrangements within the landscape as a prerequisite for mapping as a whole (Brabant, 1989). Different syntheses on the approaches to the understanding of spatial soil arrangements and their mapping, in temperate as well as in Mediterranean and tropical environments, were carried out (Ruellan et al., 1989; Jamagne and King, 1991; Jamagne et al., 1993; Ruellan and Dosso, 1993; King et al., 1994a). The concepts of a "spatial organization model" of soils and "soil functioning units" were consequently introduced by King et al. (1994b).

DEVELOPMENT IN TYPOLOGIES AND SOIL MAPPING

During the last decades, there have been tremendous advances in information technology and manipulation of spatial data. Quality control in data acquisition has been achieved by development of manuals such as the Soil Survey Manual (Soil Survey Staff, 1961; USDA-SCS, 1993), and through the recognition of the importance of the relationships between soil mapping and classification (Simonson, 1989). Concerning the processing of collected data, statistical methods and especially geo-statistics have been used more and more frequently for the drawing up of "Spatial Information Systems" (Webster, 1977; Webster and Oliver, 1990). Computer technology has become a very powerful tool that is essential for the processing of georeferenced data, especially with the introduction of the Geographic Information Systems (GIS) (Burrough, 1986). The combination of all existing statistical and computing techniques led to very powerful methods, notably those for assessing the uncertainties linked to the interpretations. Nevertheless, a judicious use of these tools is necessary (McBratney, 1992).

Mermut and Eswaran (2001) have recently synthesized the major developments in soil science that have occurred since the 1960s, highlighting the application of these new techniques in soil science: computers, statistics, remote sensing, GIS, etc. The use of DEMs (Digital Elevation Models) in the thematic domain in soil science dates from a little over 10 years ago. The variation intervals are inconsistent, going from grids of a few meters up to 1000 m. Recent contributions to the use of DEMs for modelling "Landscapes" and "Soil-Landscapes" include those of Skidmore et al. (1991), Bell et al. (1994), Odeh et al. (1994), Thompson et al. (1997), Sinowski and Auerswald (1999), King et al. (1999), Chaplot et al. (2000), and Thompson et al. (2001). These modern techniques, enabling a more dynamic approach to spatial analysis, have resulted in the progressive abandonment of traditional mapping procedures in favor of an approach that gives priority to soil arrangements in the landscape, as elaborated by Jamagne (1993), King et al. (1994), Jamagne et al. (1995), FAO-ISRIC (1995), and EC (1998).

Cartography and Classification

An important development has taken place, over the years and by the experience acquired, in the conceptions of the definition and of the contents of the mapping units. This started from the notion of "soil series" introduced by the American school of thought (Soil Survey Staff, 1961;

Simonson, 1989). The soil series is designated by the name of the place where the soil was first identified. When the series is established, it carries a description of the modal profile with information on its distribution and relations to associated soils. The original idea considered that all soils belonging to the same series constituted a taxonomic unit corresponding to a group of similar soils (polypedons). For all practical purposes, series or phases of series have similar management properties and respond similarly to management. As the basic map unit for detailed soil maps, the series therefore represents both a *taxonomic unit* and a *map unit.*

The evolution of the ideas concerning the concept of series eventually led to more precise tests of definition by selecting the most striking features enabling their differentiation: texture, drainage class, organic matter content and distribution, color, and, eventually, "diagnostic" horizons. The infinite combinations have thus enabled the definition of a very large number of soil series (currently about 22,000 in the U.S.). The higher categoric levels in Soil Taxonomy are then built up by an aggregation of the lower series, with adjustments made to the definition of the higher levels to accommodate the preconceived group of soil series.

Because of the conceptual difficulties associated with this approach, a progressive dissociation between taxonomy and cartography appeared, and the format of "typology" was developed. The typology approach attempts to differentiate the concept of mapping, in the strictest sense, from those of classification and taxonomy, which we believe are devices to structure our knowledge. Subsequent to developing the two components independently of each other, their linkage could or should occur (Baize, 1986; Jamagne, 1993; Buol et al., 1997).

Principles of classification, taxonomy, or reference may differ according to the school of thought. The basis of several worldwide soil classification systems has been presented in a number of works (Duchaufour, 1983; 1997). However, the majority of classification and taxonomic systems (CPCS, 1967; Soil Survey Staff, 1975; 1999), such as the Referential Systems (AFES-INRA, 1992; FAO-ISRIC-ISSS, 1998) or map legends (FAO, 1988) concern the profiles, sola, or pedons, and take very little account of soil-landscape arrangements. The relationships between classification and pedogenesis within the framework of ecosystems and soil-systems have nevertheless been revealed on several occasions (Buol et al., 1997).

CURRENT FRENCH IDEAS ON SPATIAL ANALYSIS IN PEDOLOGY

The "Pedological Cover" or "Soil Cover"

Soils, as we observe them today, are a result of transformations that affect the material of the earth's crust. Successive climates and biological and human activities are the agents directly responsible. Their effect depends not only on the nature of the rocks and their derived formations that have resulted from them, but also the relief and the migration of matter in solution or in suspension in water. The original arrangement of geological material disappears, leaving an entirely new arrangement of pedological origin. We are obliged to recognize that soils are frequently, if not always, a legacy of the past.

The processes of soil differentiation at the expense of a geological material are regrouped under the designation of *pedogenesis*. It is precisely this differentiation and the processes involved that place soils in different soil systems. The factors of pedogenesis are well known: parent material, climate, relief, water balance, land cover, time factor, and anthropic influences.

The Pedological Cover, or Soil Cover, is a continuum that consists of all the soils distributed throughout an area, an idea that tends to progressively overtake the more familiar concept of "soil" that appears today as a portion of this cover (Fridland, 1975; 1976). The distinction between soil covers and the original rocks or materials is not always easy to carry out. However, some criteria that enable soil to be distinguished from geological materials are the following:

- Structure with specific aggregates
- Chemical and mineralogical transformations
- Presence of organic matter and living organisms

All the components of these mantles (clods, aggregates, horizons, empty spaces, cracks, channels, diverse concretions, and concentrations) represent many volumes, sizes, and shapes which fit together or juxtapose according to an architecture whose design is a result of complex laws of distribution. This important variety in structure, differentiation, and distribution is due to the variations of the different pedogenetic factors.

The soil cover is notably the main body through which water flows, both vertically and laterally. Gradients exist as a result of many variables, including structure, calcium content, organic material, and clay content. Under the effect of these water transfers, and according to environmental conditions, a double phenomenon is observed. First, an impoverishment of the surface layers in certain constituents, which represent the "exiting" or "out-let structures." Second, a concomitant enrichment of the underlying or lateral layers play the part of "accumulative structures" for the migrant elements, whether they be in solution, colloidal, or as particles. These mechanisms impart to structures different morphologies and properties specific to the materials on which they have acted. The nature of the dynamics recorded during the pedogenesis depends a lot on the topography. The movement of soil particles and attached chemical elements generally occurs mostly vertically on soils that develop on level to gently sloping surfaces, more laterally on steeper slopes. The observed variability is not random, and it is advisable to try to identify and locate in space the factors responsible for their respective pedogenesis.

Further, each cover also has a historical context, which can be lengthy. The numerous arrangements that it presents are often the result of the action of several pedogenetic processes that have operated successively, simultaneously, or sporadically during the evolution of the soil. It is also advisable to know how to find this history, by "*going back in time*," in order to better understand the present functioning of this soil cover.

CHARACTERIZATION AND ARRANGEMENT OF SOILS

Soil formation factors are closely interdependent, and therefore act together in the process of pedogenesis. Processes, such as alteration, humification, migration, etc., which either progressively erase the original structures of the geological materials, while substituting, or at least superimposing, new structures on it, result in "pedological organizations." The examination of these arrangements is essential to the understanding of the relationships that the soil maintains with its environment, and may determine land use and management practices.

The characterization of soil mantles covers several categories of investigation. It is necessary to identify the components, as well as to understand how these components are associated according to different arrangements or organizations. The reciprocal arrangement of the constituents of the solid phase, at different scales, is considered basic to the understanding of soils.

The Different Levels of Soils Organization

Some concepts, though well accepted, could be reiterated. Generally, soil presents an internal architecture at several levels of organization. At each of these levels, we can identify volumes of similar dimensions, presenting the same morphological features that enable their identification.

Elementary arrangements and **structural aggregates.** The former arise from the direct assembly of basic soil constituents and have dimensions that start at the microscopic level (Eswaran and Banos, 1976) to reach that of the first aggregates that can be seen with the naked eye. These aggregates, whose dimensions range from a millimeter to one or several decimeters, constitute what is commonly called the soil "structure," one of the parameters essential to its "morphological description".

Pedological or **Soil horizons** have volumes that are more or less parallel to the surface of the terrain, the lateral dimensions being far larger than the vertical dimension. They show one or more types of structural aggregates, fitted together for the most part, and a certain number of associated "pedological features" such as color staining on ped faces, clay skins, iron, and manganese concretions in the mass, concentrations of organic products, etc. The upper horizon is characterized by the presence or even accumulation of organic matter.

The genetic conditions of soils result in a double differentiation: in their vertical arrangement and in their spatial distribution. The former corresponds to the common notion of soil profile, or solum, or pedon: succession of the soil horizons viewed in vertical cross-section. The latter corresponds with the lateral arrangement of different types of horizon within the landscape, thus allowing for the definition of Soil-systems.

Soil-systems or **Pedological-systems** are often large fragments of the soil cover within which we can distinguish different soil horizons organized in vertical superposition and lateral successions, at the scale of a geomorphologic relief unit. Therefore a Soil-system is described in terms of horizons and relationships between horizons, both horizontal and vertical. This constitutes the origin of our current approach to typology.

Subdivision or Partitioning of the Soil Cover

Analysis of soil mantle arrangements proceeds from an approach whose primary aim is to establish and formulate the laws of spatial distribution of soils at different levels of organization in a geographic area. Historically, this analysis has been expressed as a soil map. However, new technology has provided this expression in the form of *Spatial Soil Databases* and *Geographical Information Systems (GIS)*. These constitute spatial analytical tools and bring considerable flexibility of choice in representative modes, thus facilitating the management and processing of data.

The usual job of soil surveyors, however, remains to divide the soil mantle into approximately similar conceptual volumes to describe the contents and to locate them in space (Simonson, 1989; Wilding, 1989; Girard, 1984; Buol et al., 1997). The continuum constituted by the soil cover may be subdivided into two types of compartments (Figure 13.1):

- *Homogenous* compartments: the ***horizons***, enabling a better characterization, and corresponding to vertical subdivisions within a profile
- *Heterogeneous* compartments, constituting the territorial groups enabling the mapping: the ***typological*** and ***mapping units*** of soils, corresponding to a horizontal sequence of combinations of horizons

For the first of those divisions, we find here the well-known concept of diagnostic horizons, introduced in many taxonomies and reference systems (Soil Survey Staff, 1975; FAO, 1998). For the second, we can distinguish several types of Soil Units as shown below.

- **Soil Typological Units (STU)**
  These are made up of what we have defined as "Profile," "Solum," or "Pedon"— that is to say a volume of soil mantle showing the same superposition of horizons. These units have been the basis of the conventional soil classifications.
- **Soil Mapping Units (SMU)**
  By drawing their boundaries onto a map, each STU should, in principle, correspond to one SMU. However, the correspondence between STUs and SMUs is more complex than it would appear. Actually, due to graphic limitations, it is scarcely possible to delimit the so-called simple or pure SMUs, each corresponding to a given STU, except on large-scale topographic bases (1/5,000, 1/10,000). In most cases, a SMU has to regroup several STUs, so as to give the final document a

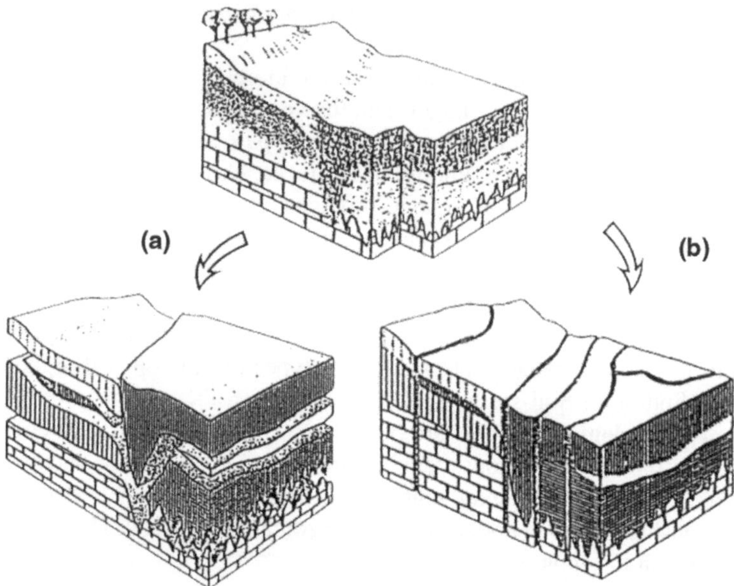

Figure 13.1 Compartments of the Pedological Cover (M.C. Girard, 1984). (a) - in homogeneous volumes, by horizontal subdivision : *Horizons.* (b) - in heterogeneous volumes, by vertical subdivision : *Solums* – « *Profiles* » - « *Pedons* » - *Typological* and *Mapping Units.*

satisfactory legibility. This is the reason for the creation of the well-known "soil associations." These SMUs thus clearly concern the numerous soil map legends elaborated over a long time. In this case we have to try to describe the spatial arrangement of the STUs within the so-called "complex" SMUs.

In this way, three main types of arrangement within these complex units are recognized (Duchaufour, 1983):

- *Soil sequence:* group of soils having a succession that is always in a given order; the reason for this consistency is the dominating influence, regularly repeated, of one of their pedogenetic factors
- *Soil "chain":* group of soils that are genetically linked, each one having received from the others, or given to the others, certain of its constituents
- *Soil juxtaposition:* group of soils whose coexistence within a map unit seems to be linked to no evident rule of distribution

- **Soil Functioning Units (SFU)**

  There is also a need for a method of subdivision that takes into account the functional relationships between typological units, essentially by means of hydric flows and transfer of constituents. Thus the concept of SFUs came about: fragments of a landscape unit for which we know the arrangement, and can therefore predict the functional dynamics. Thus all the STUs belonging to the same catchment area may be perfectly well grouped into a single SFU, in order to describe an overall functioning, for example, a generator of an erosive run-off or sub-soil transfers. (See Figures 13.2 to 13.4).

  Therefore it is possible to propose a *Spatial Organization Model of Soils* (King et al., 1994a; King et al., 1994b), which takes into account the nested combination of the structures previously described: horizons–typological units–mapping units–functioning units.

  Finally, there is the concept of Soilscape or "pedolandscape," which is defined as the soil cover, or part of cover, whose spatial arrangement results from the integration of a group of arranged soil horizons and other landscape elements. This appears to be near the concept of the so-called soil sequence, or catena (Milne, 1935; Fridland, 1976; Buol et al., 1997; Sommer and Schlichting, 1997).

Soilscapes, Soil Sequences: Necessity for a Classification

Numerous works in soil survey have shown the existence of repetitive soil organizations in many landscapes in France (INRA, 1969–2001) and many other countries (Buol et al., 1997). Soil scientists have often used this concept to define SMUs. The concept of Soilscapes and soil sequences involves morphological and pedogenetic criteria at the origin of their differentiation. Examples include scale and spatial arrangement for morphological criteria, factors of differentiation, degree of relationships, and time for pedogenetic ones. The authors recommend classifying various soil sequences according to these criteria.

Spatial Scale

Soils constituing a sequence are, geographically, generally quite close. However, this proximity is relative, and may cover landscape groups of a varied range, simple slope or extensive natural regions. Scale of investigation here is the most important element to take into account.

Interarrangement of Soils

The taking into consideration of this arrangement as "soil patterns" and the functional relationships necessitates a description of the three-dimensional spatial structure of soil types, and if possible, horizons. Methods of description and expression of the results of these spatial structures exist (Fridland, 1976; Buol et al., 1997; King et al., 1994; EU-JRC, 1998).

Differentiation Factors

Different genetic factors prove to be *dominant* in the definition of certain Soilscapes or sequences; if:

- The nature of the *parent material* is the dominant factor, there is a "lithosequence" (gentle slopes cutting across contrasted stratigraphic units, or variable substrate depth)
- It is a question of *relief*, there is a "toposequence" (transfer of dissolved products or particles from upstream to downstream)
- It is a question of *climate*, there is a "climosequence" (oceanic to continental sequences, or elevation sequences)
- It is a question of the influence of *water tables*, there is a "hydrosequence" (influence of a permanent water table, distance to the hydrographic collector)
- It is a question of *time*, there is a "chronosequence" (progressive soil differentiation of different materials, geomorphological origins of relief units such as the succession of terraces)
- *Human* action is the dominant factor, there is an "anthroposequence" (resulting from deforestation, input or out-take of material, construction of agricultural terraces)

Of course, different qualifiers may be applied to the same Soilscape. The most frequent sequences that are closely linked to the dominance of one genetic factor seem to be topographical and lithological sequences.

Relationships

The functional relationships between soils lead to classifying also the sequences, or catenas (Sommer and Schlichting, 1997):

- A catena of "transformation": vertical but no lateral exchange of matter between the soils and horizons of the sequence
- A catena of "translocation": vertical and lateral exchanges of matter between the soils and horizons of the sequence

- A catena of "loss": exportation of matter from the whole soil sequence (erosion processes)
- A catena of "gain": importation of matter for all the soils of the sequence (eolian deposits)

The age of the differentiation must also be taken into account. The sequences can be named as *paleo,* if inherited from an ancient process, or *current* if they are characteristic of an active pedogenesis. There are many chronosequences that integrate both these temporal aspects. In reality, soil formation results from a series of pedogeneses, each of which creates conditions favorable to the appearance of the following one. This leads to the concept of *sequential evolution*, associated with that of *pedogenetic phyllum.* This concept of pedogenetic chronosequence, well represented in the evolution of soils on the loessic formations of Western Europe (Jamagne, 1978), has been supported by other works reconstructing the historical alteration and progressive pedological differentiation of soils on other types of materials. Its importance has been emphasized as much in the field of pedology as in the field of pedo-geomorphology (Bornand, 1978; Vreeken, 1984; Boulet et al., 1995).

SOIL-SYSTEMS

Our concern for a typology of *Soilscapes* has led us to give initial priority to the more evident soil sequences from a spatial viewpoint. These are the toposequences where the differentiation is linked with a functioning process. This type of Soilscape we consider ***Soil-systems.***

Numerous lateral transfers affect the majority of the formations covering the slopes. In many landscapes, the landform dictates the distribution of precipitation, by surface run-off and infiltration. Flow transfer varies according to the permeability of the surface layers, and leads to a differentiation by transport of matter in solution or of particles in suspension. This differentiation which is as much vertical as lateral, is particularly explicative of the functioning of the landscape units, and their detailed analysis has often aided the comprehension of soil distribution on slopes and in watersheds. The reference relief unit is therefore made up of the catchment or watershed area. The analysis of the lateral transfers: on the surface, hypodermic, in depth, must be considered to understand the functioning of the landscape units. This understanding and analysis is accomplished through on-site monitoring.

The concept of "Soil Functioning Unit" (part of a slope or catchment that we can use to estimate the transformation kinetics), with all the significance of the mechanisms and processes acting at the interfaces of the arranged groups, particularly those of the "transformation interfaces" between horizons, is the basis (Boulet et al., 1994).

Numerous studies that focused on the analysis of Soil-systems showed factors of soil differentiation, as well as the results of this differentiation in morphological, geochemical, and mineralogical terms. The most recent and significant studies include the analysis of a hydromorphic planosols/hydromorphic soil-system (Lamotte et al., 1992); the soil types and altitude relationships on volcanic material, taking into account the amplitude of the toposequences under study (Alexander et al., 1993); the mechanical and geochemical transfers within catenas (Sommer and Stahr, 1996); upstream-downstream mineralogical evolution on a basaltic toposequence (Righi et al., 1999); geochemical evolution in a typical Luvisols/Cambisols toposequence (Brahy et al., 2000); the advantage of working on "micro-catchment areas" and catenas rather than on profiles, in the context of the biodiversity of forest regions (Thwaites, 2000); and bringing to the fore the significance of the length of the slope on different attributes of the soils in question (Applegarth and Dahms, 2001).

Soil-systems and Soilscapes in France

Elaboration of a typology involves possibilities of generalization, and we have to think about ways of defining a structure of the approach in the whole French territory. The first global information on the soils of the national territory comes from the pedological map of France at 1/1 M, published in 1966 (Dupuis, 1966). This was subsequently greatly supplemented by other studies so as to be integrated into the European Community soil map (EEC, 1985; 1986). These additions were carried out using the work of the "Soil Survey Staff of France" (SESCPF) of INRA for the drawing up of soil maps at 1/100,000, a program which is still underway (INRA-SESCPF, 1969–2001). The European program "Geographical Soil Database of Europe" subsequently harmonized and structured the whole of the collected information in the information system EUSIS: European Soil Information System (EU-JRC, 2001).

The French Soil Database contains, at an accuracy of 1/1 M, 917 STUs regrouped in 318 SMUs (Jamagne et al., 1995). These STUs belong mostly to the following Reference Soil Groups of the WRB (FAO-ISRIC, 1998): *Histosols, Anthrosols, Leptosols, Vertisols, Fluvisols, Gleysols, Andosols, Podzols, Planosols, Calcisols, Albeluvisols, Luvisols, Umbrisols, Cambisols, Arenosols, Regosols*, and to the following Orders of Soil Taxonomy (Soil Survey Staff, 1999): *Alfisols, Andisols, Entisols, Histosols, Inceptisols, Mollisols, Spodosols, Ultisols, Vertisols.*

There are a great number of soils in the French territory, and we can see quite easily the major fundamental pedogenetic processes that control the evolution of the surface formations of this territory. These dominant processes are essentially "brunification" (weathering without important transfer of matter within the pedon) and leaching, and, to a lesser degree, podzolization. However, when a dominant factor, such as significant calcium content or excess water, intervenes in the genesis, other types of evolution can occur. Among the different genetic factors that have a great significance in the evolution of French soils are, first, the *geolithological context*, then the *geomorphological context*, and lastly the *climatic and paleoclimatic influences*.

Different types of Soil-systems have been defined, cartographically delimited, and described in the course of the many works of cartographic inventory (INRA, 1969–2001), and are made up of groups of pedons or STUs linked among each other by functional processes. It is sometimes difficult to select soil systems that are sufficiently representative of large areas, and some rules had to be followed (Bui et al., 1999). The Soilscapes in question are essentially defined by the nature of the parent material and by a certain number of topographic criteria enabling the definition of these types of relief, that is to say, the "landforms" (FAO, 1995), as well as by the different soils distributed on the slopes. A first framework for a typology of the French soil-systems could therefore be based on these parameters.

The *lithological* information is provided, for areas mapped at scales ranging between 1/10,000 and 1/250,000, by the dominant parent material identified during the surveys at the level of the map units. *Geomorphology* and *topography* are of fundamental importance, and the information has, until recently, been obtained by simple analysis of topographic levels contours and specialists' descriptions. The Digital Elevation Models (DEMs) now enable higher precision in this domain, as well as the possibilities of quantification. From DEMs, it is possible to locate and delimit catchments or watershed areas of varying surfaces (IRD-ORSTOM, 1995). Independently of the numerous results obtained with reduced resolutions, 10 m to 30 m, we have been able to test the relevance of DEM at much larger grids: 250 m for the territory of France and 1000 m for Europe.

Our current approach, which is to attempt a typology of the principal soilscapes represented in the French territory, will be illustrated with three concrete examples of toposequences that are Soil-systems.

Examples of Soil-systems in France

These examples come from three representative areas that are geomorphologically and lithologically different, including the following:

- a hilly loessic region in the Paris Basin
- a region of depression on a detritic cover in the Central Region
- a medium mountainous region in the Massif Vosgien

The distribution of the main soil types of the three regions are briefly discussed.

Soils of the Paris Basin

Different geological beds show up in a number of types, belonging to Secondary formations made up of the Lias, the Jurassic, and the chalky Cretaceous. However, the most extensive cover is that of the Quatenary formations, comprising numerous sediments. The Pleistocene is essentially made up of Loess. Several types of loess of different ages are present, the most recent cover dating from the end of the last glaciation (Würm III), with quite significant original carbonate contents. The presence of loess from Würm II and I has been noted. During the Quaternary, alternations of glacial periods and thaws permitted erosive phenomena, typical of these climates, which included gelifraction, cryoturbation, and solifluction, significantly affecting the surface layers. It appeared that a loess sheet covered the whole landscape during the Upper Pleistocene. The late glacial erosive phenomena at the end of the Pleistocene left only the present day cover of the plateau, as well as outliers located on the flats and terraces, or maintained on local topographic units.

Soils of Loamy Landscapes

These materials, most often of loess origin, are favorable to a rapid differentiation by clay leaching. However, a number of them were initially calcareous and have previously undergone decarbonation, successively giving rise to *calcareous brown soils* then *calcic (Calcaric or Eutric Cambisols* [WRB]; *Typic or Rendollic Eutrudepts* [Soil Taxonomy]). The continuation of this evolution leads to *brown soils (Haplic Cambisols* [WRB]; *Typic Eutrudepts* [Soil Taxonomy]), which differentiates itself to a well-structured and colored horizon at shallow depth. The progressive desaturation of the soil by leaching then creates conditions favorable to eluviation. We then witness the differentiation of *leached brown soils* (*Leptic Luvisols* [WRB]); *Inceptic Hapludalfs* [Soil Taxonomy]), and then of *leached soils* (*Eutric or Haplic Luvisols* [WRB]*; Typic Hapludalfs* [Soil Taxonomy]), typified by a horizon impoverished in clay and iron overlying a horizon enriched in these same elements. Mobilized clay distributes itself by concentrating in the pores and on faces of the structural elements, in characteristic coatings.

Soils of Calcareous Landscapes

Soils of calcareous landscapes consist of numerous areas in chalky, marly regions, and on hard limestones. The majority of the parent materials are made up of the products of chalk alteration, most often resulting from mechanical or physical phenomena, due to freezing and to the different modes of displacement closely linked to the climatic fluctuations. The soils developed from chalk are *typical rendzinas* or *brown rendzinas* (*Rendzic, Calcaric,* or *Mollic Leptosols* [WRB]; *Lithic* or *Typic Haplrendolls*, *Typic Udorthents* [Soil Taxonomy]), *calcareous brown soils* and mostly *calcic brown soils* (*Calcaric* or *Eutric Cambisols* [WRB]; *Typic* or *Rendollic Eutrudepts* [Soil Taxonomy]). A progressive decarbonation may occur, which results in the differentiation of a structural horizon within the insoluble residual products that form in the upper part of the solum. On these deposits of chalky origin, we find allochtonous Pleistocene material of eolian, niveo-eolian or land fill origin, having a mostly silty texture.

Example from a Hilly Loessic Region of the Paris Basin

The example shown comes from a silty-chalky landscape in Picardy (Figure 13.2), in the northwest of the Paris Basin, and corresponds with one of the most representative Soil-systems in this region. A loessic cover dating from the recent Quaternary (Würm III) overlies a Cretaceous chalky substrate strongly marked by phenomena of cryoturbation dating from the last glacial period. This substrate is exposed at the slight break of slope, while the depression is characterized by the

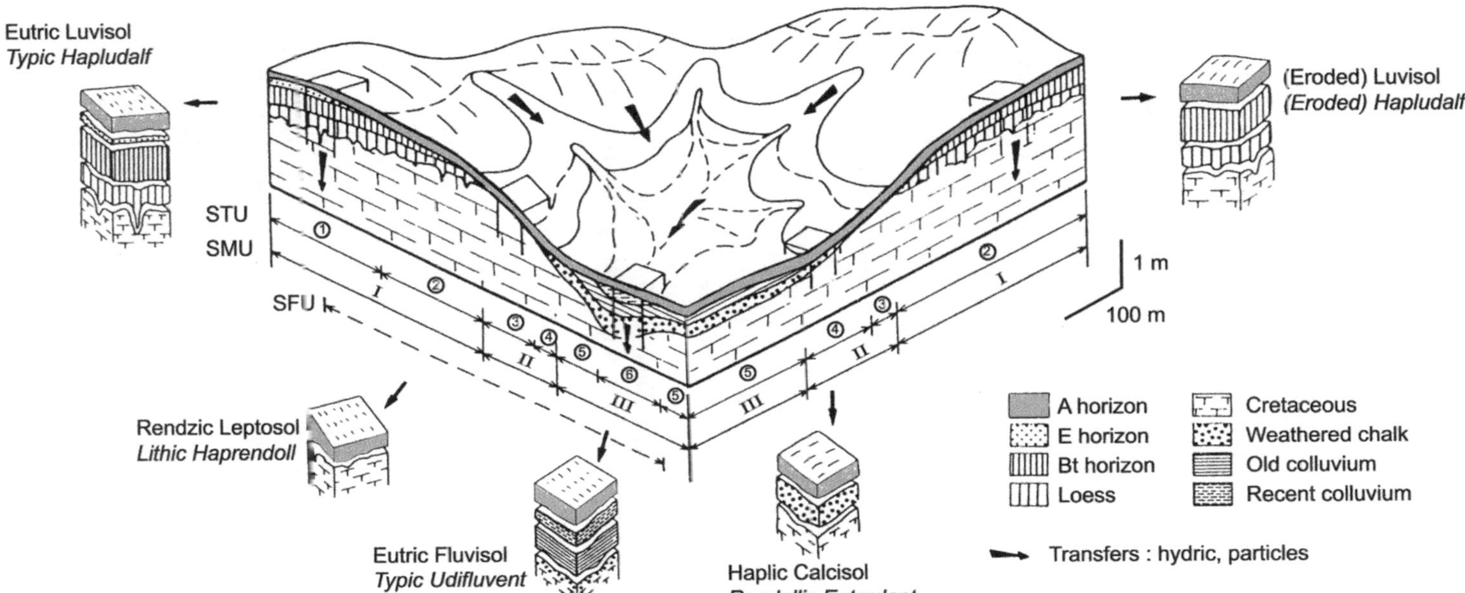

Figure 13.2 Soil-system 1 in a Loamy Landscape of the Paris Basin Region (Picardy). Diagram showing the organization of the *Soil-system* with *Soil Typological Units* (STU): 1 to 6, *Soil Mapping Units* (SMU): I to III, and the *Soil Functioning Unit* (SFU). The SFU is represented by run-off and erosion processes.

superposition of soliflucted material, apparently dating from the late glacial period and by colluviated materials, the uppermost of which are contemporary with present-day phenomena.

The typological units comprising this landscape are, from top to bottom of the hill slope, *Eutric Luvisols, Rendzic Leptosols, Haplic Calcisols,* and *Eutric Fluvisols* (WRB), corresponding successively to *Typic Hapludalfs, Lithic Haplrendolls, Rendollic Eutrudepts,* and *Typic Udifluvents* (Soil Taxonomy).

Figure 13.2 shows the arrangement of this system, comprising the STUs distributed on the slopes of the catchment area (from 1 to 6). The SMUs that could be distinguished using detailed mapping (from I to III), as well as the SFU characterizing the dynamics of this Soil-system, are also included. The dominant functioning is represented here by the phenomenon of degradation of surface layers by rain (structure breakdown and crusting/surface sealing), resulting in intense runoff and water erosion processes. The system described also corresponds with numerous landscapes of southwest France, where loamy material covers marly calcareous formations.

Soils on the Detritic Formations of the Centre Region

Large areas of the central region are covered by old weathering products of the French Massif Central. Therefore soils of the region have different origins consisting mainly of ancient detritus of the Sologne, which is frequently resting on clay sediments, and fluviatile deposits, which are linked to the hydrologic network of the Loire, and are located on terraces and along riverbanks. These formations may present varied granulometric distributions; in other words, the riverbanks show a dominance of the sandy fraction, while a large part of the recent alluvial deposits are richer in silty fractions.

Soil differentiation is extremely complex, and in it we can observe numerous transitions between the well-known evolutionary types, with the frequent intervention of a clay substrate at shallow or medium depth, creating occasionally predominant hydromorphic conditions. The Soil-systems of the type *planosols/hydromorphic soils* (*Planosols/Gleysols* [WRB]; *Aquic* or *Albaquic Hapludalfs/Aquepts* [Soil Taxonomy]) or *hydromorphic soils/podzolic soils* (*Gleysols/Spodic Gleysols* or *Entic Podzols* [WRB]); *Aqualfs* or *Aquepts/Entic Haplorthods* [Soil Taxonomy]) typify numerous landscape units. Leaching may be followed by a moderate podzolization.

Example from a Depressed Area on a Detritic Cover in the Central Region

This example typifies numerous Soilscapes of the detritic zones, located in the foothills of the mountainous and hilly massifs. They extend furthest to the approaches of the hydrologic network that drains the northern slopes of the Massif Central. These sediments constitute the accumulation of alteration products resulting from the geomorphological phenomena representative of different geological periods. The detritic materials are comprised of a sedimentary superposition of clayey sands and compact clay formations. The two main typological units present and generating this soil system are, from the interfluve to the depression, *Dystric Planosols* giving way to *Umbric Gleysols* (WRB), that is to say *Albaquic Hapludalfs,* which pass into *Typic Humaquepts* (Soil Taxonomy).

Figure 13.3 shows the arrangement of the system, which is composed of the STUs following one another on a very gentle slope (from 1 to 5). The SMUs shown by a detailed mapping (from I to III), as well as the SFU linked to the dominance of the hydric flows in this system, are also illustrated. The dominant functioning is represented by the mainly lateral orientation of the hydric transfers at the level of the particularly prominent planic contact. The system described here is also found very often in the northern foothill zones of the Massif Central.

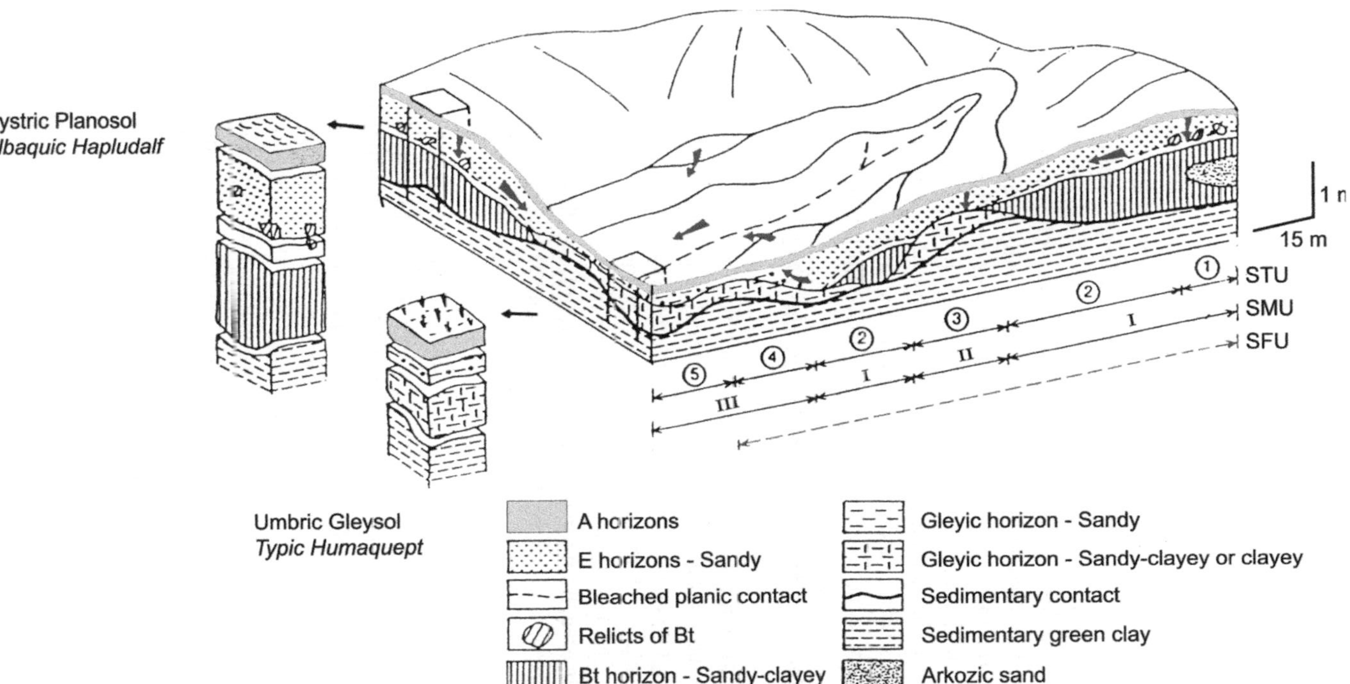

Figure 13.3 Soil-system 2 in a Detrital Landscape of Centre Region (Orleans–Loire Valley). Diagram showing the organization of the *Soil-system* with *Soil Typological Units* (STU): 1 to 5, *Soil Mapping Units* (SMU): I to III, and the *Soil Functioning Unit* (SFU). The SFU is represented by hypodermic hydric transfers along the planic contact.

Soils of the Massif Vosgien

In the Vosges, analysis of the soil cover shows two zones of podzolized soils. They are especially well represented in the zone that corresponds with the presence of the Vosgian sandstones. The second zone is that of the granitic Vosges; here, the podzolized soils occupy smaller surfaces, and are limited to the coarsest filtering quartz "arenites."

The weathered and decayed matter of the granitic rocks is, however, one of the types of material susceptible, in France, to be affected by podzolization. On the ancient crystalline massifs of the Vosges and the Massif Central, the podzolization shows up mainly at altitudes above 800 m; the climatic conditions are therefore cool and humid. The morphology of the profiles rarely shows a marked podzolization, but more often than not, expresses a moderate podzolization characterized by the presence of *podzolic ochric soils* and moderately differentiated *podzols* (*Dystric* or *Spodic Cambisols* and *Entic* or *Haplic Podzols* [WRB]; *Typic* or *Spodic Dystrudepts* [Soil Taxonomy]).

The products resulting from the alteration of crystalline, volcanic, or metamorphic rocks constitute arenites of varied granulometry according to the quantity and quality of the original weatherable minerals and to the intensity of the argilization processes. The migrations-redistributions in the form of organo-metallic complexes result in the formation, at depth, of spodic horizons. In addition, this distribution is modulated by the topography; the podzols are situated on the flats where the coarse arenites tend to accumulate and where the stability of the materials favors vertical migrations, as well as the concentration of constituents coming from uphill by lateral migration.

Example from a Medium Mountain Region of the Massif Vosgien

The toposequence shown here comes from a catchment area that is very representative of the granitic zone of the Massif Vosgien. The parent material is a granite which has been altered to a granitic arenite with a clayey-sandy texture, but whose weatherable mineral content leads to a moderate argilization. We therefore have a Soil-system that has differentiated within a lithologically homogenous material, the pedological evolution being the only process responsible for the soil distribution. We can observe, from top to bottom of the relief, the different typological units making up the Soil-system: *Dystric Cambisols, Haplic Podzols,* and *Dystric Fluvisols* (WRB), corresponding to *Typic Dystrudepts, Typic Haplorthods,* and *Typic Udifluvents* (Soil Taxonomy).

In Figure 13.4 the preferential location of the podzolization processes on the flat in the middle of the slope is clearly shown. The system is arranged from the top toward the bottom of this relatively steep slope by the succession of STUs (from 1 to 7) and the SMUs (from I to III), which have been defined by medium scale mapping. Two SFUs can be defined here; both, however, generated by the lateral hydric transfers at the level of the structural transition between the granitic arenite and the underlying rock. The former, situated on the upper part, is at the origin of the observed podzolisation. The dominant general functioning is therefore constituted by the action of the lateral flows mobilizing and displacing certain constituents in the subsoil at the level of the contact between B horizons and underlaying granite. The Massif Central presents, in its granitic domain, Soil-systems that are very similar to those that have just been described for the Massif Vosgien.

Possibilities of Generalizing the Approach

There has always been a problem with generalizing the collected information at a given level of detail over greater territories. The concepts of "downscaling approach" (starting from detailed maps for smaller scale representations over larger territories), or "upscaling approach" (going from general broad-scale knowledge to the choice of representative zones to be inventoried at a more detailed level), have led to the perfecting of diverse approaches according to environmental conditions.

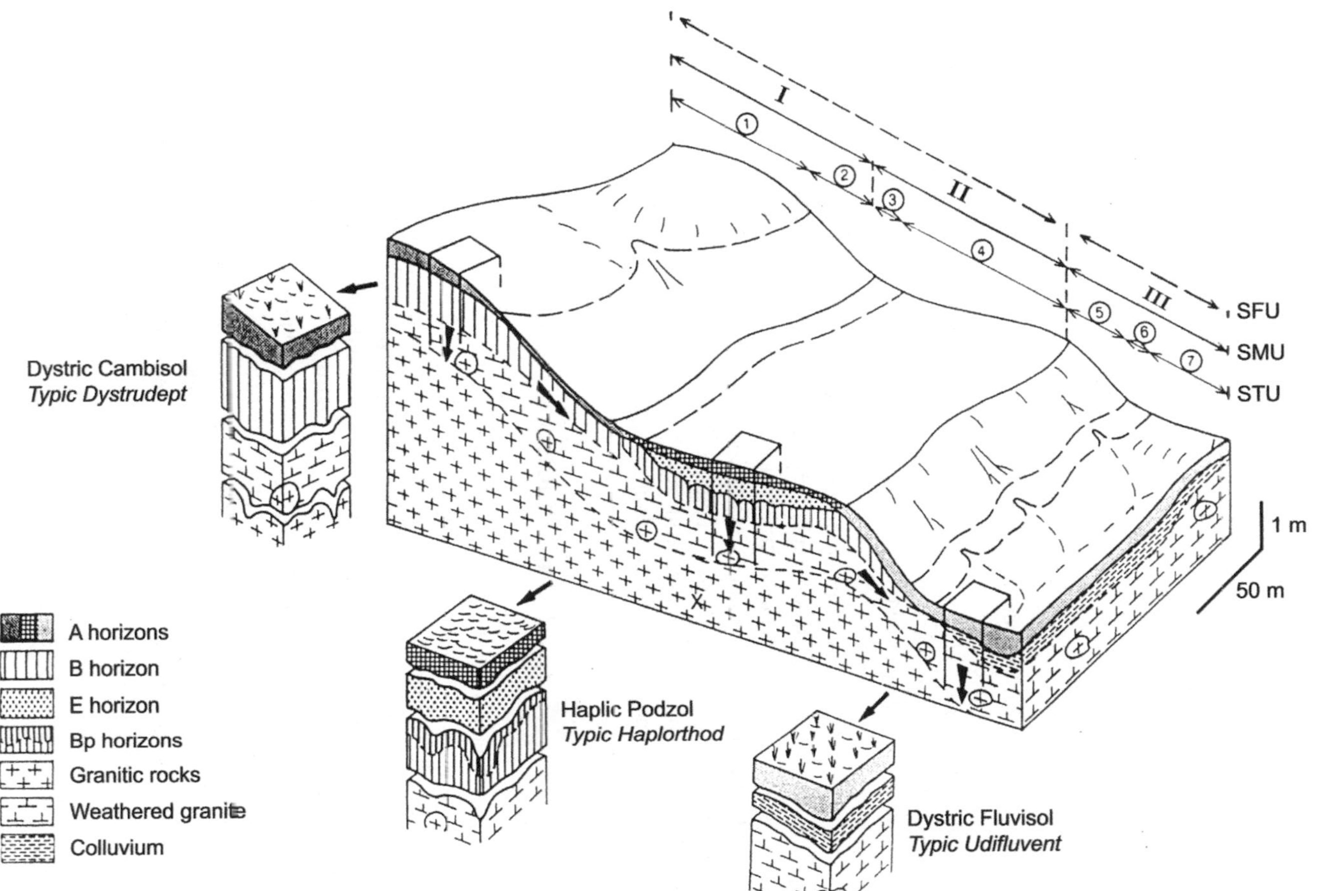

Figure 13.4 Soil-system 3 in a Granitic Landscape of the Vosges Region. Diagram showing the organization of the *Soil-system* with *Soil Typological Units* (STU): 1 to 7, *Soil Mapping Units* (SMU): I to III, and the *Soil Functioning Units* (SFU). The SFUs are represented by lateral underground fluxes transfers at B/C contact.

These difficulties have been mentioned in several contexts: the challenge of scale transfer from pedon to landforms and up to the "pedosphere" (Arnold and Wilding, 1991), structuring of the ecological and pedological diversity at different levels (Ibanez et al., 1995), and the necessity of utilizing reference areas (Lagacherie and Voltz, 2000), difficult generalizations due to the number of factors involved, particularly historical and polygenetic (Philips, 2001).

An attempt to validate the representativeness of the three Soil-systems presented here has been carried out. The basis has first been a delimitation of the watershed areas of less than 100,000 ha obtained from the DEM available for the whole French territory at 250 m intervals, associated with the hydrologic network that constitutes the main collector of these watershed areas (IRD-ORSTOM, 1995). Next, catchment areas comprising different materials and their associated soils were selected from the Soil Geographical Database of France (Jamagne et al., 1995), corresponding to a pre-eminence of the Soil-systems thus described, which allowed a zonage of the areas of extension represented in Figure 13.5.

- Watersheds of less than 100,000 ha delimitated from France DEM at 250 meters grid size
- Modelling Hydrographic Network
- Soil-system 1: on eolian loam covering cretaceous formations (secondary chalk)
- Soil-system 2: on detrital formations and old sedimentary deposits
- Soil-system 3: on crystalline rocks and migmatite

Generally, the main axes of these systems are respected, although adaptations will be necessary according to the particularities of the physiographic environments under consideration.

The examples presented are thus typical of three major Soilscapes of the French territory: North-West for System 1 on loessic formations, Central depression zones for System 2 on detritic materials, medium altitude mountain ranges of the Vosges, and the Massif Central for System 3 on granitic alteration products. The geographical databases from which the cartographic outline presented here is taken should enable a first generalization of the results obtained by means of the chosen approach.

CONCLUSION

This approach is based mainly on the analysis of spatial soil organizations, and thus complements the taxonomic references systematically attributed to the different soils that make up the landscape groups concerned. This method would allow for a better understanding and clarification of the soil distribution in the landscape than would the sole allocation of a denomination to a soil based on a classification or a referential system. We hope that it will contribute to the research of the principal processes of soil genesis and differentiation, and that it will permit a structuring of our knowledge in this field with the possibility of large transmission of that knowledge.

The test that we present here remains to be completed and validated. A similar typological test based on the principal Soilscapes belonging to the most important natural regions of the French territory is currently under way. In the future, we intend to apply this approach in zones that are particularly representative of the wide open European spaces, territories for which the information required for this structuring exists and is available.

ACKNOWLEDGMENTS

We would like to express our thanks to all our colleagues at the soil science research unit, whose ideas have been synthesized in these pages, and in particular, Odile Duval, Nicolas Saby, Sacha Desbourdes, and Alain Couturier for their help with the more technical aspects.

We also express our thanks to the scientific reviewers who have greatly helped us to improve this work.

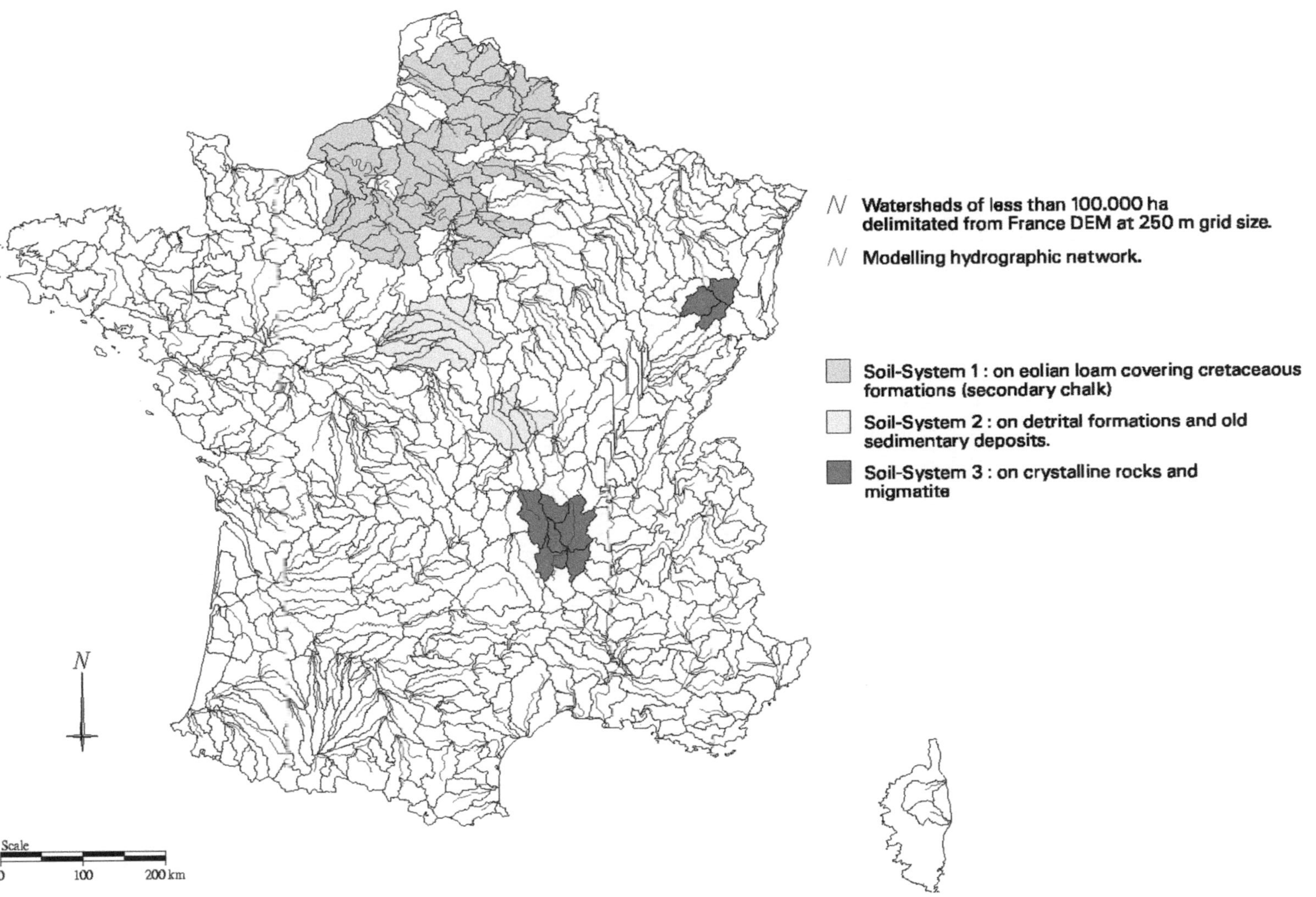

Figure 13.5 Sketch of concerned a-eas, by Watersheds, of the described Soil-systems.

REFERENCES

AFES-INRA. 1992. D. Baize and M.C. Girard, coord. Référentiel Pédologique, principaux sols d'Europe. *INRA*, Paris.

Alexander, E.B., Mallory, J.I., and Colwell, W.L. 1993. Soil-elevation relationships on a volcanic plateau in the Southern Cascade Range, North Carolina. *Catena*. 20:1–2, 113–128.

Applegarth, M.T. and Dahms, D.E. 2001. Soil catenas of calcareous tills, Whiskey Basin, Wyoming. *Catena*. 42:1, 17–38.

Arnold, R.W. and Wilding, L.P. 1991. The Need to Quantify Spatial Variability, in Spatial Variabilities of Soils and Landforms. SSSA Special Publication 28.

Baize, D. 1986. Couvertures pédologiques, cartographie et taxonomie. *Science du Sol*. 24:227–243.

Bell, J.C., Thompson, J.A., Butler, C.A., and McSweeney, K. 1994. Modeling Soil Genesis from a Landscape Perspective. 15th International Congress of Soil Science, Transactions. Acapulco, Mexico. 6a:179–195.

Bornand, M. 1978. Genèse et évolution des sols sur terrasses. Thèse ENSAM. INRA Montpellier.

Boulaine, J. 1986. La dispersion latérale dans les sols ou: de l'horizontalisme au verticalisme; essai sur la loxostasie, Cah. ORSTOM, sér. Pédol. Vol. XXII, 3:319–327.

Boulet, R., Chauvel, A., Humbel, F.X., and Lucas, Y. 1982. Analyse structurale et cartographie en pédologie. Cah. ORSTOM Pédologie. Vol. XIX, 4:309–351.

Boulet, R., Chauvel, A., and Lucas Y. 1984. Les sytèmes de transformation en pédologie, in Livre Jubilaire AFES–Paris. 167–179.

Boulet, R., Curmi, P., Pellerin, J., and Pereira de Queiroz-Neto, J. 1995. Contribution de l'analyse morphologique tridimensionnelle de la couverture pédologique à la reconstitution de l'évolution du modelé (Paulinia, Sao Paulo, Brazil). *Géomorphologie*, 1:49–59.

Brabant, P. 1989. La connaissance de l'organisation des sols dans le paysage, un préalable à la cartographie et à l'évaluation des terres. In SOLTROP 89. ORSTOM, Coll. Colloques et Séminaires, 65–85.

Brahy, V., Deckers, J., and Delvaux, B. 2000. Estimation of soil weathering stage and acid neutralizing capacity in a toposequence Luvisol–Cambisol on loess under deciduous forest in Belgium. *European J. Soil Sci*. 51:1, 1–13.

Bui, E.N., Loughead, A., and Corner, R. 1999. Extracting soil-landscape rules from previous soil surveys. *Aust. J. Soil Res*. 37:495–508.

Buol, S.W., Hole, F.D., McCracken R.J., Southard, R.J. 1997. *Soil Genesis and Classification*. Ed. 4, Iowa State University Press, Ames. XVI.

Burrough, P.A. 1986. Principles of geographical information systems for land resources assessment. Monographs on soil and resources survey 12. Oxford Science Publications.

Chaplot, V., Walter C., and Curmi, P. 2000. Improving soil hydromorphy prediction according to DEM resolution and available pedological data. *Geoderma*. 97:405–422.

C.P.C.S. 1967. Classification des Sols. INRA.

Duchaufour, P. 1983. Pédologie. T 1. *Pédogénèse et Classification*. 2nd ed. Masson, Paris.

Duchaufour, P. 1997. Abrégé de pédologie. *Sol, Végétation, Environnement*. 5th ed. Masson, Paris.

Dupuis, J. 1966. Carte Pédologique de France à 1/1 000 000. INRA. SESCPF, Paris.

ECC. 1985. Soil Map of the European Communities 1/1 000 000. CEC-DGVI, Luxembourg.

EEC–ISSS. 1986. Soil Map of Middle Europe 1:1 M. EEC publications, Luxembourg.

EC-JRC. 1998. Georeferenced Soil Database for Europe. Manual of Procedures. Version 1.0. European Soil Bureau. JRC. Ispra.

EU-JRC. 2001. Instructions. Guide for the Elaboration of the Soil Geographical Database of Euro-Mediterranean Countries. Version 4.0. Ispra.

Erhart, H. 1956. Biostasie et Rhexistasie. *CR. Acad. Sci. Paris*. 241:1218–1220.

Eswaran, H. and Banos, C. 1976. Related distribution patterns in soils and their significance. *Anal. Edafol. Agro. Espana*. XXV:33–45.

FAO. 1988. Soil Map of the World. Revised Legend. 60, Rome.

FAO–ISRIC-IUSS. 1995. Global and National Soils and Terrain Digital Databases (SOTER). Procedures Manual. UNEP-ISSS-ISRIC-FAO. *ISRIC*. Wageningen, Netherlands.

FAO–ISRIC. ISSS. 1998. World Reference Base for Soil Resources. FAO, Rome.

Fridland, V.M. 1975. Structure of the soil cover. Xe Congrès A.I.S.S., II Moscou 552–558.

Fridland, V.M. 1976. Levels of organization of the soil mantle and regularities of soil geography. 23rd Geographical Conference, Section 4. Biogeography and Soil Geography. Moscow, Russia.

Fritsch, E., Bocquier, G., Boulet, R., Dosso, M., and Humbel, F.X. 1986. Les systèmes transformants d'une couverture ferrallitique de Guyane française. Cah. ORSTOM, série Pédol., Vol. XXII, 4:361–396.

Gessler, P.E., Moore, I.D., McKenzie, N.J., and Ryan, P.J. 1995. Soil-landscape modeling and spatial prediction of soil attributes. *Int. J. Geogr. Inf. Systems*. 9(4):421–432.

Girard, M.C. 1984. Analyse spatiale de la couverture pédologique. –Cartographie.–Cartogenèse. –AFES. Ouvrage jubilaire, 153–166.

Hall, G.F., and Olson, C.G. 1991. Predicting Variability of Soils from Landscape Models. SSSA Special Publication 28.

Hewitt, A.E. 1993. Predictive modelling in soil survey. *Soils Fert.* 56 (3):305–314.

Hole, F.D. 1978. An approach to landscape analysis with emphasis on soils. *Geoderma*. 21:1–23.

Huggett, R.J. 1975. Soil landscape systems: a model of soil genesis. *Geoderma*. 13:1, 1–22.

Ibanez, J.J., De-Alba, S., Bermudez, F.F., and Garcia, A.A. 1995. Pedodiversity: concepts and measures. *Catena*. 24:3, 215–232.

INRA–SESCPF. "Cartes Pédologiques de la France." 1969–2001. Vichy L15 (1969), Perpignan L 24–25 (1970), Toulon P23 (1974), Moulins L14 (1974), Angoulème H16 (1975), Condom H21 (1975), Dijon 012 (1976), Brive J18 (1976), Privas N19 (1977), Tonnerre M10 (1978), Saint-Dié Q19 (1978), Lesparre F17 (1978), Châteaudun I9 (1978), Chartres J8 (1981), Montpellier M22 (1983), Langres 010 (1986), Saint Dizier N8 (1992), Lodève L22 (1993), Arles N22 (1994), Beaune N12 (1996), Digne P21 (2000).

IRD-ORSTOM. 1995. Notice d'instructions du logiciel DEMIURGE 2, 3, module LAMONT, Montpellier.

Jamagne, M. 1978. Les processus pédogénétiques dans une séquence évolutive progressive sur formations limoneuses en zone tempérée froide et humide. *C.R. Acad. Sci.* 286, D, 25–27.

Jamagne, M. and King, D. 1991. Mapping methods for the 1990s and beyond, in Soil survey, a basis for European soil protection. Silsoe, 11–12/12/1989. Soil and groundwater research report I. CEC. 181–196.

Jamagne, M., King, D., Girard, M.C., and Hardy, R. 1993. Quelques conceptions actuelles sur l'analyse spatiale en pédologie. *Sci. Sol.* 31(3):141–169.

Jamagne, M., 1993. Evolution dans les conceptions de la cartographie des sols. *Pédologie*, XLIII, 1:59–115.

Jamagne, M., Hardy, R., King, D., and Bornand, M. 1995. La base de données géographique des sols de France. *Etude et Gestion des Sols*. 2(3):153–172.

Kilian, J. 1974. Etude du milieu physique en vue de son aménagement. *L'Agronomie Tropicale*, XXIX (2–3):141–153.

King, D., Daroussin, J., and Jamagne, M. 1994a. Proposal for a Model of a Spatial Organization in Soil Science: Example of the European Communities Soil Map. *J. Am. Soc. Info. Sci.* 45(9):705–717.

King, D., Jamagne, M., Chrétien, J., and Hardy, R. 1994b. Soil-space organization model and soil functioning units in Geographic Information Systems. 15th International Congress of Soil Science, Transactions. Vol. 6a, Commision V, Symposia, Acapulco, Mexico, 743–757.

King, D., Bourennane, H., Isambert, M., and Macaire, J.J. 1999. Relationship of the presence of a non-calacareous clay-loam horizon to DEM attributes in a gently sloping area. *Geoderma*. 89:95–111.

Kozlovskiy, F.I. and Goryachkin, S.V. 1994. Current status and future direction of the theory of soil-mantle structure. *Eurasian Soil Sci.*. 26(11):1–18.

Lagacherie, P. and Voltz, M. 2000. Predicting soil properties over a region using sample information from a mapped reference area and digital elevation data: a conditional probability approach. *Geoderma*. 97:187–208.

Lamotte, M., Duval, O., Humbel, F.X., and Jamagne, M. 1992. Une démarche itérative pour l'identification et la cartographie d'un système pédologique en Orléanais. ORSTOM. Séminaire: "Organisation et fonctionnement des altérites et des sols." Paris, 105–116.

McBratney, A.B. 1992. On Variation, Uncertainty and Informatics in Environmental Soil Management. *Aust. J. Soil Res.* 30:913–35.

McSweeney, K., Slater, B.K., Hammer, R.D., Bell, J.C., Gessler, P.E., and Petersen, G.W. 1994. Towards a New Framework for Modeling the Soil-Landscape Continuum. SSSA Special Publication 33.

Mermut, A.R. and Eswaran, H. 2001. Some major developments in soil science since the mid-1960s. *Geoderma*. 100:403–426.

Milne, G. 1935. Some suggested units of classification and mapping. *Soil. Res.* 4.

Milne, G. 1936. A provisional soil map of East Africa. Am. Mem. no. 28; East. Afric. Agr. Res. Stn.. Tanganiyka Territory.

Moore, I.D., Grayson, R.B., and Ladson, A.R. 1991. Digital Terrain Modelling: a review. Hydrological, Geomorphological, and Biological Applications, Hydrological Processes. 5:3–30.

Moore, I.D., Gessler, P.E., Nielsen, G.A., and Peterson, G.A. 1993. Soil attribute prediction using terrain analysis. *SSSA* J. 57: 443–452.

Nachtergaele, F.O., Spaargaren, O., Deckers, J.A. and Ahrens, R. 2000. New developments in soil classification. World Reference Base for Soil Resources. *Geoderma*. 96:345–357.

Northcote, K.H. 1984. Soil-landscapes, taxonomic units and soil profiles. A personal perspective on some unresolved problems of soil survey. *Soil Surv. Evaluation*. 4(1):1–7.

Odeh, I.O.A., McBratney, A.B., and Chittleborough. D.J. 1994. Spatial prediction of soil properties from landform attributes derived from a digital elevation model. *Geoderma*. 63:197–214.

Philips, J.D. 2001. Contingency and generalization in pedology, as exemplified by texture-contrast soils. *Geoderma*. 102:347–370.

Righi, D., Terribile, F., and Petit, S. 1999. Pedogenic formation of kaolinite-smectite mixed layers in a soil toposequence developed from basaltic parent material in Sardinia (Italy). *Clays & Clay Minerals*. 47(4):505–514.

Ruellan, A., Dosso, M., and Fritsch, E. 1989. L'analyse structurale de la couverture pédologique. *Science du Sol*. 27:319–334.

Ruellan, A. and Dosso, M. 1993. Regards sur le sol. Faucher. Aupelf, Paris.

Simonson, R.W. 1989. Historical highlights of soil survey and soil classification with emphasis on the United States, 1969–1970. *ISRIC* Wageningen.

Sinowski, W. and Auerswald, K. 1999. Using relief parameters in a discriminant analysis to stratify geological areas with different spatial variability of soil properties. *Geoderma*. 89(1):113–128.

Skidmore, A.K., Ryan P.J. et al. 1991. Use of an expert system to map forest soil from a geographical information system. *Int. J. GIS*. 5(4):431–445.

Slater, B.K., McSweeney, K., Ventura, S.J., Irvin, B.J. and McBratney, A.B. 1994. A Spatial Framework for Integrating Soil-Landscape and Pedogenic Models. SSSA Special Publication 39.

Soil Survey Staff. 1961. Soil Survey Manual. Washington.

Soil Survey Staff. 1975. Soil Taxonomy. U.S. Department of Agriculture Handbook 436, Washington, DC.

Soil Survey Staff. 1999. Soil Taxonomy. 2nd Edition, Washington.

Sommer, M. and Schlichting, E. 1997. Archetypes of catenas in respect to matter—a concept for structuring and grouping catenas. *Geoderma*. 76:1–33.

Sommer, M. and Stahr, K. 1996. The use of element:clay-ratios assessing gains and losses of iron, manganese and phosphorus in soils of sedimentary rocks on a landscape scale. *Geoderma*. 71(3/4):173–200.

Thompson, J.A., Bell, J.C., and Butler, C.A. 1997. Quantitative soil-landscape modeling for estimating the areal extent of hydromorphic soils. *SSSA J*. 61:971–980.

Thompson, J.A., Bell, J.C. and Butler, C.A. 2001. Digital elevation model resolution: effects on terrain attribute calculation and quantitative soil-landscape modeling. *Geoderma*. 100:67–89.

Thwaites, R.N. 2000. From biodiversity to geodiversity and soil diversity. A spatial understanding of soil in ecological studies of the forest landscape. *J. Tropical Forest Sci*. 12(2):388–405.

USDA–SCS. 1993. Soil Survey Manual. U.S. Dept. Agric. Handbook 18. Washington, DC.

Vreeken, W.J. 1984, Soil-landscape chronograms for pedochronological analysis. *Geoderma*. 34(2):149–164.

Webster, R. 1977. *Quantitative and Numerical Methods in Soil Classification and Survey*. Clarendon Press, Oxford.

Webster, R. and Oliver, M.A. 1990. Statistical methods in soil and land resource survey. *Spatial Information Systems*. Oxford University Press.

Wilding, L.P. 1989. Improving our understanding of the composition of the soil-landscape, in *Soil Resources: Their Inventory, Analysis and Interpretation for Use in the 1900's*. Proc. International Workshop, University of Minnesota. 13–35.

CHAPTER 14

New Zealand Soil Classification—Purposes and Principles

A.E. Hewitt

CONTENTS

INTRODUCTION

The New Zealand Soil Classification was published in 1993 (Hewitt, 1993a) and replaced the New Zealand genetic soil classification (Taylor, 1948; Taylor and Pohlen, 1968) as the national system of soil classification in New Zealand. A comprehensive description of the New Zealand Soil Classification is not given in this account. Instead the reader is referred to publications in the References list. The aim here is to discuss some features of the classification (some of which are distinctive) that guided its development and application.

PURPOSES FOR SOIL CLASSIFICATION IN THE CONTEXT OF NUMERICAL ANALYSIS

The New Zealand Soil Classification was developed at a time when computing power was beginning to show its potential for rapid and convenient numerical analysis. The classification was

0-8493-1339-2/03/$0.00+$1.50

challenged as being largely unnecessary; numerical analysis of soil data, it was argued, would replace the former functions of soil classification.

In reply (Hewitt, 1987), it was noted that classification is basic to human thought processes. We observe similar objects and find it efficient to group them and provide a label as a mental key to the class. For example, we label the group of objects with a flat surface and four legs "table". Likewise the word "soils" assumes classification. A grouping of soil entities together constitutes a kind of soil. Our formal soil classification is a codification of these "soils," defined so that we can all communicate efficiently. Soil classification is, therefore, primarily designed for human cognition. Data is the currency of computers, and no matter how sophisticated the analyses, the results must be summarized and reduced to make the results comprehensible for humans. A soil classification provides a framework that greatly facilitates understanding and application of the results of numerical analysis.

With this in mind, four purposes were given priority in development of the New Zealand Soil Classification:

1. *Communication and understanding.* Soil scientists and allied specialists are able to efficiently carry on informed discussion about a set of soils when there is prior agreement on concepts and definitions. As well as facilitating specialist communication, there is a pressing need to make soil information more accessable to nonspecialists. Soil classification is the human face of soil science, and often a soil name is the first contact the nonspecialist or nonscientist will make with it. When it is learned that a soil name can be given to a soil, and what that name implies, new vistas of understanding open up.
2. *Soil survey support and application.* Soil survey is becoming more quantitative with greater focus placed on soil attributes rather than classes. However, the traditional role of soil classification to support soil survey and interpretations, as promoted by Soil Taxonomy (Soil Survey Staff, 1999), will remain as far as the need for human understanding of the process and results remain. There will always be a need to express an overview of soil entities.
3. *Key soil database field.* Soil classification provides a frequently used key when using the New Zealand National Soils Database, particularly when defining the data sets relevant to a query, summarizing the results, and checking the regional and national representativeness of data.
4. *New synthesis of knowledge.* The new classification provides a synthesis of several decades of intensive research on New Zealand soils. Providing a statement of our understanding or conception of the natural soil system would be sufficient reason alone to develop a new national soil classification. This is work that holds cultural as well as scientific values. Synthesis is extremely important in science. It is our responsibility to push our general model of the soil system so that our minds and understanding do not become reduced to mere software; reduced to match our neatly regimented data sets.

OUTLINE OF THE PRINCIPLES

To accomplish these purposes, the following principles guided development of the New Zealand Soil Classification:

1. The classification should be hierarchical, providing ascending levels of generalization.
2. The grouping of soils into classes should be based on similarity of measurable soil properties, rather than presumed genesis.
3. Classes must be designed to allow the greatest number and most precise accessory statements to be made about them, consistent with their level in the hierarchy.
4. Differentiae should be based on soil properties that can be reproducibly and precisely measured or observed.
5. Differentiae should, where possible, allow field assignment of soils to classes, either directly or by tested inferences.

6. The nomenclature of higher categories should be based, where possible, on connotative English words chosen for their acceptability to nonspecialists.
7. Where possible, continuity with successful parts of the New Zealand genetic classification should be maintained.
8. The soil classification must be valid for the main islands of New Zealand. Classes must be correlated with Soil Taxonomy (Soil Survey Staff, 1999) to support international extension.

The application of some of these principles is illustrated following an outline of the structure of the classification.

CLASSIFICATION STRUCTURE AND IDENTIFICATION OF THE SOIL INDIVIDUAL

The New Zealand Soil Classification is hierarchical, comprising four categories above the level of soil series. Classes of the order, group, and subgroup categories are defined in Hewitt (1993a). Classes of the fourth category, soilform, are defined in Clayden and Webb (1994).

There are 15 orders (Table 14.1), 76 groups, and currently 253 subgroups. Unlike the Australian Soil Classification, which was developed at a similar time, the New Zealand Soil Classification has a prescribed number of subgroups. This does not allow the flexibility of the Australian system, but aids reproducibility in assignment and greater clarity in making soil survey correlation decisions. New subgroups can be added, and a number of them were, at the time that the second edition was printed (Hewitt, 1998). The number of soilforms is not prescribed. Soilform classes are formed from a combination of parent material, particle size, and permeability profile classes (Clayden and Webb, 1994). They are applied to subgroups in the same manner as Families are applied in Soil Taxonomy.

The keys to orders, groups, and subgroups use diagnostic horizons, pans, layers, soil materials, contacts, profile forms, features, and other differentiae. These are defined primarily for efficient application of the keys.

The order of classes in the keys is determined mainly by considering potency and uniqueness of class attributes. Potent attributes such as wetness, that override other attributes in their effect on the utility of the soils, key out first. Classes with some unique character that depart clearly from the central concept of the group are keyed out next. The assignment of key soil series also affected

Table 14.1 List of Orders of the New Zealand Soil Classification (NZSC) and Their Correlation with Soil Taxonomy

NZSC order	Correlation with Soil Taxonomy
Allophanic Soils	Andisols excluding vitric soils
Anthropic Soils	Drastically disturbed, mixed, or truncated soils, or soils made from human-deposited material. Mostly correlated with the Arents
Brown Soils	Dystrudepts and Dystrustepts
Gley Soils	Aquents, Aquepts, and Aquox
Granular Soils	Aquults, Humults and Udalfs, in basic igneous parent materials
Melanic Soils	Ustolls, Aquolls, Rendolls, Vertisols
Organic Soils	Histosols
Oxidic Soils	Oxisols
Pallic Soils	Aquepts, Aqualfs, Fragiaquepts, Dystrustepts
Podzols	Aquods, Orthods
Pumice Soils	Vitraquands, Vitrands, Vitricryands
Raw Soils	Entisols or not-soil
Recent Soils	Aquents, Orthents, Psamments, Fluvents
Semiarid Soils	Aridisols (mainly Xeric subgroups)
Ultic Soils	Aquults, Humults, Udults

A more detailed correlation at soil group level is given in Hewitt, 1997.

keying-order decisions. The last class in any key is usually the class closest to the central concept of the classes in that category. For this reason, many of the final groups to be keyed out are named "orthic," in the sense of ordinary or central. Similarly, many of the final subgroups to be keyed out are named "typic."

The soil individual is the fundamental unit of soil that is assigned to classes, and the population of soil individuals may be seen as the foundation of the hierarchy. Cline (1949) defined an individual as "the smallest natural body that can be defined as a thing complete in itself." Soil Taxonomy (Soil Survey Staff, 1999) regards the polypedon as the soil individual. This definition was rejected in the New Zealand Soil Classification because, as argued by Hewitt (1982), it does not fulfill Cline's (1949) or Johnson's (1963) requirements for a soil individual.

In New Zealand, the soil individual has traditionally been the soil profile. Usually conceived as a two-dimensional section exposed by a soil pit, it is in fact a three-dimensional slice sufficiently thick to sample and examine hand specimens. With the realization that soils should be examined in successive horizontal sections as well as the vertical profile, there is increasing acceptance in New Zealand that a volume of soil the size of the pedon (Soil Survey Staff, 1975) represents a better soil individual than the soil profile slice.

Accordingly, the pedon, as defined in Soil Taxonomy (Soil Survey Staff, 1975), is applied as the soil individual for the New Zealand Soil Classification. It is understood, however, that assignments are often made from the examination of volumes of soil smaller than a complete pedon, where they are assumed representative of the pedon.

HIERARCHICAL POSITION OF THE HYDROMORPHIC SOILS

The operation of principles 2 and 3 is illustrated by treatment of soils with evidence of wetness and biochemical reduction, identified as Gley or Mottled classes. Gley classes may be defined in terms of either measures of water saturation, reduction morphology, chemical measures of reduction, or some combination of these. Problems in the use of water saturation as the basis for identification have been discussed by Childs and Clayden (1986). Probably the most convincing argument against reliance upon water saturation differentiae is that the required data are seldom available, and where they are available, extrapolations must be made to other soils by use of the reduction morphology. Emphasis in the New Zealand soil classification has been placed on reduction morphology as the alternative that enables assignment based on direct field observations in any season.

The New Zealand Soil Classification recognizes two kinds of Gley soil classes. Where features associated with strong gleying occur together with the distinct marks of other soil-forming processes of soil in defined orders, then the soils are recognized as either Perch-gley or Groundwater-gley groups of these orders. The Gley Soils order, however, was defined for soils in which such distinct marks of other soil-forming processes were absent. Gley Soils therefore comprise soils that are either too young to display the distinct marks of other soil-forming processes, or in which the marks of those soil-forming processes have been retarded or destroyed by gleying.

Two other means of classifying gley soils were considered. All gley soils could have been collected into the Gley Soils order, as in European soil classifications (e.g., Avery, 1980), or distributed among other orders, as in Soil Taxonomy. Although either approach would have produced a more symmetrical classification, they both had disadvantages for the New Zealand Soil Classification.

The collection of all gley soils into one order would have defeated principle 3 of the New Zealand Soil Classification. That is, classes must be designed to allow the greatest number and most precise accessory statements to be made about them. Such an order would have incorporated a very diverse set of soils for which the only possible accessory statements would concern wetness. It should be noted that other soil classifications that recognize gley soils at order level do not incorporate all gley soils into that one class. The Soil Classification for England and Wales (Avery,

1980) and the Canadian system of soil classification (Canada Soil Survey Committee, 1978) both place strongly gleyed podzols with Podzols, rather than with Gley Soils. Thus the precedent is established that, for some soils, features other than gleying are marked so that they determine order assignment. The New Zealand Soil Classification applies this not only to the Podzols but also to nine other orders.

FIELD ASSIGNMENT TO SOIL CLASSES

Principle 5 addresses a dilemma faced in the development of any soil classification concerning the choice of differentiae. Should they be based on high-precision laboratory-derived measurements or on lower-precision field-based measurements or observations? Laboratory-based attributes provide for more certainty and reproducibility in assignments. The disadvantage is that assignment decisions frequently cannot be made in the field and must await the results of costly laboratory procedures. As a consequence, few pedons in a survey area can be confidently assigned to the classification. Although field-based differentiae have lower precision, they enable frequent field assignment of pedons.

In the development of the New Zealand Soil Classification, it was judged that better characterization of survey areas would result from frequent assignments based on lower-precision field differentiae than from few assignments based on higher-precision laboratory differentiae. Confidence that this was possible was based on improved definition of morphological properties in New Zealand (Milne et al., 1991; Griffiths et al., 1999) that has resulted in more reproducible field measurements.

This principle was adhered to in the definition of most classes, but it was found necessary to include pH, phosphate retention, fine earth dry bulk density, and electrical conductivity as differentiae. Where possible, lab-based differentiae were used to back up field-based differentiae. For example, in the definition of allophanic soil materials, field differentiae involving stickiness, soil strength, sensitivity, and the reactive-aluminium test were backed up by an alternative: phosphate retention limit.

NOMENCLATURE

In the 1960s, the New Zealand genetic soil classification was recast as the New Zealand Technical Classification (Taylor and Pohlen, 1968). It did not endure, largely because of an unfortunate nomenclature. The elegant nomenclature of Soil Taxonomy is efficient and connotative, and suitable for an international system, but the New Zealand Technical Classification presented an unnecessary initial barrier for a national audience, which restricted effective use to the specialist.

Consequently it was decided that the nomenclature of the New Zealand Soil Classification should be acceptable to nonspecialists, and even to the general public. Names are obviously important in marketing a product for which acceptability depends on initial appeal and public exposure. An elitist scientific nomenclature was to be avoided, especially at order level. It was noted that, after all, English is now the international language of science and diplomacy. Accordingly, a system of connotative, common, English words was used in an initial draft of the classification. It was found, however, that common English words could easily create confusion between common and technical uses of the same word.

The nomenclature of the published classification represents a somewhat uneasy compromise between apparently appealing technical names and simple or modified common English words. Names such as "Brown Soils" are differentiated as nouns by capitalization. Other names such as "Allophanic Soils" are clearly not based on connotative recognizable English words. In this case, because of the prominence of allophanic soil materials in New Zealand, it was decided to introduce an academic word to help popularize knowledge of the important soils derived from them.

PRIORITY OF CLASSES OVER KEY

Principle 3 places attention on the function of classes and on class intent. It is useful to separate in our minds the intention of a class and the keying differentiae. The intention of a class and its ability to carry accessory information is the most important, and keying differentiae should be less important. If priority is fixed on the keying differentiae, the classification may force the user to make inappropriate assignments that violate principle 3.

The introduction to the New Zealand Soil Classification states that the classes are the most important part of the soil classification. The key is merely a means of assigning soils to these classes, and by its nature it is imperfect, because only a sample of all the possible soils that might potentially be assigned were used in its development. Soils will be found that are not assigned to the appropriate class by the key. This will be apparent when an assigned soil does not conform to the concept and accessory statements that can normally be made about the class. Because the key is the servant of the classes, the allocator is justified in placing the soil misfit into a more appropriate class. If this is done, however, it must be registered with the person who has responsibility for the national soil classification system, so that appropriate adjustments may be made to the key when the soil classification is next revised. An assignment contrary to the key must also be noted in any records or publication of the allocation.

DISTINCTIVE CHARACTER OF NEW ZEALAND SOILS

Many New Zealand soils are representative of major world soils, such as the Allophanic Soils and Pumice Soils, which were studied during development of the Andisols order of Soil Taxonomy (Parfitt and Clayden, 1991). However, some soils and the relationships they display in the landscape are apparently unique, and are probably related to the following characteristics:

- New Zealand loess, which is a prominent soil parent material, is predominantly acidic. Calcium carbonate is not present in the unweathered loess. The loess source rocks have low contents of calcium, compared to similar loess source basement rocks in North America and Europe (Hewitt, 1997).
- Fragipans developed in loess occur in ustic rather than udic moisture regimes. Consequently the Fragiudepts and Fragiudalfs of Soil Taxonomy are not appropriate great groups for New Zealand; rather, Fragiustept and Fragiusalf great groups are needed.
- Argillic horizons do not have the same prominence of expression and widespread occurrence in New Zealand as they have in other regions of the globe, although they might be expected from environmental relationships.
- Udepts with ochric epipedons and cambic horizons are almost exclusively dystric.

The broad-scale sequence of soils on quartzo-feldpathic parent materials, not influenced by tephra, on late Pleistocene or early Holocene land surfaces, in mesic and cryic temperature regimes, from aridic through ustic to udic and to preudic moisture regimes, is different in New Zealand from that found in North America. The expectation from North America is the sequence Aridisols–Mollisols–Alfisols–Spodosols. In New Zealand, however, the sequence is Aridisols–Usepts/Usalfs–Udepts/Udalfs–Spodosols.

The New Zealand ustic and udic soils in this sequence were recognized early in the history of pedological investigations in New Zealand, and were defined, respectively, as the yellow-grey earths and yellow-brown earths (Taylor and Pohlen, 1968). Following principles 7 and 8, development of the new New Zealand Soil Classification maintained the concepts of the soils that expressed

Table 14.2 Areas of Soil Orders in New Zealand

	Area (km^2)	% of Total NZ Soil Areas
Allophanic Soils	13,670	5
Anthropic Soils	180	<1
Brown Soils	108,450	43
Gley Soils	7,120	3
Granular Soils	2,930	1
Melanic Soils	3,220	1
Organic Soils	2,260	1
Oxidic Soils	440	<1
Pallic Soils	30,910	12
Podzols	32,870	13
Pumice Soils	17,210	7
Raw Soils	7,090	3
Recent Soils	14,020	6
Semiarid Soils	2,220	1
Ultic Soils	7,620	3

soil–environmental relationships and had proven utility, and defined them, respectively, as the Pallic Soils and Brown Soils.

Discrimination between the extensive (Table 14.2) Pallic Soils and Brown Soils orders was based on the effects of relative differences in leaching and weathering. The Pallic Soils had relatively low contents of dispersed secondary iron and aluminium oxides expressed in low phosphate retention values (Hewitt, 1993a) of less than 30%, with consequent pale colors (hence the name 'Pallic'). Fragipans are ubiquitous in soils in loess on stable sites. The Brown Soils have relatively higher content of dispersed secondary iron and aluminium oxides expressed in phosphate retention values of 30% or more. They are generally more acid with lower base saturation than the Pallic Soils.

CONCLUSION

Development of a soil classification system requires a multitude of decisions. There is no shortage of opinions offered by those involved, and there is pressure to make ad hoc decisions that may have the effect of distorting the logic of the system. A clear statement of purpose and widely agreed-upon principles, focused on at debates during the development of the New Zealand Soils Classification, greatly aided decision making. In New Zealand, we were fortunate in having a soil science community that was supportive and enthusiastic, and it was this support that has resulted in a system that shows it can achieve its purposes.

REFERENCES

Avery, B.W. 1980. Soil classification for England and Wales (higher categories). Soil Survey Technical Monograph 14. Rothamstead Experimental Station, Harpendon, UK.

Canada Soil Survey Committee. 1978. The Canadian system of soil classification. Research Branch, Canada Department of Agriculture. Publication 1646.

Childs, C.W. and Clayden, B. 1986. On the definition and identification of aquic soil moisture regimes. *Aust. J. Soil Res.* 24:311–316.

Clayden, B. and Hewitt, A.E. 1993. A new system of soil horizon notation of New Zealand soils. *Catena* 19:405–410.

Clayden, B. and Webb, T.H. 1994. Criteria for defining the soilform—the fourth category of the New Zealand Soil Classification. Landcare Research Science Series 3, Manaaki Whenua Press, Lincoln, New Zealand.

Cline, M.G. 1949. Basic principles of soil classification. *Soil Sci.* 2:81–91.

Griffiths, E., Webb, T.H., Watt, J. P.C., and Singleton, P.L. 1999. Development of soil morphological descriptors to improve field estimation of hydraulic conductivity. *Aust. J. Soil Res.* 37:1–12.

Hewitt, A.E. 1982. Decisions in the establishment of soil series. Ph.D. thesis, Agronomy Dept., Cornell University.

Hewitt, A.E. 1987. Perfume and promise. *N.Z. Soil News*. 35:240–243.

Hewitt, A.E. 1990. A new soil classification for New Zealand, Proceedings of the 14th International Congress of Soil Science, Kyoto, Japan, August 1990. p V, 22–26.

Hewitt, A.E. 1992. New Zealand soil classification. DSIR Land Resources, Scientific Report 19. Reprinted in 1993 as Landcare Research Science Series 1, Manaaki Whenua Press, Lincoln, New Zealand.

Hewitt, A.E. 1993a. Methods and rationale of the New Zealand Soil Classification. Landcare Research Science Series 2. Manaaki Whenua Press, Lincoln, New Zealand.

Hewitt, A.E. 1993b. New Zealand Soil Classification: legacy and lessons. *Aust. J. Soil Res.* 30:843–854.

Hewitt, A.E. 1995. Soil map of South Island, New Zealand Soil Classification, 1:1,000,000 scale.

Hewitt, A.E. 1996. Classificacion de los suelos de Nueva Zelanda. Spanish version of the New Zealand Soil Classification, translated by D.P. Huertas and J.L.G. Rodriguez of Programa de Edafologia, Instituto de Reecurso Naturales, Colegio de PostGraduados, Montecillo, Edo. De Mexico. C.P. 56230.

Hewitt, A.E. 1997. Are New Zealand soils distinctive? A subterranean view of NZ ecosystems. Norman Taylor lecture, November 1996. *N.Z. Soil News*. 45:7–16.

Hewitt, A.E. 1998. New Zealand soil classification. 2nd edition. Landcare Research Science Series 1. Manaaki Whenua Press, Lincoln, New Zealand.

Johnson, W.M. 1963. The pedon and the polypedon. *Soil Sci. Soc. Am. Proc.* 27:212–215.

Milne, J. D.G., Clayden, B., Singleton, P.L., Wilson, A.D. 1991. Soil description handbook. DSIR Land Resources, Lower Hutt, New Zealand.

Parfitt, R.L. and Clayden, B. 1991. Andisols—the development of a new order in Soil Taxonomy. *Geoderma.* 49:181–198.

Rijkse, W.C. and Hewitt, A.E. 1995. Soil map of North Island, New Zealand soil classification, 1,000,000 scale.

Soil Survey Staff, 1975. Soil Taxonomy: A Basic System of Soil Classification for Making and Interpreting Soil Surveys. U.S. Dept. Agric. Handbook 436. U.S. Government Printing Office, Washington, DC.

Soil Survey Staff, 1999. Soil Taxonomy: A Basic System of Soil Classification for Making and Interpreting Soil Surveys. USDA Agricultural Handbook 436. Washington, DC.

Taylor, N.H. 1948. Soil map of New Zealand, 1:2,027,520 scale. DSIR, Wellington, New Zealand.

Taylor, N.H. and Pohlen, I. 1968. Classification of New Zealand soils, in Soils of New Zealand, Part 1. Soil Bureau Bulletin 26 with 1:1,000,000 scale soil map of New Zealand. DSIR, Wellington, New Zealand, 15-33.

CHAPTER 15

Changing Concepts of Soil and Soil Classification in Russia

S.V. Goryachkin, V.D. Tonkonogov, M.I. Gerasimova, I.I. Lebedeva, L.L. Shishov, and V.O. Targulian

CONTENTS

ABSTRACT

The main concepts of soil classification in Russia advanced during the last centuries are analyzed. The early classification systems, the different approaches to soil classification, the official classification of soils of the Soviet Union, and the new classification of Russian soils that is based on the profile-genetic concept are elucidated. The prospects of the Russian soil classification development are discussed.

INTRODUCTION

During the life of Dokuchaev, the founding father of modern Pedology in Russia, the problem of soil classification has received considerable attention and become an important component of scientific research. Soil classification presents a state-of-the-art of soil science, with results from

0-8493-1339-2/03/$0.00+$1.50

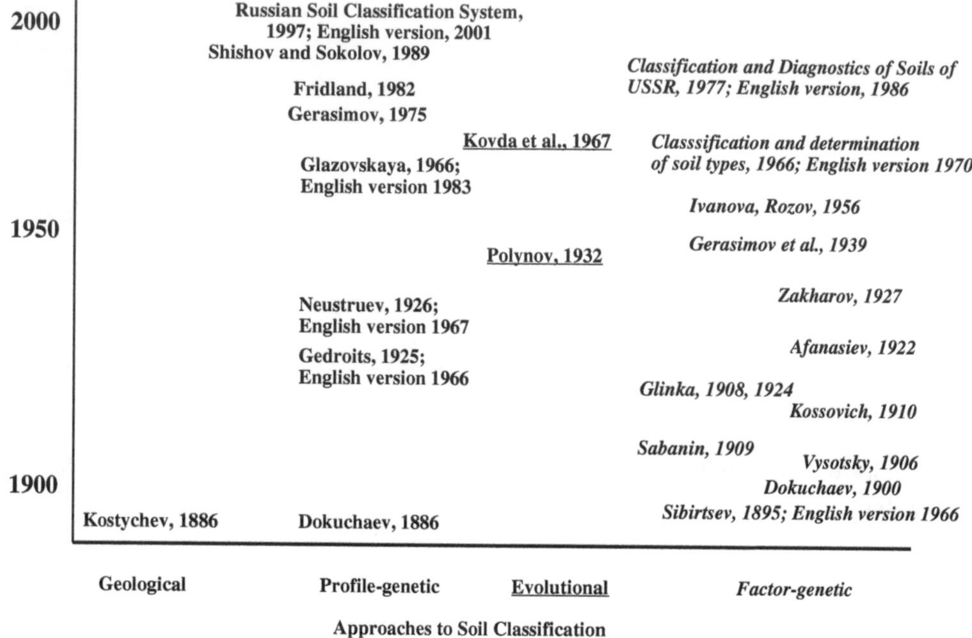

Figure 15.1 The development of different approaches to soil classification in Russia.

systematic research in the different aspects of the science, and attempts to show the linkages and interrelationships between the soil bodies as they occur on the landscape. The approaches to the different global soil classification systems stem from the stated purposes and functions they are expected to fulfill. There have been considerable discussions on this in Russia, as illustrated in the contributions of Strzemski (1975), Sokolov (1979), Gennadiev and Gerasimova (1994), Dobrovolsky and Trofimov (1996), and Tonkonogov et al. (1999).

The main goal of this paper is to elucidate the main concepts and systems of soil classification in Russia, and to analyze the change of concepts of soil classification in recent times.

CHANGING CONCEPTS

An analysis was made of the development of the system, represented in Figure 15.1. The approach was based mainly on the work of Dobrovolsky and Trofimov (1996), but modified by the authors. The different approaches to soil classification were grouped into four categories:

1. Geological, pre-Dokuchaev approach based on mineralogical, physical, and chemical characteristics of substrates
2. Factor-genetic approaches, based mainly on linkages of soils with climatic zones and other soil-forming factors, such as topography and parent materials
3. Evolutionary approaches that take into account the development of soils in time
4. Profile-genetic approaches based on soil characteristics (horizons, features) formed by pedogenesis

Figure 15.2 shows the photos of the late principals who contributed to these developments.

Figure 15.2 Russian pedologists—the authors of different classification systems.

EARLY SOIL CLASSIFICATION SYSTEMS IN RUSSIA

The pre-Dokuchaev period of soil classification in Russia was the same as in other European countries — agro-geological, chemical, and physical approaches to soil classification predominated (Dobrovolsky and Trofimov, 1996). The more advanced system was proposed by P.A. Kostychev in 1886 (cited from Sibirtsev, 1966). The principal dimensions were based on physical properties, whereas secondary groupings were chemical. The main divisions were the following:

I. Clayey soils (containing at least 50% particles smaller than 0.01 mm)
   1. Quartz or silicate
   2. Clayey
   3. Marly, calcareous, and dolomitic
   4. Humus soils
      a. bog (marsh)
      b. muck
      c. chernozem
   5. Ferruginous (ferric and ferrous oxides)
   6. Sulfate (gypsum)
   7. Solonchaks

II. Loess soils (the term "loess" being, in the physical sense, a rock or soil with a predominance of particles from 0.01 to 0.05 mm in diameter); the same seven secondary groups as in main division I above.

III. Sandy loams with a predominance (50% and more) of particles from 0.05 to 0.5 mm; the same seven secondary groups as in main division I above.

IV. Sandy soils with a predominance of particles larger than 0.5 mm
   1. Quartz or silicate
   2. Calcareous

V. Gravelly soils, with a predominance of particles larger than 2 mm in diameter
   1. Quartz or silicate
   2. Calcareous

This system could be named as a physicochemical (Dobrovolsky and Trofimov, 1996) or a geological one, as it is based mainly not on soils but on geological characteristics of substrates.

In the same year (1886), V.V. Dokuchaev proposed the first genetic classification of soils (Sibirtsev, 1966). The main scheme is as follows:

I. Normal
   A. Terrestrial-vegetal:
      1. Light gray northern
      2. Gray forest
      3. Chernozem
      4. Chestnut
      5. Brown solonetsous

(Each of these five types is subdivided as follows: sandy, sandy-loamy, light loamy, medium loamy, heavy loamy, and clayey.)

   B. Terrestrial-bog (these are soils of acid non-floodplain meadows and "chernoramens")
   C. Bog soils (peat bogs, tundras, and "plavni")

II. Transitional
   A. Eroded
   B. Terrestrial-deposited

III. Abnormal
   (Deposited-fluvial and lacustrine alluvium; aeolian deposits)

According to V.V. Dokuchaev, all soils may be subdivided into the following three groups: normal, transitional, and abnormal. This division is based on the permanence or temporary occurrence of a soil, and whether it retains those properties and features that are the product of undisturbed processes of weathering and activities of organisms, converting the bedrock to soil.

Soil, which was formed and remained *in situ*, is a normal soil. Soils of disturbed occurrence, separated from the rock from which they were formed, deposited or alluvial, were termed abnormal soils by V.V. Dokuchaev.

Thus the first classification system of V.V. Dokuchaev was not based on climatic zones, but on the real combination of pedogenic and geomorphic processes, reflected in soils, and it was more profile-genetic or substantive than the "zonal" system (Sokolov, 1979). The person who introduced the "zonal," or factor-genetic, approach to soil classification was N.M. Sibirtsev (Sokolov, 1979). Because of the contribution of N.M. Sibirtsev, the Russian pedology became "climatic" for many decades. Even V.V. Dokuchaev, under the influence of his colleague, changed his approach to soil classification to a "zonal" one in 1900. The system, proposed by N.M. Sibirtsev in 1895, was as follows (Sibirtsev, 1966):

- Class A. Zonal, melkozem (fine earth)-humus, complete soils, including the following types:
  - Laterite soils
  - Atmospheric-pulverulent or aeolian-loess soils
  - Desert soils, or soils or arid steppes
  - Chernozem soils
  - Gray forest soils
  - Sod-podzolic (or "ramen"-podzolic) soils
  - Tundra soils
- Class B. Intrazonal soil
  - Solonets soils
  - Bog (marshy) soils
  - Humus-carbonate (and other) soils
- Class C. Azonal, incomplete soils
- Subclass. Soils beyond floodplains
  - Skeletal
  - Coarse
- Subclass. Alluvial soils
  - Floodplain soils

This approach, termed the factor-genetic system, influenced the development of soil classification systems in Russia for about a century.

FACTOR-GENETIC CLASSIFICATION SYSTEMS

In 1906, G.N. Vysotsky proposed the classification system that was very similar to the system of N.M. Sibirtsev. He developed the class of "intrazonal" soils. G.N. Vysotsky divided this class into "Intrazonal soils becoming zonal ones in adjacent soil-climatic areas" (for example, podzols in chernozem zone), "Absolutely intrazonal soils" (for example, swampy soils), and "Skeletal soils."

In 1908, K.D. Glinka, while proposing a similar classification system, introduced the terms "ektodynamomorphic" and "endodynamomorphic" soils. The first term was used for the soils with a genesis mainly related to outer soil-forming agents (climate, organisms), and the second was used for soils closely connected to the specific rock, as Rendzinas for hard calcareous substrate. The second innovation was the proposal of six classes of moistening and drainage.

The close relationship of soils with vegetation type was the basis for the system proposed in 1909 by A.N. Sabanin:

- Soils of evergreen-deciduous type: 1st class—ferrugenious soils (laterites, red soils)
- Soils of coniferous-deciduous type: 2nd class—podzolic soils, 3rd class—nonpodzolic silica soils
- Soils of black-forest type: 4th class—angular-blocky soils
- Soils of meadow-forest type: 5th class—chernozems, 6th class—chestnut soils
- Soils of wormwood-grass (semidesert) type: 7th class—platy-columnar soils, 8th class—solonchaks and solonetz soils
- Soils of swampy-vegetal type: 9th class—meadow-swampy soils, 10th class—peaty dry soils, 11th class—marshy soils

The next stage in the development of approaches to soil classification was the system proposed in 1910 by P.S. Kossovich. He divided all the soils into two classes: genetically autonomous soils and genetically subordinated soils. The first class embraced the "zonal" soils and the second one, oversaturated, saline, and solonetzic soils of depressions and micro-lows of different soil-climatic zones.

In 1924, K.D. Glinka introduced a major change in the approach in which the type of pedogenesis was recognized at the highest level. This is perhaps the first instance in which soil development was considered in the classification, and this approach was soon to be adopted worldwide. There were five such types: Laterite, Podzolic, Steppe, Solonetzic, Swampy. They contain all the well-drained and poorly drained soils at the second level.

Until about 1939, the interrelationships between classes were the result of class definition, and there was no direct emphasis on priorities. The first hierarchical system, from which modern systems were developed, was developed in 1939 by I.P. Gerasimov, A.A. Zavalishin and E.N. Ivanova (Gerasimov et al., 1939). They distinguished ten genetic soil types:

1. Solonchaks
2. Solonetz
3. Serozem desert soils
4. Chestnut soils of dry steppes
5. Chernozem steppe soils
6. Gray podzolized soils of forest-steppe
7. Brown forest soils of southern forest areas
8. Sod soils of forest areas
9. Podzolic soils of northern forest areas
10. Swampy soils

These types were divided into subtypes using differences in vegetation as criteria. The lowest level of the system was species, determining "the stage of development of pedogenic processes." This is an important stage in the development of the Russian system of soil classification. It recognized the genetic aspects of soils, it provided a prioritization in the ranking of soils, it used soil properties to define the classes, it showed the linkage with geographic patterns of other factors, specifically vegetation, and it attempted to evaluate or differentiate the stage of development of the soil.

The refinement of factor-genetic approach in Russian pedology is connected with the names of E.N. Ivanova and N.N. Rozov. In the 1950s to 1960s, they proposed several schemes of classification systems, mostly reflected in the book "Classification and Determination of Soil Types"

(Ivanova and Rozov, 1970). They distinguished 110 soil types in the Soviet Union in accordance with climatic zones, drainage, and such soil characteristics as organic matter, base saturation, and salinization. The second hierarchical level is subtype. Subtypes are divided on the base of subdominant soil features (for example, calcareous or noncalcareous, saline or nonsaline chernozems). The third level is genera, which is based on the genetically important specificity of parent material, groundwater, etc. The fourth level is species. Species is distinguished by quantitative criteria — depth of humic or bleached horizons, humus content, percentage of salts. These refinements reflect the hierarchical approach and the quantification of the classification criteria, becoming the basis for the official soil classification system of the Soviet Union.

EVOLUTIONARY APPROACH TO SOIL CLASSIFICATION

Developments in the understanding of pedogenesis, as illustrated previously, were a dominant factor contributing to the modifications of the classification systems. The first classification system that considered evolution of the soil as an important factor was proposed in 1932 by B.B. Polynov. According to him pedogenesis is a stage in the process of formation of a weathering mantle. He distinguished four trends of pedogenesis:

1. Eluvial with downward movement of soil solutions
2. Lacustrine-swampy-solonchakous with weak or upward movement of soil moisture
3. Swamping and salinization
4. Draining and leaching

Trends are divided into chemical or mineralogical groups (for example, alkaline and acid), and groups are divided into types by the stages of development (for example, alkaline group is divided into primitive-alkaline, pre-chernozemic and chernozemic types). The lower levels of Polynov's systems are subtypes based on temperature; stages differentiated by quantitative parameters of the leading process; forms differentiated by the development of soil profiles; species differentiated by the presence or absence of the laterally transported material; and variations differentiated by land use. The total number of variants in Polynov's systems is 2800. So, this system was a major deviation from existing systems but it considered only the mineral composition of the soil. Further, it was considered to be too theoretical, and was ignored to some extent.

Another major development was introduced by Kovda in the 1960s. Perception of soils as results of broad geochemical mass-energy exchange processes, interacting in time in accordance with the evolution of landscapes, was the background for the world system elaborated from 1967 to 1975 by V.A. Kovda with his colleagues (Kovda et al., 1967). The first entry, or the upper category, is presented by broad soil-geochemical formations differentiated by major trends of weathering, types of humus, clay minerals, acid-base properties, and neoformations, supplemented with characteristics of climate and vegetation. The second entry is "stadial groups of soils," corresponding to hypothetical evolution stages, from submerged sites to excessively drained ones. The following stadial groups were proposed: hydroaccumulative, hydromorphic, mesohydromorphic, paleohydromorphic, proterohydromorphic, automorphic (including mountain soils), paleoautomorphic. The number of stadial groups, as well as that of soil-geochemical formations, varied in different versions of this system. Moreover, climatic zones or belts were introduced in the early ones, and climatic facies (sections) in the final version.

Such theoretical considerations in soil classification were scientifically interesting, but were considered speculative to some extent, as they were not supported by data or experimentation. They were considered to be based only on the author's hypotheses, and many of the assumed processes or trends were debatable. However, they reflected a stage in the evolution of Soviet Soil Science and even contributed to study of soils as an independent discipline. Many of the concepts that emerged were to be adopted later, and paved the way for new thinking in soil classification.

PROFILE-GENETIC APPROACHES TO SOIL CLASSIFICATION AND RELEVANT SYSTEMS

The profile-genetic approach to soil classification implies that soil horizons should be analyzed with due respect to their interdependence; in other words, the attention should be focused on the system of these horizons that make up the genetic profile of soil. Unfortunately, after Dokuchaev, this approach has never been fully realized in Russia, although it was taken into account in many other soil classifications. None of the official classifications, however, were based on the profile-genetic concept.

The first example of the elaboration of soil classification on the basis of internal soil parameters in Russia is that of K.K. Gedroits in 1925 (Gedroits, 1966). The author was a well-known specialist in physical chemistry. For decades he studied soil absorptive complex. He decided to develop a soil classification system based on the parameters of the absorptive complex. He distinguished two soil groups (with and without H^+) and four types of pedogenesis—chernozemic, solonetzic, podzolic, and latheritic. K.K. Gedroits recognized that he did not have enough information to elaborate on the working soil classification, but he believed in the future of this approach. His attempt should be evaluated as a substantive approach to soil classification; but further data on soil absorptive complex showed the severe limitation of this approach, and the impossibility of reflecting real soil diversity by only this parameter.

A more holistic approach to soil classification was proposed by S.S. Neustruev in 1926 (Neustruev, 1967). He produced a system of classification of pedogenic processes, rather than that of soils. The same criteria—"elementary pedogenic processes"—were used about 50 years later by I.P. Gerasimov (Gerasimov, 1975) for diagnostics and classification of soils. The problem of "process" approach to soil classification in both cases is lack of information and speculative character of understanding of soil processes. In later years, the processes were defined in terms of products and, consequently, the "process" concept became an integral part of classification.

The most comprehensive soil classification system in the world based on profile-genetic approach was proposed by M.A. Glazovskaya in 1966 (Glazovskaya, 1983). The system is strictly hierarchical, in that there are three above-type levels: associations, generations (classes), and families, and soil types are arranged in families in accordance with distinct rules. Lower categories were traditionally preserved, as in the above-mentioned system of Ivanova-Rozov. Thus criteria for the highest category of geochemical associations of soils were pH and soil features indicative of redox conditions; members of the second category are differentiated by broad processes of organic matter accumulation, formation of secondary minerals, translocation of substances, effects of ground water, etc. The difference in composition of pedogenetic accumulations (humus, neoformations, illuvial horizons, etc.) permit the specification of families within classes.

As is evident from this brief review, by the end of the 20th century, concepts were sufficiently well tested to develop a scientific system that also has practical value. The development of soil classification in Russia in the 1980s and 1990s is related to profile-genetic approach only. After Soil Taxonomy (Soil Survey Staff, 1975) and the FAO-UNESCO map legend, it became increasingly evident that the approach was rational, and clearly subscribed to by the global soil science community. This approach was actually deemed the basis of classification principles and schemes developed by V.M. Fridland (Fridland, 1981, 1982). The classification schemes suggested by L.L. Shishov and I.A. Sokolov (1989; 1991) are also based on soil profile morphology, along with some other substantive and dynamic characteristics of soils. This experience was used for elaborating the new soil classification system of Russia.

THE OFFICIAL CLASSIFICATION SYSTEM AND DIAGNOSTICS OF SOILS OF THE SOVIET UNION

The official system of soil classification in the Soviet Union was published in Russian in 1977. An English translation was also published in 1986. Thus this classification is well known not only in Russia, but also abroad. The classification of 1977 developed the earlier system suggested by Ivanova and Rozov (1970). The authors considered this classification as ecologic—genetic-based; it was based on the analysis of the "factors-processes-properties" triad. It was believed that the components of this triad are strictly interrelated, i.e., that a particular combination of pedogenetic agents strictly corresponds to a particular soil profile morphology and vice versa. It should be noted that, while analyzing the agents of soil formation, the authors did not consider them as equally important. To a certain extent, this was a deviation from Dokuchaev's principle of the equal importance of all soil-forming agents. This was a matter of much debate; many foreign pedologists considered this system of soil classification too climate oriented. Indeed, the highest taxa (soil types and subtypes) were differentiated with respect to bioclimatic conditions, or, more precisely, to the geographic position of soils in the system of bioclimatic zones, subzones, and facies determined by temperature criteria (influenced by the zonal concept). The degree of soil hydromorphism (drainage) was also taken into account at these taxonomic levels. This principle was strictly followed. The inner logic and consistency of this classification made it very popular in the Soviet Union. It was efficiently used in all soil survey works, including the systematic large-scale soil mapping of agricultural soils performed by regional institutes for land management (Giprozem) for more than 20 years. This classification reveals the essence of soil formation in different zones, and allows for forecasting soil development in a changing environment. The influence of this classification, its genetic principles, and its adherence to the traditional nomenclature of Russian soils can be traced in many national classifications throughout the world, in the legend to the FAO-UNESCO soil map, and in the more recent World Reference Base system.

However, with time and better understanding of soils and their distribution (resulting from the large number of soil survey and accompanying databases), the official system required modification. Information and data came not only from Russia but from all over the world. Communication among international soil scientists encouraged debates on the merits and demerits of different systems and, as a result, the last 20 years of the 20th century saw a flurry of activities to enhance national systems around the world. (This book is the first documentation of these efforts). It was determined that the 1977 Russian system did not satisfy the demands of agriculture, and did not incorporate the newly accumulated knowledge on the soils of Russia. Another important drawback of the system of 1977 was that it did not take into account the soils lying outside the zone of active agriculture; in particular, Siberian soils were not included. Thus soils occupying two-thirds of the country were not included in the official soil classification! Little attention was paid to anthropogenically transformed soils. At the same time, it contained too many virtually similar soil subtypes distinguished on the basis of exclusively hydrothermic criteria. Being based on the bioclimatic approach, this classification could not provide distinct morphological diagnostics of soil types and subtypes. Finally, this classification virtually excluded the soils that did not fit the criteria of zonal soil types and subzonal subtypes. Most of the recently described Siberian and northern soils (that were shown on small-scale soil maps of the Soviet Union) could not be placed into this classification without considerable changes in its principles and structure, including the approaches to soil diagnostics. The need for a totally new system was evident.

THE MODERN SOIL CLASSIFICATION SYSTEM OF RUSSIA

As mentioned previously, the official classification of soils of the Soviet Union could not satisfy the growing demands of the applied sciences such as agriculture, forestry, and ecology. A new classification had to be developed. Theoretically, there were three possible ways to achieve this purpose:

1. To refine and supplement the classification of 1977 without considerable changes to its basic principles
2. To adapt Soil Taxonomy (the U.S. system of soil classification) to Russian conditions
3. To develop a new classification that would have its own principles and would combine the advantages of both above-mentioned classification systems

The special commission on soil classification at the USSR Soil Science Society, which started its work in 1979, decided that the new classification should be based on the principles suggested by V.M. Fridland (Fridland, 1981). The rationale of this system is summarized as follows: the profile-genetic component of soil classification is based on the morphology of soil profiles and on soil properties that are considered the results of pedogenic processes. To realize this basic principle, the nomenclature and diagnostics of most important genetic soil horizons had to be developed. Many of these horizons were not separately distinguished in previous classification systems, and this became the first challenge.

The new ideas introduced in the Russian Soil Classification System (Shishov et al., 2001) distinguish this system from both the Classification of Soils of the Soviet Union and Soil Taxonomy.

- First, the new classification is a profile-genetic system. Therefore, in contrast to its predecessors, the highest taxonomic categories of this classification (soil orders and soil types) are distinguished on the basis of analysis of a conjugated system of genetic soil horizons that make up the soil profile. In turn, the definitions of these horizons are based on the integrity of substantive soil features and properties, dictated by pedogenic processes. Environmental agents, including climatic parameters (soil water and temperature regimes), are virtually excluded from the diagnostics of most soil taxa.
- Second, special attention is paid to anthropogenically (agrogenically and technogenically) transformed soils. These soils are considered the results of soil evolution under the impact of human activities. Several stages of this evolution are distinguished. Thus, taken together with corresponding natural soils, anthropogenically transformed soils form evolutionary sequences (from natural soils to transformed natural-anthropogenic soils and, finally, to nonsoil surface formations). The reverse trend of soil evolution after the release from anthropogenic impact is also possible in some cases (the restoration of natural soils from anthropogenically transformed ones). The classification position of anthropogenically transformed soils does not take into account the goals and character of human-induced impacts on soil and the level of soil fertility; it is fully dictated by the morphology of the soil profile (Tonkonogov and Shishov, 1990; Lebedeva et al., 1996b). The Classification of Soils of the Soviet Union considered some of such soils only in the context of the degree of soil cultivation and the character of soil transformation upon irrigation. At present, Soil Taxonomy (Soil Survey Staff, 1999) does not separate anthropogenically transformed soils into a special category, except for very strongly transformed soils (Lebedeva et al., 1996a), though the suggestions on the inclusion of the order of Anthrosols into this classification is being discussed.
- Third, the new classification system of the soils of Russia is an open system; i.e., it provides a means of introducing new soils into it, independently of their position in the system of geographical zones and subzones and of environmental conditions. In this context, it is closer to Soil Taxonomy than to the previous official classification of the Soviet Union.

In spite of the fact that the new classification system is formally limited by the territory of Russia, the profile-genetic approach to soil classification allows us to expand the system; in the future, it can be transformed into the classification of soils of the world.

The system of diagnostic soil horizons and soil properties has been specially developed for the diagnostics of soil types and subtypes. A particular combination of soil horizons serves as the basis for separation of soil orders and soil types. Qualitative peculiarities of soil horizons (genetic soil features) that are not taken into account at the type level specify the separation of soil subtypes.

The concepts of diagnostic soil horizons (including epipedons) and diagnostic soil properties are similar to U.S. Soil Taxonomy. However, according to Soil Taxonomy, all these diagnostic horizons have virtually the same weight in definitions of soil taxa. In the new classification of Russian soils, taxonomic values of different diagnostic horizons are also different. Initially, the system of diagnostic horizons was elaborated by V.M. Fridland (1982); however, the authors of the new classification have revised it considerably and introduced new quantitative and qualitative indices for soil diagnostics. The list of diagnostic horizons has been enlarged. For the first time, the group of anthropogenically transformed horizons has been distinguished; these horizons differ from natural soil horizons by specific structural status, substantive composition, morphology, and some other indices.

The definitions of diagnostic soil horizons in the new classification differ much from the system of strictly defined quantitative characteristics that are used for soil diagnostics in Soil Taxonomy. Mainly qualitative morphological and analytical characteristics are used in these definitions. In most cases, quantitative indices are rather diffuse, which is conditioned by the continual nature of soil bodies that usually have no distinct boundaries in nature. Strictly defined quantitative criteria are used only in exceptional cases. The absence of overly long and complicated keys to soil diagnostics with numerous quantitatively defined parameters (as in Soil Taxonomy) simplifies the procedure of soil correlation during a fieldwork.

The new classification retains most of the soil names that have been traditionally used in Russia, including the nomenclature decisions of the previous (1977) classification. The names of taxonomic categories (soil types, subtypes, genus, species, etc.) are also retained. However, all soil names that had certain indications of environmental conditions (e.g., forest soils, meadow soils, hydromorphic soils, etc.) are excluded from the new soil classification.

GOALS AND CHALLENGES FACING THE DEVELOPMENT OF SOIL CLASSIFICATION IN RUSSIA

The recently published "Russian Soil Classification System" is considered the first version of a comprehensive (basic) soil classification. The most important challenge for the near future is the improvement of the diagnostics of soil horizons and other criteria; quantitative indices that are used for soil classification at low taxonomic levels should also be refined. The changes in diagnostic criteria will result in some changes in the system of soil taxa. The soils that currently are not included in the classification because of the complexity of their diagnostics should find their adequate place in the next version of this classification.

The basic classification of soils is a unified language of scientific communication, the means to explain the character of soil formation, and the means to encode information on the genetic diversity and properties of soils. It should serve as the basis for soil mapping, evaluation of land resources, environmental monitoring of the soil cover, and other applied purposes. In this context, the problem of revealing not only stable features, but also dynamic characteristics of soils (including their water [moisture] and temperature regimes, the status of soil biota, and the current conditions of soil functioning that control the level of soil fertility and many other practically important soil features), emerges. The authors of the previous official classification of soils (1977) directly introduced these important characteristics in their classification; moreover, the environmental approach to soil grouping prevailed over the substantive approach in this classification. From our point of view, the combination of both substantive (stable soil features) and environmental (information on environmental factors) diagnostic indices in the integral system of soil classification is not an appropriate decision. Fridland (1982) and Sokolov (1991) suggested that the profile-genetic component of the basic soil classification should be supplemented by additional and independent components bearing information on soil regimes (dynamic characteristics of soils) and soil mineralogical and textural features. However, we should admit that the currently available data on soil temperature and soil moisture regimes are insufficient in Russia for elaboration of their separate classification. As for the second mineralogical (or lithological) component of soil classification, the main problem is to distinguish between proper soil (pedogenic) and lithogenic features. From our point of view, the more promising way is to create a special ecological-landscape classification on the basis of the profile-genetic classification of soils. The ecological classification of landscapes should take into account not only climatic and lithological indices, but also the geomorphic position of soils, the character of natural and anthropogenic vegetation, the presence of geochemical barriers in soils, etc. (Lebedeva et al., 2000).

CONCLUSIONS

1. Russia has more than 100 years history of soil classification. The first Dokuchaev's soil classification system of 1886 was mainly a profile-genetic one.
2. The approaches to soil classification in Russia were changing in the following sequence: geological → factor-genetic → evolutionary → profile-genetic systems.
3. The current and new Russian soil classification system is based on substantive approach taking into account genetically significant soil features. It includes both natural and human-modified soils (that are specially classified) and it preserves old traditional names of soils.
4. The reasonable way to include soil climatic parameters and other important nonpedogenic features into soil classification systems is to develop a special soil-ecological-landscape system that is an adjunct to the soil classification system.

REFERENCES

Anon. 1986. *Classification and Diagnostics of Soils of the USSR*. Oxonian Press Pvt. Ltd., New Delhi.

Dobrovolsky, G.V. and Trofimov, S.Ya. 1996. *Soil Systematrics and Classification (History and Up-to-date Status)*. Moscow State University Press, Moscow (in Russian).

Fridland, V.M. 1981. Fundamentals of the profile-genetic component of the basic soil classification. *Pochvovedenie*. 6:106–118 (in Russian).

Fridland, V.M. 1982. Main Principles and Elements of the Basic Soil Classification and the Working Program for Its Realization. Moscow (in Russian).

Gedroits, K.K. 1966. Genetic soil classification based on the absorptive soil complex and absorbed soil cations. Israel Program for Scientific Translation. Jerusalem.

Gennadiev, A.N. and Gerasimova, M.I. 1994. The Evolution of Approaches to Soil Classification in Russia and United States: Divergence and Convergence Stages. *Pochvovedenie*. 6:34–40 (in Russian).

Gerasimov, I.P. 1975. Elementary pedogenic processes as a basis for genetic diagnostics of soils. *Pochvovedeniye*. 5:3–9 (in Russian).

Gerasimov, I.P., Zavalishin, A.A., and Ivanova, E.N. 1939. A new pattern of the general classification of soils in the Soviet Union. *Pochvovedenie*. 7:10–43 (in Russian).

Glazovskaya, M.A. 1983. *Soils of the World. Vol. 1. Soil Families and Soil Types*. Amerind Publishing Co. Pvt. Ltd., New Delhi.

Ivanova, E.N. and N.N. Rozov, Eds. 1970. Classification and determination of soil types. Nos. 1–5. Israel Program for Scientific Translation, Jerusalem.

Kossovich, P.S. 1910. Pedogenic processes as the basis of the genetic soil classification, ZJL *Opytn. Agron.*,11 (in Russian).

Kovda, V.A., Lobova, E.V., and Rozanov, B.G. 1967. Problem of world soils classification. *Pochvovedenie*. 7:3–15 (in Russian).

Lebedeva, I.I., Tonkonogov, V.D., and Gerasimova, M.I. 1996a. Anthropogenically transformed soils in world soil classifications, *Eurasian Soil Sci*. 29(8):894–899.

Lebedeva, I.I., Tonkonogov, V.D., and Gerasimova, M.I. 2000. An experience in developing the factor-based classification of soils. *Eurasian Soil Sci*. 33(2):127–136.

Lebedeva, I.I., Tonkonogov, V.D., and Shishov, L.L. 1996b. Agrogenically transformed soils: Evolution and taxonomy. *Eurasian Soil Sci.* 29(3):311–317.

Neustruev, S.S. 1967. A tentative classification of soil-forming processes as related to soil genesis. Israel Program for Scientific Translation. Jerusalem.

Shishov, L.L. and Sokolov, I.A. 1989. Genetic classification of soils of the Soviet Union, *Pochvovedenie.* 3:112–120 (in Russian).

Shishov, L.L., Tonkonogov, V.D., Lebedeva, I.I., and Gerasimova, M.I. 2001. Russian Soil Classification System. Moscow.

Sibirtsev, N.M. 1966. Selected Works. Vol.1. Soil Science. Israel Program for Scientific Translation. Jerusalem.

Soil Survey Staff. 1975. Soil Taxonomy: A Basic System of Soil Classification for Making and Interpreting Soil Surveys. U.S. Dept. Agric. Handbook 436. U.S. Government Printing Office, Washington, DC.

Soil Survey Staff. 1999. Soil Taxonomy: A Basic System of Soil Classification for Making and Interpreting Soil Surveys. 2nd edition. U.S. Dept. Agric. Handbook 436. U.S. Government Printing Office, Washington, DC.

Sokolov, I.A. 1979. Problem of Classification in Soil Science. Results of Science and Technique. VINITI Moscow (in Russian).

Sokolov, I.A. 1991. The basic substantive-genetic classification of soils. *Pochvovedenie.* 3:107–121 (in Russian).

Strzemski, M. 1975. Ideas Underlying Soil Systematics. Warsaw, Poland.

Tonkonogov, V.D. and Shishov, L.L. 1990. Classification of anthropogenically transformed soils. *Soviet Soil Sci.* 22(6):52–59.

Tonkonogov, V.D., Lebedeva, I.I., and Gerasimova, M.I. 1999. The development and current status of soil classification. *Eurasian Soil Sci.* 32(1):37–42.

CHAPTER 16

Advances in the South African Soil Classification System

Michiel C. Laker

CONTENTS

ABSTRACT

South Africa is part of the relatively unknown "Third Major Soil Region" of the world. The dominant soils of this major soil region differ significantly from those of the other two major soil regions, namely, the soils of the high latitude big continents of the northern hemisphere and the soils of the humid tropics. Development of a systematic soil classification system for South Africa started about 1960 and culminated in the publication of "Soil Classification: A Binomial System for South Africa" in 1977. The system had two categories: the soil form and the soil series. The form was the higher category. The system was based on a number of well-defined diagnostic horizons. Easy to determine, mainly morphological criteria were used to define diagnostic horizons. Each soil form had a specific diagnostic horizon sequence. A revised version of the system, "Soil Classification: A Taxonomic System for South Africa," was published in 1991. The soil family became the lower category, instead of the series. A series category is supposed to be developed below the family, but this has become totally bogged down, and thus no progress is being made because of a decision that series should represent natural soil bodies and not pragmatic entities. In the interest of efficient land use planning and sustainable land use,

0-8493-1339-2/03/$0.00+$1.50

it is imperative that series should be defined as soon as possible, especially for the higher potential soils.

INTRODUCTION: THE PIONEER AND LEADERS OF SOIL CLASSIFICATION IN SOUTH AFRICA

Before embarking on a discussion of soil classification in South Africa, it is fit to pay tribute to the pioneer and subsequent leaders of soil classification in the country. There is no doubt that Dr. C.R. van der Merwe was the pioneer of soil classification in South Africa. In the foreword to the report on South African soil series by MacVicar et al., (1965), it is stated:

> Others have made important contributions from time to time, but in the field of soil classification and soil genesis in South Africa, he stands supreme. ... Concepts will change with time, old ideas will be discarded, names will be forgotten, but not the name of C.R. van der Merwe.

When the "7th Approximation" (Soil Survey Staff, 1960) confronted the world with a new approach to soil classification and brought a new impetus to soil classification, Drs. C.N. MacVicar, R.F. Loxton, and J.M. de Villiers became the leaders of a new movement to develop a systematic soil classification system, according to the new lines of thinking, for South Africa. Their strength was a combination of brilliant intellectual understanding of pedology coupled with extensive field experience. One of their biggest challenges was to develop a useful soil classification system for a country that forms part of the internationally little understood "Third Major Soil Region" of the world.

THE "THIRD MAJOR SOIL REGION"

When I became involved with international soil classification, as a member of the World Reference Base for Soil Resources (WRB) Working Group of the International Soil Science Society (now the International Union of Soil Sciences), I battled with the question of why so many of the most important soils in Southern Africa (and Australia) are not catered for adequately (or at all) in the two major international soil classification systems, namely, the USA's Soil Taxonomy (Soil Survey Staff, 1975) and the FAO system (FAO, 1988). Looking at a world map, it then dawned upon me that the world can be divided into three broad different "major soil regions." These are the following:

1. The soils of the relatively high to high latitudes of the big continents of the northern hemisphere, i.e., those at latitudes 35°N and higher. These include, *inter alia*, the whole of Europe, and nearly the whole of North America and the former Soviet Union. These soils have been well studied and their classification has been developed to a high degree of refinement. In the southern hemisphere, there is almost no land at these latitudes, not forgetting the ice-covered Antarctica, of course. Habitable land at these latitudes in the southern hemisphere includes virtually only New Zealand and minute parts of Australia and South America.
2. The soils of the humid and subhumid tropical areas. These soils have been fairly well studied, documented, and classified by soil scientists from Europe, and to some extent from the U.S. Between the two aforementioned major soil regions is found, in both hemispheres, the internationally quite unknown and relatively poorly understood "third major soil region" of the world.
3. In the southern hemisphere, it lies between approximately 20° and 35°S latitudes. It includes the whole of Southern Africa and almost the whole of Australia. These include large desert and semi-desert areas. Where it rains somewhat more, potential evapotranspiration greatly exceeds rainfall. Rainfall is also strongly seasonal and unreliable. Moist periods and periods of intense desiccation

alternate. The parent materials are predominantly weathering hard rock. These factors have a profound influence on the nature of the soils. In the northern hemisphere, this major soil region starts closer to the equator than in the southern hemisphere, but also extends to about 35°N latitude. Here it includes all the big deserts of the world.

HISTORICAL REVIEW OF SOIL CLASSIFICATION IN SOUTH AFRICA

MacVicar et al. (1965), MacVicar (1978), and Turner (2000) have given comprehensive historical reviews of soil classification in South Africa. A start was made, with studies of South Africa's soils in the early 1890s, with Juritz being the pioneer in this field (MacVicar et al., 1965). By 1910, Vipond pointed out that chemical analyses alone were not adequate and suggested soil surveys, recommending that these should be based on parent material (MacVicar et al., 1965). After that, a few loose-standing small soil surveys were done by different individuals, without any coordination or development of a soil classification system. During the same period, the Soil Survey Section of the Division of Chemistry of the Department of Agriculture conducted various detailed soil surveys for the development of irrigation schemes. The soils were classified into types, based on their morphology as observed in profile pits and some chemical analyses. Specific attention was given to soil properties that would determine the behavior of soils under irrigation (MacVicar et al., 1965).

During these irrigation surveys, C.R. van der Merwe, as Officer-in-charge of irrigation surveys, inspected the work of field teams in various parts of the country. He identified characteristic soils, describing and sampling them for further analysis. On the basis of these studies, Van der Merwe (1941; 1962) "classified the soils of South Africa into soil groups and subgroups according to their characteristic morphology as influenced by soil forming factors" (MacVicar et al., 1965). In this classification system, Van der Merwe embraced the zonality concept under the influence of the Russian pedologists and Marbut, and upon "observing such large tracts of uniform soil as the red Kalahari sands, the black clays on norite and the highly weathered soils of the eastern escarpment" (MacVicar, 1978). In many areas his system placed dissimilar soils in the same class or put similar soils in different classes (MacVicar, (1978). Because of the overriding effects of other factors, especially parent material, the zonality approach did not work in many areas. "Its failure to group soils according to similarity in properties was the downfall of this classification system. The units display the character of map units rather than taxonomic units" (MacVicar, 1978). Despite these criticisms, the following must be kept in mind:

1. This was the first attempt at a systematic classification of the soils of the country.
2. It was done in line with the international views on classification at that time.
3. The delineations of present-day generalized small-scale soil maps of the country do not differ dramatically from those of Van der Merwe. Basically, only the terminology differs.

During the late 1930s, Rosenstrauch conducted surveys of the soils of the "Sugar Belt" in the KwaZulu-Natal lowlands, and in the process was the first to make broad series classifications of soils in South Africa (MacVicar et al., 1965). Beater continued this work and published the first series definitions of soils in South Africa (Beater, 1944). His refinement of the series definitions of the soils of the Sugar Belt was published in three volumes, covering the three main sugar production regions of that time (Beater, 1957; 1959; 1962). Parent material was the dominant factor that determined classes in Beater's classification. "Since the effect of drift is minimal, each rock type has its own suite of soils, due mainly to drainage differences. ... These geological groupings were in fact well defined and hence useful map units (associations of soils)" (MacVicar, 1978). They were not taxonomic units, as most contained dissimilar soils because of position in the landscape (MacVicar, 1978).

MacVicar et al. (1965) pointed out that "outside the sugar industry and apart from soil surveys for irrigation purposes, hardly any progress was made with systematic soil survey and classification between 1941 and 1958" in South Africa.

DEVELOPMENT OF "SOIL CLASSIFICATION: A BINOMIAL SYSTEM FOR SOUTH AFRICA"

In 1958, the Natal Town and Regional Planning Commission initiated a number of natural resource surveys of the Tugela River catchment, including a systematic reconnaissance soil survey, to serve as a basis for sound regional planning (MacVicar et al., 1965). The survey was completed in 1962 (MacVicar, 1978), and a comprehensive outline of the soil studies conducted during the survey, including the soil classification system that was developed, was published in 1969 (Van der Eyk et al., 1969). In the process, provisional definitions of soil series were developed by De Villiers (1962) and MacVicar (1962). MacVicar et al. (1965) pointed out that the Tugela basin survey "was the first systematic soil survey executed in South Africa using modern concepts of series definition and classification."

The Tugela basin survey and an upsurge in soil surveying in various other parts of South Africa coincided with the publication of the 7th Approximation (Soil Survey Staff, 1960), which brought a new dimension and new interest in soil classification worldwide. The possibility to use the 7th Approximation as a means to classify South Africa's soil series was considered very strongly (MacVicar, 1978). South Africa had only a small number of soil scientists, and any classification had to be easy to use by nonsoil scientists if it was to have any practical value. It was believed that the 7th Approximation, with its "strange nomenclature," would not be generally accepted (MacVicar, 1978). I have always believed that this notion that the 7th Approximation and later Soil Taxonomy (Soil Survey Staff, 1975; 1999) had a strange or difficult nomenclature is unfounded and unfair. It is very simple and easy to use by teaching oneself to break up the name of a soil into its small individual pieces, and to look at what each little piece tells you about the soil. After analyzing the name of the soil in this way, you can then "synthesize" the soil to get a very clear picture of the actual properties and characteristics of the soil.

The other reason given by MacVicar (1978) for not using the 7th Approximation is valid, namely, that it often separated soils that local soil scientists (for good reasons) wanted in one class into different classes, or, conversely, grouped soils together that local soil scientists wanted in separate classes. The most important reason why the 7th Approximation could not be used—that it does not cater to a number of our most important soils—is not listed by MacVicar. This is also true for Soil Taxonomy and the FAO system, as will be pointed out later.

The 7th Approximation did have a very positive effect on soil classification in South Africa, however, for the following reasons:

1. It stimulated new interest in systematic taxonomic soil classification.
2. It emphasized that soil attributes should be used to classify soils.
3. It established the principle of using well-defined diagnostic horizons as the key building blocks for soil classification.

Points 2 and 3 became cornerstones of the South African soil classification system that was developed from about 1960 onward.

During reconnaissance soil surveys in the then Eastern Transvaal Highveld (now called the Mpumalanga Highveld) by AOC, the soil scientists were "confronted with the problem of conveying to their clients the nature of the very large number of soils in this area" (MacVicar, 1978). This led MacVicar and Loxton to group the soil series into soil forms, each form having a specific diagnostic horizon sequence (MacVicar, 1978). The two-tier form-series approach was also used for the classification of soils in the Tugela basin survey (Van der Eyk et al., 1969).

"The Tugela Basin classification met with considerable success. It gave planners, soil scientists, agronomists and farmers a good understanding of the soils of the area, non-soil scientists found soil identification easy, and its terminology was rapidly and widely taken into use" (MacVicar, 1978). This success led to the establishment of a soil classification working group in the early 1970s under the chairmanship of first De Villiers, then MacVicar, with the purpose of developing a form-series classification for the whole country. This culminated in the publication of *Soil Classification: A Binomial System for South Africa* (MacVicar et al., 1977).

SOIL CLASSIFICATION: A BINOMIAL SYSTEM FOR SOUTH AFRICA

As its name implies, the system was meant for soil classification in South Africa, from national level right down to farm level. It therefore did not need a large number of categories (tiers) of classification, in contrast to the case with Soil Taxonomy (Soil Survey Staff, 1975; 1999). Because it had to be used in detailed surveys at farm level, its categories had to be low ones. The system was therefore constructed to have two categories in the lowest tiers. In line with the approach elsewhere, the soil series is the lowest category in the system. Related series are grouped into soil forms.

Each soil form has a characteristic diagnostic horizon sequence (MacVicar et al., 1977). In some cases, soils with similar diagnostic horizon sequences are grouped into different forms on the basis of the presence of other materials in the soil profile. Thus the Valsrivier and Swartland forms both have Orthic A horizons over Pedocutanic B horizons, but in the case of the Valsrivier the underlying material is "unconsolidated material" (drift), while in the case of the Swartland it is saprolite.

Well-defined diagnostic horizons form the basis of the classification system. Diagnostic horizons are all designated in terms of classical traditional master horizon coding, i.e., each diagnostic horizon is indicated as being an A, E, B, or G horizon, e.g., Orthic A, Melanic A, Red apedal B, Prismacutanic B, etc. In the classification book, it is clearly stated that, in the procedure for the identification of soils according to this system, the first step is the demarcation of the master horizons present in the profile. This is in direct contrast to the big international soil classification systems, such as Soil Taxonomy (Soil Survey Staff, 1975; 1999) and the FAO system (FAO, 1990), which have steered away completely from giving diagnostic horizons any master horizon labels.

The system distinguished five diagnostic topsoil horizons: Organic O, Humic A, Melanic A, Vertic A, and Orthic A. The Humic A, Melanic A, and Orthic A are similar in concept to the umbric, mollic, and ochric epipedons/horizons, respectively, of Soil Taxonomy (Soil Survey Staff, 1999) and the FAO/WRB systems (FAO, 1988; WRB Working Group, 1998). The latter names were not adopted, because in the case of Humic/umbric and Melanic/mollic, the respective concepts are similar, but the horizons are not identical and the definitions differ somewhat. "Orthic" is thought to be a better description of the "ordinary" ("orthodox") topsoil horizons, which do not have any special features as do the others, because these horizons are not all "pale" ("ochric").

The Vertic A horizon was a special feature of the system. Although the 7th Approximation, Soil Taxonomy, and the FAO system recognize Vertisols, they do not have "vertic" diagnostic horizons. The WRB system (WRB Working Group, 1998) has a vertic diagnostic horizon, but defines it as a clayey *subsurface* horizon. I believe the latter is a mistake, and that the South African approach is the correct one. Surely the classical concept of a Vertisol is a soil in which the *topsoil* has strong swell-shrink properties, which affect the land use suitabilities and management requirements of these soils, setting them aside from other soils which may have clayey topsoils. This aspect will be debated further later in this chapter.

Thirteen diagnostic subsurface horizons and two diagnostic materials (regic sand and stratified alluvium) were defined (MacVicar et al., 1977). The thirteen subsurface horizons included eleven diagnostic B horizons, plus an E and a G horizon. For each horizon, a set of specifications was given, followed by a general discussion explaining the underlying concept of the horizon and its linkages with other horizons. To be deemed diagnostic, the upper limit of a horizon had to be within 120 cm (the normal length of a soil auger) from the surface. Morphological features, which could be easily observed in the field, were almost exclusively used for the identification of different diagnostic horizons. Most of these implied certain land qualities. This characteristic of the system greatly contributed to its success.

Two of the most important features used for the identification of diagnostic subsurface horizons were soil color and structure. Extensive color specifications were given, especially for the so-called Red apedal B, Yellow-brown apedal B, and Red structured B horizons, in terms of permissible hues, values, and chromas. In addition to the color specifications, the colors of these horizons had to be "substantially uniform." Horizons with significant color variegation were not permitted. This was to ensure that only perfectly drained and aerated horizons were included under these brightly colored subsoils, which included the best subsoils from a crop production viewpoint in the country. A consequence of this color emphasis was that a soil form such as the Hutton form, having an Orthic A horizon over a Red apedal B horizon, included soils as diverse as an almost pure eutrophic sand (with less than 6% clay) at the one extreme and a highly weathered dystrophic clay at the other extreme. That is, it spanned several orders in the U.S. Soil Taxonomy. This has drawn heavy international criticism. It should be kept in mind that the soils in a form were separated very neatly into several texture and weathering/leaching class combinations at series level. The most important aspect is that it worked successfully in practice.

Soil structure was another important feature used for separating certain diagnostic subsoil horizons. Again, as in the case with the "chromic" horizons discussed above, it has important implications with regard to land suitability for rainfed cropping or irrigation. The B horizons with moderate to strong blocky structure have poor qualities (unfavorable for root development; highly erodable) and must therefore be separated from their more weakly structured counterparts, which normally have much better qualities.

An important example with regard to differentiation on the basis of structure, not only from a land suitability point of view but also in terms of understanding pedogenesis, is the distinction between the Neocutanic B horizon and the Pedocutanic B horizon. The concept of the Neocutanic B horizon is that it is a young (neo = new) horizon which has developed in transported material. The Pedocutanic B horizon is a "mature" horizon which has developed from clay illuviation over adequate time. The difference between the two is that the Neocutanic B horizon has structure that is weaker than moderate, while the Pedocutanic B horizon has at least moderate to strong structure. It is easy to perceive that in most cases Neocutanic B horizons, over a long period of time, will reach the stage at which it can become a Pedocutanic B horizon. This sequence is clearly seen on some narrow river terraces in South Africa, e.g., along the Tyumie River, near Alice in the Eastern Cape. The soil on the youngest terrace on the river bank has alluvial stratifications throughout the profile below the A horizon. The subsoil on the second (older) terrace has a typical Neocutanic B horizon–alluvial stratifications have been obliterated, but no structure has developed. The subsoil on the third (oldest) terrace is clay-enriched and has strong course blocky, and in places even prismatic, structure.

Conceptually, the Neocutanic B horizon is similar to the cambic horizon of Soil Taxonomy and the FAO/WRB system, while the Pedocutanic B horizon could be expected to resemble the majority of the argillic/argic horizons of the latter systems. In these international systems, however, minimal structure requirements are reversed: The cambic horizon must have structure "which is at least moderately developed," whereas the argic horizon has no minimal structure requirement (WRB Working Group, 1998). For our purposes, I believe that the South African approach toward differentiation between the ("young") Neocutanic B horizon and the ("mature") Pedocutanic B horizon is correct.

A case in which the morphological approach of the South African system was (and still is) far superior to the two big international systems is that of the B horizon of the "solonetzic" soils. In the South African system, the two key requirements for a "Prismacutanic B horizon" are:

- An abrupt transition from the overlying horizon in terms of texture, structure, and consistency (or at least two of them)
- Prismatic or columnar structure

There are numerous soils with typical solonetz morphology and, more importantly, solonetz behavior, in which the B horizons do not qualify as natric horizons because they do not have ESP values of 15% or more at some depth in the profile. Many have the "ancillary" requirement of lopsided Ca:Mg ratios, but some do not even have this. "Low sodium solonetzes" are also found in other countries in the "third major soil region," such as Australia and Zimbabwe (Nyamapfene, 1991), and have also been reported by Soviet soil scientists to be "common" in parts of the former Soviet Union (Laker, 1997). The subsoils of soils with typical solonetz morphology are notorious for their dispersivity and extreme susceptibility to erosion, especially gully erosion and piping (tunneling). It is, therefore, very important to classify these soils together, irrespective of the ESP of the Prismacutanic B horizon. Nyamapfene (1991) pointed out that "rigorously defined morphological requirements" would be needed in such a case. This is exactly what was done in the South African classification system: Structured B horizons that occur under an abrupt transition, but do not have prismatic or columnar structure, are not classified as Prismacutanic B horizons but as Pedocutanic B horizons. Likewise, structured B horizons that have prismatic structure, but are not under abrupt transitions, are classified as Pedocutanic B horizons, not as Prismacutanic B horizons.

Two horizons that deserve special attention, because of their uniqueness, are the *Soft plinthic B* and *Lithocutanic B* horizons.

The *Soft plinthic B* horizon is a horizon that is characterized by the presence of abundant *vesicular high chroma* (bright red or reddish brown and/or bright yellow or yellow-brown) mottles in a matrix that has at least some gray areas in it. The cores of the mottles are often black, and occasionally some of the mottles may be black, indicating the presence of manganese oxides. Some mottles may be hard in the dry state and some individual mottles might have hardened into iron or iron-manganese concretions. The presence of gray colors in the matrix and the localization of ferric oxides (indicated by the vesicular nature and bright red or yellow colors of the mottles) indicate that this horizon is formed in a zone of the soil profile that is subject to a fluctuating water table (alternate wetting and drying).

Although the Soft plinthic B horizons form under the same conditions as the plinthite of Soil Taxonomy (Soil Survey Staff, 1999) or plinthic horizon of the FAO/WRB systems (FAO, 1988; WRB Working Group, 1998), these horizons do not qualify as plinthite or plinthic horizons because they do not harden irreversibly upon alternative wetting and drying. It was initially (in 1993) extremely difficult to convince the members of the WRB Working Group that such horizons actually exist, although we South African soil scientists knew that we had several million hectares of soils with Soft plinthic B horizons. It was only when they saw these horizons in the field during the 1996 WRB workshop that the WRB members had to agree that it was different from plinthite, and made provision for it as a "paraplinthic" horizon (WRB Working Group, 1998). None of the South African soil scientists that I contacted have ever found plinthite, i.e., "soft plinthite" that hardens irreversibly, in South Africa. These include P.A.L. le Roux, who did his Ph.D. on plinthic soils (Le Roux, 1996). It should be pointed out that large areas with "Hard plinthic B horizons" (petroplinthite) are found in South Africa.

From a land suitability and land use planning point of view, it is very important to identify Soft plinthic B horizons correctly in soils, especially for rainfed cropping. For example, in the drier western part of the maize quadrangle, maize yields (summer crop/summer rainfall) are, during normal years and especially during low-rainfall years, higher (and more reliable) on soils with Soft plinthic B horizons than on other soils. This horizon indicates a fluctuating water table, i.e., a much higher plant-extractable water reservoir in the deeper soil layers. In the higher-rainfall eastern parts of the quadrangle, maize does not perform as well on the soils with Soft plinthic B horizons as on the better drained soils, because the former tend to be too wet during the growing season. In these areas, the soils with the Soft plinthic B horizons are the best wheat (winter crop grown totally on stored water) soils, however, and outyield the "better soils." From a land suitability evaluation point, it is very important to distinguish correctly between Soft plinthic B horizons and the following situations, which do not qualify as Soft plinthic B horizons:

- Horizons with abundant low chroma (dull) yellow, usually streaky and not vesicular, mottles. These indicate unfavorable, excessively wet conditions.
- Uniformly red horizons with abundant iron or iron-manganese concretions. Absence of gray colors indicates that these horizons are not under the influence of fluctuating water tables, and thus much drier than Soft plinthic B horizons.
- Light gray horizons with abundant iron or iron-manganese concretions that have formed from the weathering of ferricretes (hard plinthite; petroplinthite). These horizons have not formed under the influence of a fluctuating water table. They have extremely poor physical properties (being hard-setting when dry) and are very infertile. Hubble et al., (1983) also refer to such infertile "young soils formed in lateritic detritus and exposed remnants of laterite profile following dissection."

The concept of the *Lithocutanic B* horizon "is one of minimal development of an illuvial B horizon in weathering rock" (MacVicar et al., 1977). It is typically characterized by prominent tonguing of soil into partly weathered rock. It often shows clear signs of illuviation in the form of clay skins, which are often deposited around a small stone. In a country where shallow soils predominate, it is important to distinguish soils with these horizons from other shallow soils, because these horizons offer at least some room for root development and for additional storage of plant-extractable water. This type of material is classified as a Lithocutanic B horizon only if it occurs directly under an Orthic A, Humic A, or Melanic A or E horizon. Soils of the Glenrosa form (Orthic A over a Lithocutanic B) are possibly the most widespread soils which are found in South Africa. We are consequently so used to them that they are not regarded as special. We were, therefore, very surprised when Hari Eswaran, at the time Chairman of Commission V of the then ISSS, in 1993 insisted that we must include it when an international soil classification workshop is held "because no provision is made for it in the international classification systems." An example of a Glenrosa soil likewise attracted intense interest and debate during the 1996 WRB workshop in South Africa. During the workshop we noted that the WRB Working Group classified a non-calcareous series of our Mayo form, which has a Melanic A (mollic horizon) over a Lithocutanic B horizon as a Glossic Phaeozem, the "Glossic" indicating the tonguing nature of the horizon.

Once all the diagnostic horizons in a soil profile have been identified, the diagnostic horizon sequence of the soil can be determined. Once the diagnostic horizon sequence has been determined, the soil form is determined by using the "Key to the soil forms" in the classification book (Table 16.1). Thus no reading through a descriptive key is required to determine the soil form. Each soil form was given the name of a place, normally where the first good example of such soil was identified, e.g., Hutton, Clovelly, etc.

The key to soil forms also indicates the page on which a matrix table with the series for each form is given, with a color photograph of a typical profile for that form on the opposite page. The three types of soil properties most widely used as series differentiae in the South African binomial system were texture, degree of leaching/weathering, and presence or absence of lime. It must be kept in mind that these were not the only properties used and that they were not used in all soil forms.

Two textural parameters were used, namely, clay content and sand grade. The clay content classes were 0–6%, 6–15%, 15–35%, 35–55%, and >55%. Not all the classes were used in all

Table 16.1 Key to the Soil Forms in Soil Classification: A Binomial System for South Africa*

UNDERLYING DIAGNOSTIC HORIZONS/ MATERIALS	DIAGNOSTIC TOPSOIL HORIZONS				
	Organic	Humic	Vertic	Melanic	Orthic
Gleyed material	Champagne (34)				
G horizon			Rensburg (44)	Willowbrook(48)	Katspruit (60)
Pedocutanic B/ saprolite					Swartland (62)
Pedocutanic B/ unconsolidated material					Valsrivier (64)
Pedocutanic B				Bonheim (50)	
Prismacutanic B					Sterkspruit (66)
E horizon/ Prismacutanic B					Estcourt (68)
E horizon/ Gleycutanic B					Kroonstad (70)
E/Yellow-brown apedal B					Constantia (72)
E/Red apedal B					Shepstone (74)
E/Neocutanic B					Vilafontes (76)
E/Ferrihumic B/ saprolite					Houwhoek (78)
E/Ferrihumic B/ unconsolidated material					Lamotte (80)
E/Lithocutanic B					Cartref (82)
E horizon/Hard plinthic B					Wasbank (84)
E horizon/Soft plinthic B					Longlands (86)
Soft plinthic B				Tambankulu (52)	Westleigh (88)
Yellow-brown apedal B/Soft plinthic B					Avalon (90)
Yellow-brown apedal B/Hard plinthic B					Glencoe (92)
Yellow-brown apedal B/ Gleycutanic B					Pinedene (94)
Yellow-brown apedal B/Red apedal B		Kranskop (36)			Griffen (96)
Yellow-brown apedal B		Magwa (38)			Clovelly (98)
Red apedal B/Soft plinthic B					Bainsvlei (100)
Red apedal B		Inanda (40)			Hutton (102)
Red structured B					Shortlands(104)
Neocutanic B				Inhoek (54)	Oakleaf (106)
Regic sand					Fernwood (108)
Stratified alluvium				Inhoek (54)	Dundee (110)
Lithocutanic B		Nomanci (42)		Mayo (56)	Glenrosa (112)
Hard rock, etc.				Milkwood (58)	Mispah (114)
Not specified			Arcadia (46)		

* Pages for matrix tables in brackets. Adapted from MacVicar et al., 1977.

forms. In some, the sandy classes (0–6% and 6–15% clay) were excluded, because no soils in the form had such low clay content. In other forms, the high clay classes were excluded, because no soils in the form had such high clay content. Three sand grade classes—fine, medium, and coarse—were used. Sand grades were derived from a sand grade triangular diagram similar in concept to the well-known texture triangles (MacVicar et al., 1977). Sand grade subclasses were distinguished only in the relatively low clay classes. This is because in these classes, the sand grade has major effects on soil physical conditions, such as plant-available water storage capacity and susceptibility to soil compaction or hard-setting. Where the B21 (upper B) horizon was the series classification control horizon, sand grades were used in the 0–6% and 6–15% clay classes only. Where an Orthic A or E horizon was the control horizon for series classification, sand grades were used in the 0–6%, 6–15%, and 15–35% clay classes.

Although the concept of "control horizons" was nowhere mentioned in the classification system, it was in fact applied. Some examples are the following:

- In *all* soils with Red apedal B or Yellow-brown apedal B horizons, the B21 (upper B) horizon was the horizon which had to meet certain requirements with regard to the two textural parameters and degree of leaching/weathering (where the latter was applicable).
- In the case of *all* soils with E horizons, the E horizon had to meet certain requirements with regard to the two textural parameters and was the control horizon, although in some cases certain properties of the underlying horizon were also used in series classification.
- In the case of *all* soils with Melanic A (mollic) horizons, properties of the A horizon, including clay content classes, were used for series classification.

Three leaching/weathering classes were distinguished. These were defined in terms of the amount of exchangeable (i.e., not including soluble) bases (Ca, Mg, K, and Na) *per unit clay*, as expressed in me/100 g clay, i.e., the system did not use percentage base saturation, as is done in Soil Taxonomy and the FAO/WRB systems. The following three classes were distinguished:

- Dystrophic: < 5 me exchangeable bases/100 g clay
- Mesotrophic: 5–15 me exchangeable bases/100 g clay
- Eutrophic: > 15 me exchangeable bases/100 g clay

These criteria were applied to the upper parts (B21 horizons) of the "high quality" B horizons, namely, the Red apedal B, Yellow-brown apedal B, and Red structured B horizons.

Table 16.2 gives the series classification matrix for the Hutton form, the form generally considered to include the highest potential soils in the country. The diagnostic horizon sequence for this form is an Orthic A over a Red apedal B horizon.

The system quickly found wide acceptance in the country, not only among soil scientists, but also among the users of soil information. This can be attributed to the following:

- Its simple, straightforward structure
- Its emphasis on morphological features, making it easy to identify soils
- The fact that only a few simple routine laboratory analyses are required
- The fact that most of the parameters used in the classification can be directly related to some practical land use implication
- The fact that no "difficult" terminology was used

An advantage of the system was its "open" nature, i.e., if new diagnostic horizon sequences were found in nature, these could be added as new forms without disrupting the existing classification. The same would be true if the need arose for defining additional diagnostic horizons.

Table 16.2 Series Matrix Table for the Hutton form* in "Soil Classification: A Binomial System for South Africa"

Clay Content of B21 Horizon (%)	Grade of Sand in B21 Horizon	Noncalcareous in B Horizon: Dystrophic in B21 Horizon		Mesotrophic in B21 Horizon		Eutrophic in B21 Horizon		Calcareous in B Horizon	
0 – 6	Fine	Alloway	10	Whithorn	20	Roodepoort	30	Lowlands	40
	Medium	Arnot	11	Joubertina	21	Gaudam	31	Nyala	41
	Coarse	Stonelaw	12	Chester	22	Moriah	32	Quaggafontein	42
6 – 15	Fine	Wakefield	13	Lichtenburg	23	Mangano	34	Maitengwe	43
	Medium	Middelburg	14	Clansthal	24	Zwartfontein	35	Malonga	44
	Coarse	Kyalami	15	Bontberg	25	Portsmouth	36	Vergenoeg	45
15 – 35	Undifferentiated	Hutton	16	Msinga	26	Shorrocks	37	Shigalo	46
35 – 55	Undifferentiated	Farningham	17	Doveton	27	Makatini	38	Hardap	47

* Adapted from MacVicar et al., 1977.

The main weaknesses of the system were:

- Each soil form had a series (with a much narrower range of properties) with the same name as the form, e.g., the Hutton form included a Hutton series (Table 16.2). This often led to confusion when people just talked about "Hutton soils," for example.
- Knowledge about the soils of the vast semiarid and arid parts of the country was very poor when the classification system was constructed, and the system did not cater well for the classification of these soils.

DEVELOPMENT OF "SOIL CLASSIFICATION: A TAXONOMIC SYSTEM FOR SOUTH AFRICA"

In 1981 a new classification working group, still under chairmanship of MacVicar, started with revision of "Soil Classification: A Binomial System for South Africa." Aspects initially identified for special attention included, *inter alia,* the following:

- Possible creation of new soil forms for soils with diagnostic horizon sequences for which provision has not been made in the 1977 system
- Improved definitions for some existing diagnostic horizons according to new information that has become available
- More subdivisions of the podzolic soils. This was requested by the people in forestry because of the importance of this in suitability evaluations of such soils for different *Pinus* species
- Classification of the soils of the semiarid and arid regions
- Creation of a soil family category to fit in between the form and series categories

At the second meeting of the Working Group in 1983, two new (interrelated) issues came on the agenda and would not only become controversial, but would bog down progress with regard to important classification issues. Neither of these has thus far been resolved. The issues are these:

- A proposal that soil series should be natural soil bodies and not artificial classes. The soil textural classes of the 1977 system were identified as the "culprit" in this regard.
- How provision should be made for silt in the use of soil texture as classification criterion. In the 1977 series classification, silt was ignored. This was because the silt content of the vast majority of South African soils is low to very low, and can be ignored. There are soils with enough silt (20% or more) to significantly affect soil properties, such as plant-available water capacity and the susceptibility of the soil to compaction. Special concern was expressed with regard to the

identification of "hard-setting" soils and the possible role of silt, in combination with fine sand, in these subsoils (Turner, 2000).

The work of the Working Group culminated in the publication of "Soil Classification: A Taxonomic System for South Africa" in 1991 to replace the 1977 binomial system as South Africa's soil classification system (Soil Classification Working Group, 1991).

SOIL CLASSIFICATION: A TAXONOMIC SYSTEM FOR SOUTH AFRICA

Soil Classification: A Taxonomic System for South Africa is again a system with two categorical levels (Soil Classification Working Group, 1991). The soil form is again the higher category, as in the 1977 binomial system, but now the soil family is the lower category, instead of the soil series. The system was supposed to have had the soil series as lowest (third) category, but since no series could be defined, this category fell away. This is because the majority decision, with which I totally disagree, was that series should represent natural soil bodies in terms of their textural ranges. Series classification could not be done on this basis, because the data required to do it were simply not available, and are still not available.

In the 1991 taxonomic system, the upper boundary of a horizon had to be within 150 cm from the soil surface to qualify as a diagnostic horizon, and not within 120 cm, as in the 1977 binomial system. Master horizon designations were changed to fit in with the newer international trend. The new designations were as follows (with the old equivalents used in the 1977 system in parentheses): A(A1), E(E or A2), AB(A3), BA(B1), B(B2), BC(B3). As in the 1977 binomial system, each soil form still had a characteristic diagnostic horizon (or diagnostic material) sequence. In the 1991 taxonomic system, there are 73 soil forms, as opposed to 41 in the 1977 binomial system.

With one exception, i.e., the Gleycutanic B horizon, which was merged with the G horizon, all the diagnostic horizons and materials of the 1977 system were retained. Slight changes to the definitions of some were made, however. The following are some of the most important changes:

- Plasticity index was made one of the criteria that can (on its own) be used to distinguish between Vertic A and Melanic A horizons.
- In the 1977 system, the E horizon had to meet the "gray" color criteria in either the dry or the moist state. In the 1991 system, it must meet these criteria in the dry state.
- "Youthful, cross-bedded aeolian sands, usually in the form of dunes, with color that is predominantly red or yellow, no longer qualify as red or yellow-brown apedal B horizons; provision is made for these regic sands in the Namib form" (Soil Classification Working Group, 1991). Meanwhile, the deep light gray sands, with or without lamellae, of the Fernwood form are now considered to be E horizons and not regic sand, as was the case in the 1977 classification.

A number of new diagnostic horizons and materials have been added. Of these, the following deserve mentioning:

- The placic pan, found in some Podzol B horizons.
- The Neocarbonate B horizon, including all the calcareous *apedal B horizons* that were formerly classified as calcareous Red or Yellow-brown apedal B or Neocutanic B horizons.
- The Soft carbonate (hypercalcic) and Hardpan carbonate (petrocalcic) horizons. These are horizons of which the morphology is totally dominated by the lime. From a practical land-use point, in marginal rainfall areas it is very important to distinguish between the two. It has been found that large quantities of plant-extractable water are stored in the Soft carbonate horizons. Because of this, much more consistent, and in the long term considerably higher average, maize yields are obtained under these rainfall conditions on these soils than on the "better" deep soils in the same

areas. The Hardpan carbonate horizons do not have this advantage because they are impenetrable to water and roots.

- Dorbank (Afrikaans: dor = very dry; bank = pan). The definition of the dorbank is like that of the duripan of Soil Taxonomy and the FAO/WRB systems. Its name indicates that it is associated with extremely arid conditions, typically drier areas than where Soft carbonate or Hardpan carbonate horizons are found.
- Man-made soil deposits.

Laker (2000) gives very short descriptions of the basic concepts of all the diagnostic horizons and materials distinguished in the 1991 South African system. (He also discusses the distribution, main properties, and land-use suitabilities of the main South African soils.) The key to the identification of soil forms in the 1991 system (Table 16.3) is structured differently from that of the 1977 system (Table 16.1), and is more logical and user-friendly than the latter.

With the exception of the texture criteria, which were eliminated, all the soil properties used for series differentiation in the 1977 system are used for family differentiation in the 1991 system. Twelve new criteria were added for family differentiation, i.e., the number of family criteria is more than double the original number of series criteria. Some of the most noteworthy new criteria are the following:

- Differentiation between bleached and nonbleached *Orthic A horizons* in some soil forms. Some Orthic A horizons have bleached (light gray as defined for E horizons) colors. These are not E horizons that have been exposed at the surface. They are found mainly in relatively dry areas, with gray mudstones or shales as parent materials. The bleached A horizons have very poor physical properties, being especially prone to severe surface sealing (crusting). For this reason, it is important to recognize them and to classify these soils separately from Orthic A horizons that are not bleached. This has since also been recognized in the WRB system (WRB Working Group, 1998).
- Gray and *yellow E horizons*. Some E horizons assume uniform high chroma yellow colors, as defined for the Yellow-brown apedal B horizon, in the moist state. These horizons are classified as E horizons on the basis of their light gray colors in the dry state. At family level, a distinction is made between those that remain gray in the moist state and those that become yellow in the moist state. After the WRB workshop in South Africa in 1996, this phenomenon has also been recognized in the definition of the albic horizon in the WRB system (WRB Working Group, 1998).
- *Luvic B horizons*. Some apedal B horizons (i.e., some Red apedal B, Yellow-brown apedal B, Neocutanic B, and Neocarbonate B horizons) have markedly more clay in B horizon than in the overlying A or E horizon. The luvic criterion was introduced to distinguish these horizons. To qualify as luvic, the following conditions must be met (quoted from Soil Classification Working Group, 1991):
  - When any part of the A or E horizon has 15% clay or less, the B1 horizon must contain at least 5% more clay than the A or E.
  - When any part of the A or E has more than 15% clay, the ratio of clay percentage in the B1 to that in the A or E must be 1:3 or greater.
- *Subangular/fine angular and medium/coarse angular structure* in Pedocutanic B and Red structured B horizons. By definition the blocky structure in these horizons must be at least moderately developed in the moist state. Horizons with medium to coarse angular blocky structure are more common than those with subangular/fine angular blocky structure, and are much less suitable for crop production than the latter. In fact, soils with such horizons are usually rated unsuitable for crop production, both rainfed and under irrigation. Soils with Pedocutanic B horizons with medium to coarse blocky structure are also highly erodable.
- *Hard and not hard Lithocutanic B horizons* (and saprolite). Because they are found in such shallow soils, Lithocutanic B horizons are important for root development and storage of plant-available water. Its capacity for water storage and its space for root development are determined by how much actual soil the horizon has. If more than 70% by volume of a Lithocutanic B horizon "is bedrock, fresh or party weathered, with at least hard consistence in the dry, moist and wet states," the horizon is classified as "hard" (Soil Classification Working Group, 1991). Because of the smaller rooting volume and lower water storage capacity, the "hard" Lithocutanic B horizons are less favorable for plant growth than the "not hard" ones.

Table 16.3 First Approximately Half of the Key to the Identification of Soil Forms in "Soil Classification: A Taxonomic System for South Africa"

Diagnostic Horizons and Materials					
Topsoil	**Subsoil**			**Soil Form**	**Page**
Organic	Unspecified			Champagne	48
Humic	Yellow-brown apedal B	Red apedal B		Kranskop	50
Humic	Yellow-brown apedal B	Unspecified		Magwa	52
Humic	Red apedal B	Unspecified		Inanda	54
Humic	Pedocutanic B	Unspecified		Lusiki	56
Humic	Neocutanic B	Unspecified		Sweetwater	58
Humic	Lithocutanic B			Nomanci	60
Vertic	G horizon			Rensburg	62
Vertic	Unspecified			Arcadia	64
Melanic	G horizon			Willowbrook	66
Melanic	Pedocutanic B	Unspecified		Bonheim	68
Melanic	Soft carbonate			Steendal	70
Melanic	Hardpan carbonte			Immerpan	72
Melanic	Lithocutanic B			Mayo	74
Melanic	Hard rock			Milkwood	76
Melanic	Unspecified			Inhoek	78
Orthic	G horizon			Katspruit	80
Orthic	E	G horizon		Kroonstad	82
Orthic	E	Soft plinthic B		Longlands	84
Orthic	E	Hard plinthic B		Wasbank	86
Orthic	E	Yellow-brown apedal B		Constantia	88
Orthic	E	Podzol B + placic pan		Tsitsikamma	90
Orthic	E	Podzol B	Unconsolidated material with signs of wetness	Lamotte	92
Orthic	E	Podzol B	Unconsolidated material without signs of wetness	Concordia	94
Orthic	E	Podzol B	Saprolite	Houwhoek	96
Orthic	E	Prismacutanic B		Estcourt	98
Orthic	E	Pedocutanic B		Klapmuts	100
Orthic	E	Neocutanic B		Vilafontes	102
Orthic	E	Neocarbonate B		Kinkelbos	104
Orthic	E	Lithocutanic B		Cartref	106
Orthic	E	Unspecified		Fernwood	108
Orthic	Soft plinthic B			Westleigh	110
Orthic	Hard plinthic B			Dresden	112
Orthic	Yellow-brown apedal B	Soft plinthic B		Avalon	114
Orthic	Yellow-brown apedal B	Hard plinthic B		Glencoe	116
Orthic	Yellow-brown apedal B	Unspecified material with signs of wetness		Pinedene	118
Orthic	Yellow-brown apedal B	Red apedal B		Griffin	120
Orthic	Yellow-brown apedal B	Soft carbonate		Molopo	122
Orthic	Yellow-brown apedal B	Hardpan carbonate		Askham	124
Orthic	Yellow-brown apedal B	Unspecified		Clovelly	126

Adated from Soil Classification Working Group, 1991.

For each soil form there is a matrix table for family identification. The structure of the family tables is different from those of the series tables of the 1977 system. (Compare Tables 16.4 and 16.5 for the Glenrosa form.) The new family tables allow a very simple and easy stepwise "keying out" of families until the right one is found. New names were created for all of the families. None of the former series names was retained.

Table 16.4 Series Matrix Table for the Glenrosa Form* in "Soil Classification: A Binomial System for South Africa"

Clay content of A horizon (%)	Sand grade of A horizon	B horizon and directly below it non-calcareous		B horizon or directly below it calcareous	
0 – 6	Fine	Martindale	10	Malgas	20
	Medium	Oribi	11	Majeng	21
	Coarse	Paardeberg	12	Knapdaar	22
6 – 15	Fine	Kanonkop	13	Southfield	23
	Medium	Platt	14	Dunvegan	24
	Coarse	Glenrosa	15	Lomondo	25
15 – 35	Fine	Williamson	16	Lekfontein	26
	Medium	Trevanian	17	Dothole	27
	Coarse	Robmore	18	Achterdam	28
Above 35	Undifferentiated	Saitfaiths	19	Ponda	29

Diagnostic horizon sequence for the Glenrosa form: Orthic A/Lithocutanic B.

Adapted from MacVicar et al.,1977.

Table 16.5 Family Matrix for the Glenrosa form in "Soil Classification: A Taxonomic System for South Africa"

1000 A horizon not bleached		
1100 B1 horizon not hard		
1110 No signs of wetness in B1 horizon		
Noncalcareous in B horizon	1111	DUMISA
Calcareous B horizon	1112	KEURKLOOF
1120 Signs of wetness in B1 horizon		
Noncalcareous in B horizon	1121	KILSPINDIE
Calcareous in B horizon	1122	KAMMIEVLEI
1200 B1 horizon hard		
1210 No signs of wetness in B1 horizon		
Noncalcareous B horizon	1211	TSENDE
Calcareous B horizon	1212	BERGSIG
1220 Signs of wetness in B1 horizon		
Noncalcareous B horizon	1221	MARINGO
Calcareous B horizon	1222	WHEATLAND
2000 A horizon bleached		
2100 B1 horizon not hard		
2110 No signs of wetness in B1 horizon		
Noncalcareous B horizon	2111	OVERBERG
Calcareous B horizon	2112	INVERDOORN
2120 Signs of wetness in B1 horizon		
Noncalcareous B horizon	2121	BOTRIVIER
Calcareous B horizon	2122	TEVIOT
2200 B1 horizon hard		
2210 No signs of wetness in B1 horizon		
Noncalcareous B horizon	2211	BISHO
Calcareous B horizon	2212	KAKAMAS
2220 Signs of wetness in B1 horizon		
Noncalcareous in B horizon	2221	SOLITUDE
Calcareous in B horizon	2222	MERWESPONT

Adapted from Soil Classification Working Group, 1991.

Some of the definite improvements of the 1991 classification system over the 1977 system include the following:

- Good provision for the classification of most soils of the semiarid and arid regions. From a practical land-use point, the clear separation between soils with Soft carbonate and Hardpan carbonate horizons, respectively, is very good. There are still some outstanding issues regarding soils of the semiarid and arid regions, however, as will be pointed out later.
- The much finer subdivision of soils with Podzol B horizons. Although these soils are of limited geographic extent and of little agricultural value, they are very important for the forestry industry.
- Redefinition of the E horizon to force soil surveyors to determine their color in the dry state.
- Some important new family criteria, such as recognition of the yellow E horizons, provision for the separation of bleached A horizons from nonbleached A horizons, provision for thin Humic A horizons, and the distinction between structured B horizons with subangular/fine angular blocky structure and those with medium/coarse angular blocky structure.
- Improved structures for the key to soil forms and the family matrix tables to make them easier to use.

The 1991 system has a number of deficiencies, some minor and some serious. These are a few of these deficiencies:

- The change in definition of the Vertic A horizon. In the 1977 system, it had to have one or more of the following: clearly visible slickensides or cracks wider than 25 cm throughout at least half of the thickness of the horizon when it is dry, or self-mulching properties at the surface. In the 1991 system, it must have at least one of the following: clearly visible slickensides or a plasticity index greater than 32 (using the standard SA Casagrande cup to determine liquid limit) or greater than 36 (using the British Standard cone to determine liquid limit). According to P.A.L. Le Roux (University of the Free State, personal communication), there are many soils with higher plasticity values than these that are not Vertic A horizons. Van der Merwe (2000) found clearly that plasticity index cannot be used as criterion to differentiate between Vertic A and Melanic A horizons. She also showed that the statement in the discussion on Melanic A horizons in the 1991 classification book, that the difference between Melanic A and Vertic A horizons in regard to swell-shrink properties can mainly be ascribed to differences in clay mineralogy, is not true (Van der Merwe et al., 2001).
- The biggest deficiency of the 1991 system is that the definition of new soil series was left hanging in the air. For many soil forms this is not a problem, because their family definitions cover just about everything that is of practical importance that can be used in taxonomic classification. The remaining criteria that can be used in soil surveying are those that are normally used as phase criteria (Van Wambeke and Forbes, 1986). In these forms, there is actually no "space" for series below the soil family. Unfortunately, these include almost exclusively the "low quality" soil forms, which are unsuitable, or at best only marginally suitable for crop production, either rainfed or irrigated. The shallow Glenrosa form had 20 in series in the 1977 system, and has 16 well-defined families in the 1991 system (Tables 16.4 and 16.5). The problem lies with the "good quality" soils, for which the family definitions are very wide and do not allow good land suitability evaluation, especially with regard to management requirements, as will be pointed out. The Hutton form, for example, had 27 noncalcareous series in the 1977 system, but has only six families in the 1991 system (Tables 16.2 and 16.6).

ADVANCES IN SOIL CLASSIFICATION IN SOUTH AFRICA SINCE 1991

From 1991 to 1996, nothing much happened with regard to soil classification in South Africa. The main emphasis during this period was supposed to be on the development of new soil series. For various reasons, this became bogged down completely. One reason was that the majority feeling in 1983 was already that series should be defined as natural soil bodies regarding soil texture, and not as artificial entities, subdivided into arbitrary textural classes. But it was never

Table 16.6 Family Matrix for the Hutton Form in "Soil Classification: A Taxonomic System for South Africa"

1000 Dystrophic B1 horizon		
Nonluvic B1 horizon	1100	LILLIEBURN
Luvic B1 horizon	1200	KELVIN
2000 Mesotrophic B1 horizon		
Nonluvic B1 horizon	2100	HAYFIELD
Luvic B1 horizon	2200	SUURBEKOM
3000 Eutrophic B1 horizon		
Nonluvic B1 horizon	3100	STELLA
Luvic B1 horizon	3200	VENTERSDORP

Adapted from Soil Classification Working Group, 1991.

made clear what exactly was meant with natural bodies, and no guidelines were given about how to find and define these natural bodies. Furthermore, no institution/organization was given the responsibility or the funding to give the necessary guidance or to do the coordination and correlation needed to avoid chaos. The consequence is that from 1983 to 2001, no progress was made on this issue. At the 1992 Congress of the Soil Science Society of South Africa, Duvenhage et al., (1992) already highlighted the problem that there were no guidelines in the 1991 classification system to guide scientists on how to define series. They emphasized that the information provided by series was essential for any land user, and therefore guidelines for series differentiation should be a high priority.

Yet it was only at the meeting of the new soil classification working group in January 2001 (nearly 10 years later), under the agenda point "Soil series/Natural soil bodies," that the chairman (Lambrechts) "invited discussion by asking the question: How should this issue be tackled?" (Soil Classification Working Group, 2001). He further added: "The issue of unclear concepts is still not resolved. We have a responsibility to define what we want to establish locally as a series concept." It was decided that "We need a concept definition of what we want soil series to be in South Africa. It should be understandable by all and within the context of the Blue Book." (The "Blue Book" is the 1991 South African soil classification system.) Turner was given the task of writing the first draft, because he recently completed a comprehensive study on the recognition of natural soil bodies in two of South Africa's nine provinces under my supervision, studying data of over 4000 profiles in the process (Turner, 2000).

Ever since the natural soil body issue was raised in 1983, I could not agree with the idea that series should be defined in terms of natural soil bodies. I agree with the series concept explained in Soil Taxonomy (Soil Survey Staff, 1975):

> The function of the soil series is pragmatic, and differences within a family that are important to use the soil should be considered in classifying the soil series. Differences in particle size, texture, mineralogy, amount of organic matter, structure, and so on that are not family differentiae should be considered at series level.

Duvenhage et al. (1992) compared soils of the Avalon form (Orthic A/Yellow-brown apedal B/Soft plinthic B), one of the most important soils for rainfed cropping in the maize quadrangle, in two districts in the somewhat drier western part of the quadrangle. In the 1977 system, they were placed in two different series on the basis of clay content of the B21 (now B1) horizon, one falling in the 6–15% clay range, and the other in the 15–35% clay range. The management techniques applied by farmers differ in at least nine respects between the two series: crop selection, type of cultivation, depth of cultivation, tractor power used, planting density, planting depth, row width, weed control measures, and fertilizer placement. The fertilizer advisors, extension officers, farmers, etc., knew the names of the series and the management requirements of each. In the natural soil body approach, the big question will be how to handle this type of practical requirement in natural bodies with wide textural ranges, as in some cases found by Turner (2000). Another question is how confusing it is going to be to the users of soil information if series are not mutually exclusive and overlap partially in regard to something such as texture.

Reactivation of a soil classification working group was started toward the end of 1996 by the Soil Science Society of South Africa (SSSSA). This was stimulated by the WRB workshop in South Africa in September 1996. A small steering committee met in February 1997 to set things in motion. The two most important recommendations by the steering committee from an international perspective are these:

- That the other countries from the Southern African Development Community (SADC) should also be included, and that the activities of the working group should not be limited to South Africa. Unfortunately this has not yet materialized.
- That greater liaison and coordination with the World Reference Base for Soil Resources (WRB) Working Group must be encouraged.

It was emphasized that the working group must strive to satisfy the needs of our clients (land users, students, consultants, etc.) and not alienate them. The working group met just after the SSSSA congress in July 1999 to get some points on the agenda, and held a fruitful workshop just before the SSSSA congress in January 2001. Progress is slow because there is no funding for the working group, and no institution has been given the responsibility to drive its activities. The 2001 workshop was encouraging because a number of definite proposals were discussed in depth. These included the creation of new forms for soils with new diagnostic horizon sequences that have been found. At last someone has been assigned the task of coming up with a definite proposal for a concept definition for what we want soil series to be, as previously discussed. From an international perspective, the two most relevant points handled at the workshop are:

- The definition and classification of soils with vertic properties. The WRB's decision to define the vertic horizon as a subsurface horizon caused quite a stir, and the question was whether South Africa should go the same way, instead of having a Vertic A horizon. In South Africa, there are many Prismacutanic B and Pedocutanic B horizons with vertic properties, i.e., strong swell-shrink properties and perfect slickensides. The vast majority of soils with these horizons have relatively sandy topsoils with no swell-shrink properties and no structure. It was decided to keep the Vertic A horizon, because the classical Vertisol is not only a soil with a clayey topsoil, but which is distinguished from other structured clayey topsoils (Melanic A horizons) on the basis of its strong swell-shrink properties. The swell-shrink properties of the topsoil cause it to have specific land-use problems and to have special management requirements. From an engineering point in particular, it is important to map out soils with vertic subsoils. Thus it was decided to also recognize vertic properties in subsoils, and to probably make provision for these at family level in the soils with Prismacutanic B and Pedocutanic B and G horizons.
- How *gypsic* and *salic* diagnostic horizons/properties should be handled in the classification system. No provision is made for these in the 1991 classification system. The need for attention to them was expressed by WRB Working Group members during the 1996 workshop in South Africa. Soils with gypsiferous horizons are found in the extreme aridic western part of the country (Ellis, 1988). Salic soils are found in various parts of the semiarid and arid regions. Ellis was given the task of drawing up proposals for their handling in the classification system. My proposal was that the handling of these in the WRB system should be used as a starting point, and adapted for South African conditions.

THE ROAD AHEAD

It seems that the approach and structure for the definition of diagnostic horizons and the classification of soil forms seem well accepted and do not have to be changed. New diagnostic horizons (e.g., gypsic and salic) and new forms can be added as new information becomes available. Minor adjustments can also be made as required. The main challenge will be to solve the deadlock regarding series definition as soon as possible so that useful series can be defined again, especially for the higher potential soils, as was the case in the old 1977 classification. This is absolutely essential to ensure efficient land suitability evaluation and sustainable use of the country's limited soil resources.

It is hoped that the proposal to include the whole SADC region in the further development of the classification system will as soon as possible become a reality. This would greatly enhance efficient cross-border technology transfer among the different countries. Optimum, sustainable land use would be promoted by this. It is foreseen that there will be closer liaison between the South African soil classification working group and the WRB working group, and that greater efforts will be made to fit in with the WRB system. During the January 2001 workshop of the South African working group, it was striking how much reference was made to relevant aspects in the WRB system by various members of the working group. The 1996 WRB workshop in South Africa brought about a big turnaround regarding the attitude of the South African pedologists. For the first time, they are really looking at interfacing with an international system while improving their own system. Much of this is due to the open-minded approach of the WRB working group members during the 1996 South African workshop. They acknowledged that there are important soils that have been continuously ignored in international systems, and started making provision for them in the WRB system. These include the following, *inter alia*:

- Acknowledging that "soft plinthite," i.e., "non-hardening plinthite," actually exists and making provision for it as "paraplinthite" (WRB Working Group, 1998)
- Making provision for the "yellow E horizons"
- Making provision for the "bleached" A horizons.
- Creating the "Durisol" reference group, for the soils of extreme aridic regions with duripans, acknowledging that these duripans are associated with extreme aridity and not with volcanic activity

Unfortunately, reaching the above ideals is hamstrung by lack of funding of systematic soil classification programs and the pedological research required to support them. Funding problems increased when the Agricultural Research Council (ARC) was formed some years ago, and all the specialist research institutes of the Department of Agriculture were transferred to it. These included the Institute for Soil Climate and Water, the country's national soil research institute, which earlier formed the backbone for the development of the 1977 and 1991 classification systems. Now the institute is simply struggling for survival.

The situation regarding soil classification in the country is well illustrated by the fact that Jan Lambrechts, present Chairman of the Soil Classification Working Group, gave me the following three points in response to my request for a list of the things that he would like me to highlight in this paper:

- Such a working group can be active and productive only if it has official/statutory status, e.g., a project registered and funded by the ARC. (Note by MCL: Unfortunately, the ARC is not a funding body and is battling to find funds from outside, especially for something such as a national soil classification system.)
- A "loose" group of volunteers, as is the case now, has no obligation to complete/finalize any task.
- The arid regions are neglected because there is no "loose" funding available to conduct intensive pedological work in such areas.

REFERENCES

Beater, B.E. 1944. The soils of the Sugar Belt. A classification and a review. *Proc. S. Afr. Sugar Technol. Assoc.* 18:25–37.

Beater, B.E. 1957. *Soils of the Sugar Belt.I. Natal North Coast.* Oxford University Press, Cape Town.

Beater, B.E. 1959. *Soils of the Sugar Belt. II. Natal South Coast.* Oxford University Press, Cape Town.

Beater, B.E. 1962. Soils of the Sugar Belt. III. Zululand. Oxford University Press, Cape Town.

De Villiers, J.M. 1962. A study of soil formation in Natal. Ph.D. thesis, University of Natal.

Duvenhage A.J., Laker, M.C., and Turner, D.P. 1992. 'n Vergelyking tussen twee gronde van die Avalon grondvorm met die oog op seriedifferensiasie. *(A comparison between two soils of the Avalon soil form with a view to series differentiation),* Proc. 17th Congress of the Soil Science Society of South Africa, 26.1–26.6.

Ellis, F. 1988. Gronde van die Karoo. *(Soils of the Karoo)* Ph.D. thesis, University of Stellenbosch.

FAO. 1988. FAO-UNESCO Soil Map of the World. Revised legend. FAO World Soil Resources Report No. 60. FAO, Rome.

Laker, M.C. 1997. Definition and classification of Solonetz. *Commun. Austrian Soil Sci. Soc.* 55:213–216.

Laker, M.C. 2000. Soil resources: distribution, utilization, and degradation, in R. Fox and K. Rowntree, Eds. *The Geography of South Africa in a Changing World.* Oxford University Press, Cape Town.

Le Roux, P.A.L. 1996. Die aard, verpreiding en genese van geselekteerde redoksmorfe gronde in Suid-Afrika. *(The nature, distribution and genesis of selected redoxmorphic soils in South Africa)* Ph.D. thesis, University of the Orange Free State.

MacVicar, C.N. 1962. Soil studies in the Tugela Basin. Ph.D. thesis, University of Natal.

MacVicar, C.N. 1978. Advances in soil classification and genesis in Southern Africa. Proc. 8th National Congress of the Soil Science Society of South Africa. Tech. Comm. 165:22–40. Dept. Agric. Tech. Services, Pretoria.

MacVicar, C.N., De Villiers, J.M., Loxton, R.F., Verster, E., Lambrechts, J.J.N., Merryweather, F.R., Le Roux, J., Van Rooyen, T.H., and Harmse, H.J. von M. 1977. Soil Classification: A Binomial System for South Africa. Dept. Agric. Tech. Services, Pretoria.

MacVicar, C.N., Loxton, R.F., and Van der Eyk, J.J. 1965. South African soil series. Part I. Definitions and key. Dept. Agric. Tech. Services, Pretoria.

Nyamapfene, K. 1991. Soils of Zimbabwe. Nehanda Publishers, Harare.

Soil Classification Working Group. 1991. Soil Classification: A Taxonomic System for South Africa. Dept. Agric. Dev., Pretoria.

Soil Classification Working Group. 2001. Unpublished minutes of workshop held in Pretoria, January 2001. Soil Sci. Soc. S. Afr., Pretoria.

Soil Survey Staff. 1960. Soil Classification. A Comprehensive System. 7th Approximation. USDA SCS, Washington, DC.

Soil Survey Staff. 1975. Soil Taxonomy. A Basic System of Soil Classification for Making and Interpreting Soil Surveys. Agriculture Handbook No. 436. USDA SCS, Washington, DC.

Soil Survey Staff. 1999. Soil Taxonomy. A Basic System of Soil Classification for Making and Interpreting Soil Surveys. 2nd Edition. Agriculture Handbook No. 436. USDA NRCS, Washington, DC.

Turner, D.P. 2000. Soils of KwaZulu-Natal and Mpumalanga: Recognition of Natural Soil Bodies. Ph.D. thesis, University Pretoria.

Van der Eyk, J.J., MacVicar, C.N., and De Villiers, J.M. 1969. Soils of the Tugela Basin. A study in subtropical Africa. Natal Town and Regional Planning Reports Vol. 15. Natal Town and Regional Planning Commission, Pietermaritzburg.

Van der Merwe, C.R. 1941. Soil groups and subgroups of South Africa. *Science Bull. 231, Chemistry Series No. 165.* Dept. Agric. and Forestry, Pretoria.

Van der Merwe, C.R. 1962. Soil groups and subgroups of South Africa. *Science Bull. 356. Chemistry Series No. 165.* Dept. Agric. Tech. Services, Pretoria.

Van der Merwe, G.M.E. 2000. Melanic soils in South Africa: Compositional characteristics and parameters that govern their formation. M.Sc. dissertation, University Pretoria.

Van der Merwe, G.M.E., Laker, M.C., and Bühmann, C. 2002. Clay mineral associations in melanic soils of South Africa. *Aust. J. Soil Res.* 40:115–126.

Van Wambeke, A. and Forbes, T. 1986. Guidelines for using Soil Taxonomy in the names of soil map units. SMSS Tech. Monogr. No. 10. USDA SCS SMSS, Washington, DC.

WRB Working Group. 1998. World Reference Base for Soil Resources. FAO, Rome.

CHAPTER 17

Soil Taxonomy and Soil Survey

C.A. Ditzler, R.J. Engel, and Robert J. Ahrens

CONTENTS

ABSTRACT

Soil taxonomy was developed primarily for the practical purpose of supporting the National Cooperative Soil Survey (NCSS) Program in the United States. It was adopted nearly 40 years ago by all of the NCSS partners, and is recognized as one of our most important standards. The adoption of the classification system had several important impacts on the soil survey program. The emphasis on observable diagnostic horizons and features for defining classes tended to make all competent soil scientists, regardless of experience and rank, equally capable of accurately and consistently classifying soils. By focusing attention on qualitative class differentiae, the quantity of field data collection has increased and the quality has improved. Property ranges of soil series and their geographic distribution have generally been narrowed over time, allowing us to make more precise interpretations. Soil Taxonomy has benefited the soil correlation process by grouping the nearly 22,000 series currently established in the United States in ways that allow us to efficiently compare and differentiate competing soil series, and coordinate their use among survey areas.

One area that has presented difficulty and confusion from the beginning of Soil Taxonomy's use has been reconciling the difference between map units and taxonomic units. Our soil maps are an attempt to depict our understanding of how natural soil bodies occur within the landscape. The delineated boundaries reflect the constraints of map scale and the conceptual landscape model of the surveyor (Hudson, 1992). We do not attempt to map taxonomic concepts. Rather, we use the taxonomy to classify the soil bodies we have mapped. Concepts such as pedons, polypedons, series, taxadjuncts, ranges in characteristics, map unit components, similar soils, dissimilar soils, and multi-taxa map units take significant effort to be mastered by soil surveyors, and are little understood outside of our profession. We must always remember that the taxonomy is simply a tool to help us organize our knowledge and transfer our experience and technology from place to place in the landscape. Our primary goal is to help individuals and society understand the soil resource by showing them where the soils are and interpreting, in as simple a manner as possible, their suitability and limitations for intended uses.

INTRODUCTION

The coupling of Soil Science, Soil Classification, and Soil Survey provides a powerful resource for the benefit of humankind. Soil science provides the foundation for our understanding of the physical, chemical, and biological properties of the soils we depend on to grow crops, sustain forests and grasslands, and support our homes and society's structures. Soil surveys put our knowledge into a spatial context so that we know the geographic distribution of the soils. Soil classification helps to organize our knowledge, facilitates the transfer of experience and technology from one place to another, and helps us to compare soil properties. It provides a link between soil science and soil survey.

A major force driving the development of Soil Taxonomy was the practical need to support the National Cooperative Soil Survey (NCSS). Dr. Guy Smith wrote

> "The system is being developed by the Soil Survey Staff to facilitate the soil survey of the United States, a cooperative work involving more than fourteen hundred soil scientists working for more than fifty different institutions. Soil maps are being made at a rate of more than sixty million acres each year. … The classification, therefore, is being developed to serve a program that has a practical objective" (Smith, 1963, p. 6).

The leaders of the NCSS program recognized the need for a classification system that could be applied by a cadre of soil scientists with varying education and experience, in a uniform manner. To do this, they devised a system that used objective criteria that focused on the properties of the soil itself, rather than on theories of its genesis, as was required by the previous classification system (Smith, 1963; Cline, 1963). One of the more ingenious devices of the system is the use of observable, quantitative diagnostic horizons and features, which reflect our understanding about soil genesis. By recognizing these diagnostic horizons and features, and by observing other key differentiating characteristics, a competent soil scientist is able to objectively observe the soil and place it into appropriate taxa.

Since 1965, Soil Taxonomy has been a standard recognized by all members of the NCSS. It has been taught in our universities, and is recognized by many scientific journals as the appropriate way to communicate about soils in scientific research.

IMPACT OF THE USE OF QUANTITATIVE CRITERIA

The adoption of Soil Taxonomy impacted soil survey operations in a number of ways. These impacts were primarily the result of the shift from a qualitative to a quantitative emphasis in observing, describing, and classifying soils (Cline, 1980). Before the adoption of Soil Taxonomy, the concept of each category in the classification scheme focused on a central concept, or typifying individual. The limits of the classes were not well defined, thus there was much latitude for soil scientists to make classification decisions, based on their judgment as to the degree of similarity to one central concept as opposed to another. Soil Taxonomy put a greater focus on observable and measurable class limits defining diagnostic horizons, features, and other differentiating characteristics. This shift from a qualitative to a quantitative emphasis had several practical effects.

Parity and Consistency in Classifying Soils

The use of Soil Taxonomy has effectively placed all soil scientists on an equal footing with regard to their ability to classify a soil. Smith (1963) explained that the choice of differentiating criteria was intended to group soils with similar genesis, but genesis itself was not in the definitions. Instead, it was one step removed from the definitions. This allowed soil scientists to focus on soil properties, and to classify soils rather than processes. This leveled the playing field for all soil scientists, regardless of status within the NCSS Program. No longer did one have to theorize about the genesis of the soil in order to classify it, something that must have been a serious problem when there were either competing theories regarding a soil's genesis, or when the genesis was simply not known by those attempting to classify the soil. With a good morphological description and key laboratory data, a junior field soil scientist was on equal footing with the most senior correlator. As a result, Soil Taxonomy could be applied universally and consistently by any competent soil scientist.

Quantity and Quality of Data Collection

The attention of field scientists became focused on those characteristics selected as class differentiae, thus influencing the kind of information being recorded. It increased the quantity and quality of data gathering by encouraging the recording of greater detail in soil descriptions, and by encouraging laboratory analyses to document properties not readily observable in the field but required for determining taxonomic placement. The kind of information obtained was directly influenced by the need to determine the presence of diagnostic horizons and features, and to document other differentiating characteristics used as class limits. A potential downside to this focus on the properties used as differentiae was recorded in the report of the committee on the "New Soil Classification System" to the 1963 NCSS National Work Planning Conference (Soil Survey Division, 1999, p. 27–28). Guy Smith is reported as commenting

> As Dr. Simonson has pointed out again and again, it is possible to become a prisoner of one's classification. The differentiae that are used in the 7th Approximation will get very great emphasis and those that have not been used can easily be overlooked. Yet, in defining a soil, one must think of all of its properties and not just the ones that have been used in differentiae. We have been able to use only a very few soil properties to define our taxa. Soils have a great many other properties, and we must be on our guard to be sure that these others, which actually may be more important from some viewpoints than those we have used, do not get overlooked.

This caution is as valid today as it was then. As observers of soils, we tend to see what we are taught to see, and may not strive to observe more once we know enough to complete the task of classifying a soil. When the need for change in the system is recognized, however, adding or changing differentiae at the higher categories can be (and has been) done, but it requires that a proposal be developed, reviewed widely, and approved, and this can take considerable time. However, at the series level of classification, there are no predetermined differentiating properties, allowing considerable freedom to define a series based on virtually any soil property observed within the series control section. This approach, however, is only useful for separating soils that the classification system has grouped into the same family. It cannot be used to group soils that taxonomy has set apart at a higher level.

Refinement of Soil Series Concepts

Another consequence of the shift to a quantitative system was a change in the range of characteristics, geographic distribution, and precision of interpretation for some series. The soil series evolved from an early concept based on parent rock and geographic province (USDA, 1903), to one that included the recognition of common genetic horizons and morphology (Kellogg, 1937), and finally to becoming the lowest level of the classification system, thus sharing limits with each of the higher taxa to which it belongs (Soil Survey Staff, 1975). With the use of class limits for particle size, mineralogy, temperature, and other family criteria, along with soil property limits in the higher categories (including moisture class), some series ranges had to be trimmed and geographic distributions limited. As a result, some new series were established. Bailey (1978) described the impact of the implementation of Soil Taxonomy on the Miami series. Before the adoption of Soil Taxonomy in 1962, the concept of the Miami series included pedons with subsoil clay contents straddling the family particle-size class limit of 35%. Since a series could not range beyond the limit of the family, the Miamian series (fine, mixed, mesic, Typic Hapludalfs) was split out of the Miami series (fine-loamy, mixed, mesic, Typic Hapludalfs) in 1969. As Soil Taxonomy continues to be amended by revising or adding new differentiating criteria (Soil Survey Staff, 1999), series concepts continue to be refined. For example, in 1992 the International Committee on Aquic Soils introduced the Oxyaquic subgroup within several great groups to recognize soils with seasonally high water tables within 100 cm, but not meeting the criteria for higher aquic taxa. In 1996, the International Committee on Families introduced additional family classes for cation-exchange activity classes (CEC). Today, the family to which the Miami soils were originally classified has been subdivided into three (semiactive, active, and superactive) of the four CEC classes, and the new Oxyaquic subgroup. These subdivisions resulted in further narrowing the range of the series and recognition of six families, where previously there had been one. Today, the Miami series is a member of the family of fine-loamy, mixed, active, mesic Oxyaquic Hapludalfs.

The use of quantitative criteria defining class limits, with the resultant refinement of series ranges, has allowed for the development of increasingly precise interpretations. For example, the introduction of the Oxyaquic subgroup to separate soils with water tables of short duration within 100 cm of the soil surface (such as Miami) from otherwise similar soils in the typic subgroup of Hapludalfs (such as Amanda) allows us to better describe the suitability of these classes of soils for homes with basements or on-site sewage disposal. We must remember, however, that providing interpretations for soil series (a conceptual taxonomic class) is not the same as interpreting a soil map unit. Map units contain not only the soil(s) for which they are named, but also similar and dissimilar inclusions (Soil Survey Division Staff, 1993).

Refinements to the concept and distribution of individual soil series, while providing the benefit of improved interpretations, also present some problems. Coordination between survey areas mapped at different times requires careful correlation to achieve adequate joining of yesterday's and today's series concepts on each side of the survey boundary. This has been especially evident in recent efforts to accelerate the digitizing of soil surveys that were mapped over a period of 30 years or more in the United States. Unless adequately coordinated and joined, GIS users cannot easily achieve effective analysis across survey area boundaries. Changing series concepts are also a burden for others who use soil survey information, because they must periodically learn new names for the soils they have become familiar with. They don't understand why the soils "change."

Improvement in Soil Correlation

Soil correlation is the process the NCSS uses to define, map, name, classify, and interpret soils and to join soil map units consistently within and among soil survey areas (Simonson, 1963; Soil Survey Staff, 2001a). It is our most important quality control process. The need to provide comprehensive field descriptions and associated laboratory data to adequately classify the soils tended to result in more uniform descriptions than in the earlier periods of soil survey (Simonson, 1963). One manifestation of this improvement in the detail and precision of soil descriptions was the development and adoption of the standard form (SCS-SOI-232) for recording soil descriptions. Although this was officially a form used within the Soil Conservation Service, it was commonly used or adapted by other NCSS partners. These standard field description templates had the beneficial effect of reminding soil scientists to record many of the important properties of the soil, such as moist and dry color, percentage of coarse fragments, mottles, ped coatings, pH, horizon boundaries, etc.; and to identify diagnostic horizons and features that are present. However, Dr. Simonson (then Director, Soil Classification and Correlation in the Soil Conservation Service) expressed the concern that while it tended to set a standard for the minimum set of data to be recorded, it may also have had the unintended consequence of setting the maximum (Simonson, 1987).

Another important way that Soil Taxonomy has improved the soil correlation process is by grouping soils with similar properties into classes, thereby making it easier to understand how they relate to one another. The number of soil series recognized in the United States has grown from about 10,500 in 1975 to nearly 22,000 today, clearly too many for the human mind to organize and remember. With the aid of the classification to group the soils with similar properties, and computer technology (Soil Survey Staff, 2001c) to query our soil database and quickly deliver series descriptions for comparison, it is fairly easy for today's soil scientists to find the names of all the soil series in a given class, or closely similar class, and to coordinate the naming and interpreting of soil map units from one survey project to another.

Some Problems with Quantitative Limits

The use of quantitative class limits has also had some drawbacks. Webster (1968) expressed several objections regarding the system. A major problem with the approach is that while the class limits are fixed, there is inherent uncertainty in the observation and measurement of many of the properties, such as recording color by comparison to color charts, or determining rupture resistance class by attempting to crush peds in the field. Laboratory measurement of properties such as particle size, base saturation percentage, or organic matter content also contain inherent measurement error. As a result of these uncertainties, data obtained by repeating a field observation or laboratory analysis could result in a different classification for the soil. Also, with no provision for overlapping of class limits, some soil bodies that have natural distributions of one or more properties that straddle artificial class boundaries will be forced into separate taxa, even though they clearly form a natural cluster in their landscape setting. As Soil Taxonomy has evolved, it has become increasingly dependent on the need for laboratory data for supplying the required quantitative values that successfully classify a soil (e.g., spodic materials, andic soil properties, cation exchange activity classes, etc.). While this reliance on laboratory measurements can be a positive addition to the quality of the data in the soil survey, obtaining it can be time-consuming and expensive, thus hindering our ability to classify pedons with confidence at the time they are being observed. Also, it is an impediment to the effective use of Soil Taxonomy in places where analytical laboratory services are not readily available.

TAXONOMIC UNITS AND MAP UNITS

Cline (1977, p. 253), in an attempt to anticipate future developments in soil survey, wrote

> At the lowest level of the system, we will have to acknowledge the differences between taxonomic soil series and mapping units that bear the same name and will probably have to rectify the confusion

this causes. It is conceivable that soil families could become the lowest category of taxonomy, but some ingenious person may find a better solution.

Soil Individuals

The concepts of the pedon and polypedon were introduced by Johnson (1963) to relate map units depicted in soil surveys to the new classification system. These were intended to bridge the gap between the conceptual taxonomic categories of soil series with the soil bodies delineated on maps. The pedon was conceived to be the basic soil unit consisting of a volume just large enough to depict the horizons present and their relationships to one another. The pedon is likened to the unit cell of a crystal (Soil Survey Staff, 1975). It is generally considered to be the entity that we describe and sample in the field. Inherent weaknesses of the pedon concept are that while its relatively small size is convenient for study, it is too small to exhibit the full range of properties for a series, and it cannot show the nature of the boundary with adjacent soils in the landscape, the surface shape, or other site characteristics of the soil. Additionally, rarely if ever have soil scientists truly identified, described, and sampled the three-dimensional pedon in the field. It has been argued that we more commonly describe and sample soil profiles rather than pedons (Holmgren, 1988).

The concept of the polypedon was introduced to overcome some of the weaknesses inherent in the pedon concept. Johnson (1963, p. 215) described it as a real soil body consisting of contiguous pedons "all falling within the range of a single soil series." It was conceived to provide a link between pedons and the taxonomy on one hand, and to relate taxonomic units to map units on the other hand. The polypedon was considered to be the individual we classify, and, in Johnson's words, "comparable to individual pine trees, individual fish, and individual men."

These concepts presented practical difficulties. First, the relationship between real soil bodies (i.e., polypedons) and the conceptual taxonomic class of the soil series presented a serious dilemma. Soil properties within a three-dimensional soil body are not mindful of arbitrary class limits of our taxonomy. The report of the Committee on the Application of the New Classification System recorded in the 1965 NCSS Conference Proceedings (Soil Survey Division, 1999) includes a discussion of "guidelines for allowable tolerances in the stretching of family class limits by series class limits." The debate was whether the range of characteristics for a series must be within the limits of the family. Alternatively, only the typical pedon itself would be required to fit within the family, thus allowing the range of characteristics to stretch beyond the family range. It was agreed that series, as the lowest level of the taxonomy, must have ranges no wider than the family to which they belong. Having made this decision, however, the NCSS leaders wanted to avoid the possible proliferation of new series simply to cover pedons and polypedons slightly beyond the rigid family class limits. They decided to study the problem further. Two years later at the 1967 NCSS conference (Soil Survey Division, 1999) there was discussion of *taxonomic inclusions, plesioseries,* and *taxal deviants* as devices to classify pedons close to, but outside the range of, a given series. These concepts evolved into the *taxadjunct* and *variant*. While we no longer correlate variants (we establish new series for these), we frequently use the taxadjunct to this day when the pedon used to typify a series in a survey area is outside the limits of the family to which the series belongs.

The difficulties presented by the concept of pedons and polypedons, and the constraints on allowable ranges for soil series, have been debated ever since that time. Soon after Johnson's introduction of the pedon and polypedon, Knox (1965, p. 83) pointed out that polypedons have no real existence apart from series concepts. He stated that "*their significance as individuals seems less than the significance of individual pine trees, individual fish, and individual men.*" Webster (1968) suggested that in order to define and maintain effective series concepts, there is a need for an "unconformity" between the series and higher taxonomic classes. Guthrie (1982) recognized that the use of the correlation devices of taxadjuncts and variants are a manifestation of the problem. He suggested two alternatives when a pedon is chosen to typify the central concept of a series that is slightly beyond the concept of a taxonomic class. Either the soil can be classified at the lowest taxonomic level that it fits, or it can be correlated to the closest available series and any properties that are outside the range of the selected series can simply be disregarded in the taxonomic description (presumably to be handled as inclusions in mapping). In most instances, soil survey practitioners have followed the second alternative, although not always. For example, the soil survey of the Flathead National Forest (Martinson and Basko, 1998) correlated the soils at the family level, and constructed map unit names using phases of higher categories for brevity. Two examples are *Fluvents, alluvial fans* and *Typic Eutroboralfs, silty till substratum, hilly.* Despite the passage of 18 years since the decision was made to restrict series ranges to the limits of higher categories, Alexander (1983) and Borst (1983), each reacted to Guthrie by arguing that series should be allowed to encompass a range reflecting the properties observed in the field, even if they cross higher class limits. Nettleton et al. (1991), describing what they called the "taxadjunct problem," reported that about half of the pedons analyzed by the National Soil Survey Laboratory have one or more properties outside the family limits for the series identified in the field. They suggested that we classify the central concept of the soil and allow the described range in characteristics to cross the boundaries of higher taxa (one of the original alternatives discussed 28 years earlier at the NCSS conference). Holmgren (1988) suggested that an operational definition should have been developed to depict the procedure for describing a pedon in a spatial context. He contends it was a mistake to develop an operational definition for the pedon itself. While he did not propose an operational definition, it seems that he envisioned a set of multiple observations, related to each other geographically (similar to "satellite sampling"), to characterize the pedon. In this way, there would be multiple possible outcomes that could be realized as one observes a pedon repeatedly. A polypedon, then, would have some "probability" or "expectation" of meeting a particular classification.

Today we still talk of pedons, although rarely do we truly identify, describe, and sample this three-dimensional body. More often than not, the pedon becomes a spoil pile next to the pit and we describe a two-dimensional soil profile. Perhaps we need a procedure to "dissect a pedon" rather than simply "digging a pit." The polypedon concept has been largely ignored because it is too difficult to locate it's boundary in the field, and it's very concept is contradictory and relies on circular reasoning (Soil Survey Division Staff, 1993).

Taxonomic Purity of Map Units

A number of pedologists have looked at the taxadjunct problem in the context of the composition of map unit delineations to estimate the proportion of pedons not fitting within the taxonomic class for which the map unit is named. McCormack and Wilding (1969) studied a portion of northwestern Ohio that had been mapped before the introduction of Soil Taxonomy. It was considered to have very complex soil patterns, resulting in a high degree of variability. Overall, they found that 74% of the more than 200 pedons observed fit the expected Order, while just 17% fit the expected Family and Series. An additional 26%, however, were close enough to be considered taxadjuncts. Despite the low taxonomic purity, they concluded that only a few of the delineations observed were improperly mapped.

Nordt et al. (1991) studied the taxonomic makeup of four map units in east Texas. The soils observed were Crockett (fine, montmorillonitic, thermic Udertic Paleustalfs), Rader (fine-loamy, mixed, thermic Aquic Arenic Paleustalf), Robco (loamy, mixed, thermic Aquic Arenic Paleustalfs), and Spiller (fine, mixed, thermic Udic Paleustalfs). The taxonomic purity of the map units, following a pattern similar to that found by McCormack and Wilding (1969), is highest at the Order, Suborder, and Great group levels, and drops with progressively lower taxa, with the percent of pedons classifying in the named series ranging from just 11% for Rader to 49% for Crockett. In addition to the taxonomic purity, they also considered the interpretive purity of the map units. When soils considered to be similar to the named series were added, the percentage of "interpretive purity" was estimated to be 48 for Rader, 52 for Robco, 81 for Spiller, and 86 for Crockett. The Rader and Robco map units, therefore, do not meet the definition of a consociation type of map unit as defined in the Soil Survey Manual (Soil Survey Division Staff, 1993), and the authors recommended that a multitaxa unit be correlated. If done, the two map units would presumably have significantly higher taxonomic and interpretive purities.

Other authors have conducted assessments of the taxonomic purity of map units, and have reported results similar to the two studies described above. Edmonds et al. (1982) suggested that taxonomic criteria that require laboratory methods to be determined often cannot be reliably inferred in the field, and therefore are not easily mapped. They showed that for an area studied in Virginia, variability that gives rise to mixed and oxidic mineralogy classes could be found within a distance of 7 m. In this landscape, then, not only is it impractical to map this class difference at scales commonly used in soil survey, soil volumes as small as individual pedons will likely contain properties of both classes. Ransom et al. (1981), Edmonds et al. (1985), and Edmonds and Lentner (1986) all point out that because soil series have property ranges limited by taxonomic boundaries, and interpretations have generally been generated based on series properties, our ability to interpret map units adequately is hampered. They suggest that we need to be able to record property ranges for map units themselves, and then interpret the map units.

Today, by taking advantage of improved computer technology, we have begun to move from interpreting soil series in our survey reports to interpreting components of map units. Soil series are conceptual soil classes defined by limits of key diagnostic properties. Components of map units are natural bodies of soils in a particular landscape. Whereas the soil series is the lowest class in our taxonomy and is constrained by higher class limits, a soil component is that portion of a map unit either fitting within the concept of the series or is close enough in its properties to interpret in essentially the same way for most uses. In the past, interpretations were generated and stored for each series nationally, and then used locally for the soil survey. While they could be adjusted, this was a cumbersome process. Today we can store soil property values for individual map unit components and generate interpretations based on the property ranges recorded for that specific soil survey area. While our Official Series Descriptions remain the lowest level of Soil Taxonomy, and are therefore required to have ranges extending no further than higher taxa limits, the National Soil Information System (NASIS) allows soil scientists to record soil property values for components of map units that more closely reflect ranges of properties in the field. Low and high values depicting the range, along with a "representative value" (RV), are recorded. The RV is required to be within the range of the taxonomic class (generally series), but the low and/or high value may *"extend beyond the established limits of the taxon from which the component gets its name, but only to the extent that interpretations do not change"* (Soil Survey Staff, 2001b). Thus in effect, the soil survey has developed a procedure to effectively accomplish what was debated by the NCSS leaders at the 1963 conference. Rather than allowing the series to "stretch family class limits," we have allowed the map unit component to stretch the series class limits. Interpretive criteria are then applied to the component data.

REFERENCES

Alexander, E.B. 1983. Comment on "The relationship between Soil Taxonomy and soil mapping." *Soil Surv. Horiz.* 24:15–16.

Bailey, G.D. 1978. A brief history of the Miami soil series. *Soil Surv. Horiz.* 19:9–14.

Borst, G. 1983. Taxonomy and soil mapping units: A reply to R.L. Guthrie., *Soil Surv. Horiz.* 23:16–19.

Cline, M.G. 1963. Logic of the new system of soil classification. *Soil Sci.* 96:17–22.

Cline, M.G. 1977. Historical highlights in soil genesis, morphology, and classification. *Soil Sci. Soc. Am. Proc.* 41:250–254.

Cline, M.G. 1980. Experience with soil taxonomy of the United States, *Adv. Agron.* 33:193–226.

Edmonds, W.J., Iyengar, S.S., Zelazny, L.W., Lentner, M., and Peacock, C.D. 1982. Variability in Family differentia of soils in a second-order soil survey mapping unit. *Soil Sci. Soc. Am. J.* 46:88–93.

Edmonds, W.J., Baker, J.C., and Simpson, T.W. 1985. Variance and scale influences on classifying and interpreting soil map units. *Soil Sci. Soc. Am. J.* 49:57–961.

Edmonds, W.J. and Lentner, M. 1986. Statistical evaluation of the taxonomic composition of three soil map units in Virginia. *Soil Sci. Soc. Am. J.* 50:997–1001.

Guthrie, R.L. 1982. The relationship between soil taxonomy and soil mapping. *Soil Surv. Horiz.* 3:5–9.

Holmgren, G.S. 1988. The point representation of soil. *Soil Sci. Soc. Am. J.* 52:712–716.

Hudson, B.D. 1992. The soil survey as a paradigm-based science. *Soil Sci. Soc. Am. J.* 56:836–841.

Johnson, W.M. 1963. The pedon and the polypedon. *Soil Sci. Soc. Am. Proc.* 27:212–215.

Kellogg, C.E. 1937. Soil Survey Manual. USDA Misc. Publ. 274, Washington, DC.

Knox, E.G., 1965. Soil individuals and soil classification. *Soil Sci. Soc. Am. Proc.* 29:79–83.

Martinson, A.H. and Basko, W.J. 1998. Soil Survey of Flathead National Forest Area, Montana. USDA Forest Service and Natural Resources Conservation Service.

McCormack, D.E. and Wilding, L.P. 1969. Variation of soil properties within mapping units of soils with contrasting substrata in Northwestern Ohio. *Soil Sci. Soc. Am. Proc.* 33:587–593.

Nettleton, W.D., Brasher, B.R., and Borst, G. 1991. The taxadjunct problem. *Soil Sci. Soc. Am. J.* 55:421–427.

Nordt, L.C., Jacob, J.S., and Wilding, L.P. 1991. Quantifying map unit composition for quality control in soil survey, in M.J. Mausbach and L.P. Wilding, Eds. Spatial Variability of Soils and Landforms. SSSA Spec. Publ. No. 28, Soil Sci. Soc. Am., Inc., Madison, WI.

Ransom, M.D., Phillips, W.W., and Rutledge, E.M. 1981. Suitability for septic tank filter fields and taxonomic composition of three soil mapping units in Arkansas. *Soil Sci. Soc. Am. J.* 45:357–361.

Simonson, R.W. 1963. Soil correlation and the new classification system. *Soil Sci.* 96:23–30.

Simonson, R.W. 1987. Historical aspects of soil survey and soil classification. Part VI. 1951–1960. *Soil Surv. Horiz.* 28:39–46.

Smith, G.D. 1963. Objectives and basic assumptions of the new soil classification system. *Soil Sci.* 96:6–16.

Soil Survey Division. 1999. Proceedings of the National Cooperative Soil Survey conferences and other soil survey documents 1963–1997. CD-ROM computer file. National Soil Survey Center, Lincoln, NE.

Soil Survey Division Staff. 1993. Soil Survey Manual. USDA-SCS Agric. Handbook 18, U.S. Government Printing Office, Washington, DC.

Soil Survey Staff. 1975. Soil Taxonomy: A basic system of soil classification for making and interpreting soil surveys. USDA-SCS Agric. Handbook. 436, U.S. Government Printing Office, Washington, DC.

Soil Survey Staff. 1999. Soil Taxonomy: A basic system of soil classification for making and interpreting soil surveys, Second Edition. USDA-NRCS Agric. Handbook 436, U.S. Government Printing Office, Washington, DC.

Soil Survey Staff. 2001a. National Soil Survey Handbook. Soil Correlation (Part 609.06). Online at *http://soils.usda.gov/procedures/handbook/main.htm* (verified October 8, 2002).

Soil Survey Staff. 2001b. National Soil Survey Handbook. Soil Properties and Qualities (Part 618.03). Online at *http://soils.usda.gov/procedures/handbook/main.htm* (verified October 8, 2002).

Soil Survey Staff. 2001c. Official Soil Series Descriptions. Online at *http://ortho.ftw.nrcs.usda.gov/osd/osd.html* (verified October 8, 2002).

USDA. Bureau of Soils. 1903. Instructions to field parties and descriptions of soil types. Washington, DC.

Webster, R. 1968. Fundamental objections to the 7th Approximation. *J. Soil Sci.* 19:354–366.

CHAPTER 18

Classification of Soils of the Tropics: A Reassessment of Soil Taxonomy

Friedrich H. Beinroth and Hari Eswaran

CONTENTS

ABSTRACT

We briefly review the history of soil classification in the tropics and trace the development of Soil Taxonomy relative to tropical soils. Although significant progress has been made in this regard, there remain several areas that need improvement. These include the following issues that are addressed in this chapter: (1) the undocumented change in the definition of the "iso-" soil temperate regimes, which increased the delta T from 5°C to 6°C; (2) the questionable criteria for the spodic horizon, which disagree with classical concepts of pedology; (3) the case of Mollisols with an aridic soil moisture regime, which are illogically classified as Ustolls; (4) the problem of extremely acid, wet Vertisols, some of which are now classified as Aquerts, thus being recognized for their

0-8493-1339-2/03/$0.00+$1.50

most limiting characteristic only at the subgroup level; (5) the ambiguity of "Pale" great groups, which include soils with thick argillic horizons and soils in which the subsoil high clay content is inherited from the parent material; (6) the dilemma of paddy soils, whose classification fails to capture the particular conditions of these soils; and (7) the predicament of the kandic horizon, whose introduction may not have been warranted. We examine these issues in some depth, and present possible solutions.

INTRODUCTION

Documenting the soil resources of the world is a formidable task, and modern classification systems have made great strides toward this goal. In the United States, efforts to build a modern classification commenced in the early 1950s, and an operational system called the 7th Approximation (Soil Survey Staff, 1960) became available in 1960. Major revisions since then led to the first edition of Soil Taxonomy (Soil Survey Staff, 1975), and eventually the second edition in 1999 (Soil Survey Staff, 1999). Between the 1975 and 1999 editions, major revisions were made, and many of the changes were directed at classifying the soils of the tropics. This was made possible through a project (Soil Management Support Services–SMSS) funded by the U.S. Agency for International Development (Eswaran et al., 1987) under which benchmark soils in the tropical regions were sampled, and studies and international committees were initiated to help refine Soil Taxonomy. SMSS established a process and protocol for evaluating, testing, and validating proposals to change Soil Taxonomy emanating from outside the U.S. With the termination of SMSS in 1994, this process was no longer operational, and the information delivery and feedback mechanism from international sources declined.

The focus during the SMSS program was on major issues articulated by international experts. However, there are many other issues embedded in Soil Taxonomy that affect the classification of soils of the tropics, which were not considered, and there still remain problem areas waiting for resolution. In the interim period between 1975 and 1999, SMSS and later the Natural Resources Conservation Service (NRCS) issued a biannual publication called "Keys to Soil Taxonomy" that incorporated all accepted changes and modifications to the system. The Keys served a very important function by permitting persons to field-test the changes, thereby validating the system. With the termination of the SMSS process, validation from outside the U.S. has become an *ad hoc* activity. The consequences of this are observed in some of the changes recently introduced in Soil Taxonomy. It should be pointed out at the outset that the issues presented in this chapter do not detract from the excellent quality of Soil Taxonomy, and that Soil Taxonomy still remains the most advanced soil classification in the world.

Having been intimately involved with and having contributed to the process of enhancing Soil Taxonomy, the authors of are keenly aware of the difficulties of obtaining a consensus, even on scientific concepts. Thus in identifying the issues and suggesting solutions, we have attempted to make them in the context of the guiding principles and concepts of the system (see Arnold and Eswaran, this publication). We are also cognizant of the fact that the classification of natural systems must necessarily be, to some extent, arbitrary and qualitative, and that subdividing a natural continuum is always artificial. Hallberg (1984) states that "the best of classification schemes, at any given time, are merely an index of the evolution of our knowledge. Such schemes must be tested, and must be capable of growth and change as our knowledge changes." It is in the spirit of the last statement that we attempt to evaluate the classification of soils of the tropics according to Soil Taxonomy.

EVOLUTION OF CONCEPTS

The Dawn of Soil Classification

When Dokuchaev (1896) formalized his concept of soils and the Russian classification, it became a benchmark event in soil science, marking the starting point of a new science. The basic concepts he enunciated influenced developments in soil classification, and elements of his concepts and principles are integral parts of modern classification efforts. The importance of soil moisture in determining soil attributes was emphasized by Vysotskii (1906), and the notion of soil zonality was born. The role of climate was further elaborated by Lang (1922), who also considered temperature, and perhaps is one of the first to recognize soils in the tropics as a separate set of soils. This was another major step in the development as aspects of this principle became incorporated into other classification systems, including Soil Taxonomy (Soil Survey Staff, 1999).

By the time of the first International Congress of Soil Science in 1924 in Washington, DC, the climatic aspects were refined. Because of the difficulties of acceptable climatic parameters, soil cover was used as differentiae (Afanasev, 1928). By this time, two large groups of soils in the tropics were known: the red soils and the black soils. The red colors and the deep weathering profiles of many soils in the tropics are distinctive, and most of the earlier classification systems in the world differentiate these soils from the soils of temperate regions, which have good horizonation and generally do not exceed a meter in thickness. Traditionally, there has also been a tendency to view the soils of the tropics differently from the colder temperate region soils. The vegetation and land use were additional reasons for making this distinction.

The highly weathered soils of the tropics were grouped under red soil, red loam, or red earth until about the middle of the last century. Among others, the work of Kubiëna (1953) influenced the thinking of the period. Kubiëna was already working with soil micromorphological analysis and developed his "Erde and Lehm" concepts, which did not have much acceptance among the soil scientists who were more familiar with physical and chemical properties. These concepts were the basis for the later differentiation between argillic and cambic horizons. With the description of laterites by Buchanan (Buchanan, 1807), the term *laterite* (and lateritic soils) became prevalent in the soils literature (Alexander and Cady, 1962). With more studies, other adjectives were added to the term *lateritic soils.* In 1949, a group of scientists created the term *Latosol* (Cline, 1975) that soon became very popular (Bennema et al., 1959) despite the fact that it was not defined in rigid terms. In the soil survey of Hawaii, the terms Humic Latosols, Hydrol-humic Latosols, etc. were used (Cline et al., 1955). This was also the period when the concept of "podzolic soils" was being accepted (Baldwin, 1928; Baldwin et al., 1938) and the Latosols (Kellogg, 1949) were considered to be a distinct group of soils. In the United States, Thorp and Smith (1949) formalized this state of soil classification after the war, but difficulties in the use of the system and the rapid increase in information required the development of new concepts and approaches. Dr. Guy D. Smith was charged by the Soil Conservation Service of USDA to develop a new system, which today is known as Soil Taxonomy. In the 1950s and 1960s, several other classification systems became in vogue, and other terms to describe these highly weathered soils of the tropics were coined.

By about the mid-1950s, the concept of Latosols as highly weathered or low negatively charged soils became firmly established, and paved the way for the modern concept of Oxisols. The decade of the 1950s also saw great advances in mineralogical studies, and with increasing studies on soils of the tropics, the processes responsible for Latosol formation were better understood. The 1950s perhaps mark the second renaissance of pedology, launched by several events or factors. First were aspirations to develop national classification systems. More important, however, was the decision of the Food and Agricultural Organization of the United Nations to embark on a "Soil Map of the World" program. During this period, the term *Ferralitic Soils* appeared in the French (Aubert, 1958) and Portuguese (Botelho da Costa, 1954; 1959) literature. The Belgians introduced the term *Kaolisols* (Sys et al., 1961; Tavernier and Sys, 1965). Charter (1958), working in Ghana, used the terms *Oxysols* and *Ochrasols*. Though many descriptive terms were introduced to describe the highly weathered soils of the tropics, most were not accompanied by quantitative definitions. The transitions of the Podsolics and the Latosols remained subjective and unresolved.

The term *Oxisol* was created around 1954, during the development of Soil Taxonomy (Soil Survey Staff, 1960; 1975), and the definition was based largely on the Latosol concept. Thus the class of Oxisols included many of the present-day Andisols, some Inceptisols, and some of the Alfisols and Ultisols. A more comprehensive definition of the term Oxisols was published in 1960 by the USDA (Soil Survey Staff, 1960), and the concept, definition, and classification were gradually modified and refined until the publication of Soil Taxonomy in 1975 (Soil Survey Staff, 1975). A comprehensive review of Oxisols, including their use and management, is provided by Van Wambeke (1992) and by Buol and Eswaran (2000).

A Consensus for Modern Classification

By the early 1970s, soil classification systems that had been initiated in the 1950s were maturing. Soil Taxonomy was published in 1975 (Soil Survey Staff, 1975). But Dr. Guy D. Smith, the principal contributor to Soil Taxonomy, felt that the knowledge and thus the classification of soils of the tropics was incomplete, and indicated that these taxa and others in Soil Taxonomy should be considerably modified. FAO-UNESCO (1971–1976) began to publish the Soil Map of the World. The legend of the soil map was developed and tested throughout the duration of the program, and the final global soil map is good evidence that the legend served its purpose. The soil map of the world, however, had areas where it was acknowledged that adequate information was not available, and many such areas occur in the tropics. After publication of the global soil maps by FAO and Soil Taxonomy by USDA-NRCS, an atmosphere of *fait accompli* descended on the richer countries, and there was a visible decline in enthusiasm for refinements or other activities. Soil survey and soil research in the tropics were also reduced, except in countries such as Brazil or in countries that still maintained strong interest in the tropics, such as France and Belgium.

The cold war, ironically, turned out to be a boon for soil science in the tropics. The principal global adversaries provided generous financial and technical support for Third World countries in an attempt to garner political influence. Some of the donor largesse trickled down to the technical level and resulted, *inter alia*, in increased soil resource inventory and pedological research. When the relationships thawed during the decade of the last century, the richer countries lost interest in the tropics and the tropical countries could not afford to support a similar research effort. In the U.S., the international politics of the 1970s resulted in a major effort to improve Soil Taxonomy for better application in the tropics. With funding from the U.S. Agency for International Development, the Soil Management Support Services (SMSS) was created to enable the USDA Soil Conservation Service (now the Natural Resources Conservation Service–NRCS) to work in the tropics (Eswaran et al., 1987).

Guy Smith, in collaboration with soil scientists from the U.S. and Europe, identified many areas in Soil Taxonomy that required priority attention. Several international committees (ICOMs) were established by SMSS (Table 18.1) to rally soil scientists around the world to help improve Soil Taxonomy. Simultaneously, there was an effort to develop a database of soils of the tropics to backstop the ICOMs. In addition, the University of Puerto Rico, in collaboration with SMSS, organized workshops (Beinroth, 1978a; 1978b; 1983; 1986) at which scientists could meet and discuss the classification issues. Many of these activities are summarized by Eswaran et al. (1987).

The second edition of *Soil Taxonomy* (Soil Survey Staff, 1999) has incorporated many of the changes proposed by the ICOM as well as other changes submitted to NRCS from outside the U.S. or suggested by soil scientists working in the U.S. Because of the investments involved and the time available, not all soils in the tropics could be scrutinized in the manner done by the ICOMs

Table 18.1 International Committees Organized to Revise Soil Taxonomy

Year	Name of International Committee	Chairperson/s
1976	International Committee on Classification of Soils with Low Activity Clays (ICOMLAC)	F.R. Moormann Netherlands
1978	International Committee on Oxisols (ICOMOX)	S.W. Buol N. Carolina
1978	International Committee on Soil Moisture Regimes (ICOMMORT)	Van Wambeke New York
1980	International Committee on Andisols (ICOMAND)	M.L. Leamy New Zealand
1980	International Committee on Aridisols (ICOMID)	Osman Syria
1981	International Committee on Vertisols (ICOMERT)	J. Comerma Venezuela
1982	International Committee on Classification of Soils with Aquic Soil Water Regimes (ICOMAQ)	J. Bouma Netherlands
1982	International Committee on Spodosols (ICOMOD)	R. Rourke Maine
1987	International Committee on Soil Families (ICOMFAM)	Hajek Alabama

(Table 18.1) and by the process developed by SMSS. Such activities were phased out beginning in 1990, and further changes to the system only resulted from *ad hoc* international meetings. The process of consensus-seeking based on data and validation is critical for Soil Taxonomy to reduce personal biases and to ensure that basic tenets of the system are not corrupted. Only an institution can sustain such a process, and the continued success of Soil Taxonomy is because the Natural Resources Conservation Service (NRCS) nurtures it. In contrast, classification systems developed by individuals become academic exercises after the demise of the author.

The 1999 edition of Soil Taxonomy made some important changes based largely on the U.S. experience. Most of the changes are by any measure significant improvements to the system. There are, however, a few that have not been validated by wide testing (as had happened in the past, through the SMSS program), and gave rise to reasonable questions when applied to the tropical context. One of the continuing shortcomings is the inability of soil scientists working in the tropics to provide the kinds of documentation needed to support their case for changes in Soil Taxonomy. As a result, many proposals from tropical countries are ignored, and consequently, although the understanding of soils and their management has improved during the last decade, changes in Soil Taxonomy have not accommodated these. At the beginning of the new millennium, there was a general consensus that the system has improved considerably, thereby realizing the dream of Guy Smith. But there were still areas for improvement. One such area is soils of the tropics.

PURPOSE AND PRINCIPLES OF SOIL TAXONOMY

The purpose and principles of Soil Taxonomy are well enunciated by Smith (1963; 1965) and Soil Survey Staff (1999). These have been strictly adhered to in many of the changes to the system that took place in the past. Recently, however, the process appears to be less stringent, perhaps because of the haste to publish the second edition by the end of the century. As indicated earlier, apart from the mandated areas of the ICOMs, there were few investigations and critical assessments of the needs of soils of the tropics. Further, in the last decade, there have been minimal changes to the system. As a result, there were few opportunities to address the many questions or inconsistencies related to soils of the tropics. As there were few opportunities to initiate such discussions in formal meetings, they have remained unanswered. Some of these are presented here, with our opinion, to initiate discussion, research, and eventual changes to the system.

Classification systems portray the state of knowledge of the science, but as Smith (1965) has warned, they can become a potent factor limiting the possibilities of new experience. Smith states:

> If its criteria are theories without some device for constant and inescapable scrutiny in relation to fact, the concepts in the system become accepted as fact. Such acceptance can mold research experience into patterns of the past and can limit understanding of even new experience to concepts established in the past.

As Soil Taxonomy evolved from previous systems, it inherited many useful notions, but also became entrapped in accepted concepts of a previous era from which it has become extremely difficult to dissociate itself. Guy Smith recognized this difficulty, as he was confronted with formidable obstacles when he tried to make paradigm shifts. At the end of his career and in retrospect, he confided to the authors that, to some extent, he had failed to make those important changes that he deemed were necessary, but rested with the confidence that he had helped develop a system that will be capable of modifications with new knowledge. He indicated in his interviews (Forbes, 1986) that if a system is not flexible when challenged by new knowledge, it is a mark of failure of the system. In this respect, it is evident that soil classification systems are a series of approximations that must be continuously evaluated and modified for a changing knowledge base and the needs of society.

ISSUES OF CONCERN

More than 25 years have passed since the first edition of Soil Taxonomy. With the advent of Soil Taxonomy and the acceptance of many of its concepts and terminology, modern soil surveys have been conducted with a better scientific basis. In the last quarter-century, many countries have published national soil maps with supporting data, and many have also commenced detailed mapping supported with pedological studies. The cumulative effect of this progress is that some concepts have been questioned and many new proposals have been formulated. These have been reported at scientific meetings or as technical papers in journals. The authors have also had the opportunity to discuss them with scientists in their countries. Some of the major issues are presented here.

Inconsistencies in the system may be inherent or introduced consequent to changes made to parts of the system. A change in one part of the system may have repercussions across the system. These require time and effort to locate and correct, and are an ongoing task of the Soil Survey Staff. The change in definition of the Vertisols and the introduction of a suborder of Aquerts are examples that are discussed later. Those inherent in the system are usually those inherited during the development of the system, and there is a general reluctance to change them. Mollisols with aridic soil moisture regime is an example.

Some changes were introduced without adequate testing and evaluation of the consequences. The change in the definition of the "iso" soil temperature regime is a case in question, and reflects the situation in which the process for accepting changes was violated. Other examples of changes that were introduced result from compromises among well-meaning scientists. The introduction of the "kandi concept" is an example. The concept, as elaborated later, not only violated the principles of the Soil Taxonomy (Cline, 1949, 1963), but was also not a major improvement to the system.

Finally, there are examples of situations in which an absence of detailed studies has prevented an acceptable consensus on the subject. Rice-growing soils of the tropics is a good example. Such soils were not included in the effort of SMSS, and little has been done by NRCS since the termination of SMSS. The problems of seasonally flooded soils, as elaborated later, are not only one of classification but have relevance to larger questions of soil management and global warming.

The Unsubstantiated Change in the Definition of the "iso" Limit

The isohypothermic, isothermic, isomesic, and isofrigid soil temperature regimes were introduced in the first edition of Soil Taxonomy (Soil Survey Staff, 1975). Their common characteristic was mean summer and mean winter soil temperatures that differ by less than 5°C at a depth of 50 cm. This value was selected because the area having such temperature fluctuations corresponds closely to the intertropical region between the Tropic of Capricorn and the Tropic of Cancer.

In the second edition of Soil Taxonomy (Soil Survey Staff, 1999), the limit for the "iso" condition was increased to 6°C. No rationale or reasons to substantiate the change were provided. Apparently, the change was made to accommodate small areas with ustic or aridic soil moisture regimes on the islands of Hawaii and Maui in the State of Hawaii, so that all of both islands could be considered "tropical." Classifying these areas as "non-iso" would have indicated their different soil temperature attributes and would have served the same purpose for use and management. But this was not considered. Although there was no widely circulated proposal, the change was accepted by the classification team of the NRCS and incorporated into the system.

We analyzed the climatic data for 55 stations in Hawaii and Puerto Rico, and calculated the difference between mean summer and winter air temperatures. The delta T values ranged from 1.7 to 4.0°C. The amplitude in soils is normally smaller, and in the Newhall (1972) model, a factor of 0.6 is used to reduce the delta T in the soil. As the islands of the State of Hawaii are located near the Tropic of Cancer, there may well exist small areas where delta T exceeds 5°C. It appears, however, that extrapolating this situation to the rest of the world is not warranted.

The ΔT 5°C break to differentiate the iso or tropical conditions from the non-iso was based on general land use conditions prevailing in the tropics. In the tropics, though winter months could become cool (at transition zones to the high latitudes), there is no frost to affect plant growth.

Although the change has worldwide implications, its effect has not been documented and there appear to be no factual global data to substantiate the need for change. However, we understand that the issue is now being reexamined by the NRCS to study the impact and implication of the new definition. Unless extensive testing produces convincing evidence that the original 5°C limit was flawed, and that the new delta T of 6°C constitutes a significant improvement, we consider the change unnecessary and indeed unfortunate. We further believe that changes in Soil Taxonomy of such magnitude and consequence should not be made in an impromptu fashion on the basis of isolated, *ad hoc* information.

The Flawed Definition of Spodic Materials

Spodic materials, the indispensable characteristic of the spodic horizon and Spodosols, are defined in Soil Taxonomy (Soil Survey Staff, 1999) as "mineral soil materials [that] are dominated by active amorphous materials that are illuvial and composed of organic matter and aluminum, with or without iron." Yet a soil material may qualify for spodic solely on the basis of pH, organic carbon content, and color. In the northern coastal plain of Puerto Rico, there are clayey deposits that are overlain by almost pure quartz sand sediments of varying thickness. At the abrupt lithological discontinuity, a black horizon has formed, likely due to lateral seepage from the surrounding lagoons with which they are geographically associated. This horizon qualifies as a spodic horizon by current definition, and the soils are therefore Spodosols.

Similar soils are extensive elsewhere in the tropics, notably in Southeast Asia, in the southern part of the Congo Basin and in the Amazon Basin of South America. A major river draining the basin, the Rio Negro, is loaded with dissolved organic matter making the water black (hence the name). Along some coastal plains of Southeast Asia, there is a back-swamp separating the inland from the raised coastal dune sands. Peat or muck occurs in the swamp and the dissolved organic matter laterally permeates the sandy beach deposits. A transect from the swamp to the sea shows that the horizon of accumulation reduces in thickness toward the sea. Above this horizon is a layer of bleached sand. The soils have a very poor vegetative cover, further indicating that there is an inadequate supply of organic matter to enrich the "spodic horizon" and develop the giant podzols of the Amazon described by Jenny (1941). On the raised coastal plains around Miri, Sarawak, further accumulation of organic matter is terminated because of the uplift. Since the uplift, the material has hardened to form a rock-hard pan, for which there is no adequate term.

We believe that the present definition of the spodic horizon can result in Spodosols that fail to conform to the genetic concept of Spodosols. This is a situation in which similar morphology results from distinct and different processes, and this must be alluded to in the system. The process described earlier is not exclusive to the tropics, and could occur in other similar landscape conditions. There are many ways in which the definition and concept could be modified to reduce the ambiguity. Regarding large areas of soils that now, perhaps illogically, key out as Spodosols, we propose that the definition of spodic materials be revised to require the presence of active amorphous materials. This is the unifying feature of all Spodosols. In view of the tropical evidence, it should also be examined if the spodic horizon should always be illuvial. Subsequent to geomorphic and laboratory studies of such soils, it would be useful to differentiate them from the classical Spodosols of the temperate regions.

The Case of Aridic Mollisols

Soils that have a mollic epipedon, an isomesic or warmer iso soil temperature regime, and a drier than ustic soil moisture regime are currently placed in aridic, torrertic, torripsammentic, torriorthentic, or torroxic subgroups of Ustolls. The soil moisture regime of these subgroups requires them to be moist in some or all parts for fewer than 90 days, and dry in some or all parts for six tenths of a year (i.e., 220 days) or more of cumulative days. This regime is similar but not identical to the aridic soil moisture regime, as the control section in soils with an aridic regime must be dry in all parts for more than 180 cumulative days. The mentioned subgroups, therefore, may be more or less dry than soils with an aridic regime. Also, there may not be too many such subgroups, as item 8 of the definition of the mollic epipedon requires that "some part of the epipedon is moist for 90 days or more (cumulative) in normal years" (Soil Survey Staff, 1999).

We are not familiar with the rationales that support these definitions, but we find them somewhat confusing and difficult to establish. In the absence of extensive field measurements, which would allow a positive placement in one of the subgroups, there is the possibility that many soils are misclassified, leading to erroneous interpretations. Further, the soils are indicated as Ustolls or Xerolls on small-scale maps, although the actual moisture regime is aridic.

In order to reduce the margin of error in taxonomy and interpretation, we propose that item 8 be deleted from the definition of the mollic epipedon, that the mentioned subgroups be eliminated, and that a suborder of Torrolls for Mollisols is established with an aridic soil moisture regime. The proposed change would have a precedent in the Torrand, Torrert, and Torrox suborders and in the Torriarent, Torrifluvent, Torrifolist, Torriorthent, and Torripsamment great groups. The change would make the system logical and would maintain the integrity of the system.

The Problem of Extremely Acid Vertisols

The suborder of Aquert was recently introduced based on extensive occurrence of such soils. However, though the need was established and rationale was defensible, the consequences of the change were not anticipated. There are soils with shrink-swell potential that are continuously wet and as they did not crack, they were classified as Inceptisols in the previous versions of Soil Taxonomy. Those with a sulfuric horizon were classified as Sulfaquepts, or the actual acid sulfate soils. Some with only sulfidic materials (the potential acid sulfate soils) are the Sulfaquents.

With the change in definition of Vertisols and the introduction of Aquerts, some of these soils are currently classified as Haplaquerts or Dystraquerts, and at the subgroup level, as Sulfaqueptic

Dystraquerts. In central Thailand and elsewhere in Southeast Asia, there are extensive areas of hydromorphic Vertisols that have a sulfuric horizon, and are therefore extremely acid. Their classification as Sulfaqueptic Dystraquerts implies, at the great group level, that the soils have a pH value of 4.5 or less within 50 cm depth, and, at the subgroup level, jarosite concentrations and a pH value of 4.0 or less within 100 cm depth.

The most constraining property of the soil is its extreme acidity, and this attribute is only considered at the subgroup level. The Asian pedologists argue that the shrink-swell nature of the soil is less important to the use and management of the soil. As the soil is classified as a Vertisol, it is also not shown on small-scale soil maps at the great group level, and hence misleads potential interpretations. They believe these soils should be classified as Vertic Sulfaquepts. This classification would indicate, at the great group level, that these soils have a pH value of 3.5 or less within 50 cm of the mineral soil surface.

This taxonomic change could be accomplished by amending item F in the Key to Soil Orders to define Vertisols (Soil Survey Staff, 1999) as "Other soils that do not have a sulfuric horizon or sulfidic materials within 1 m of the mineral soil surface." It would further require establishing and defining a Vertic subgroup of Sulfaquepts.

The Ambiguity of "Pale" Great Groups

Soil Taxonomy recognizes "pale" great groups in the Alfisols, Aridisol, Mollisol, and Ultisol orders. Most of the 14 great groups have an argillic horizon, although some Palecryalfs, Paleudolls, and Palexerolls qualify on the basis of other differentiae. The common feature of the pale great groups with argillic horizons is "a vertical clay distribution in which the clay content does not decrease by as much as 20 percent from the maximum clay content within a vertical distance of 150 cm below the soil surface" (Soil Survey Staff, 1999). In the Palexeralfs and Palexerolls, this clay distribution must occur within the argillic horizon; in all other pale great groups only depth is diagnostic. Consequently, pale great groups may have argillic horizons that do not extend to 150 cm below the soil surface. This situation is common throughout the tropics where Ultisols and Alfisols are developed in clayey parent materials. In these soils, the clay distribution is a lithological rather than pedogenetic feature. In Puerto Rico, for example, all of the 13 series of Paleudalfs, Paleudults, and Paleustalfs are developed in clayey sediments on late-Pleistocene or younger surfaces, and have not been found on autochthonous parent materials.

According to Soil Taxonomy (Soil Survey Staff, 1999), "pale" is derived from the Greek word "paleos," meaning old, and has the connotation of "excessive development." The narrative descriptions in Soil Taxonomy also refer to old and stable land surfaces of mid-Pleistocene or older age, and occasionally mention thick argillic horizons.

The current definitions of pale great groups result in a pedogenetically heterogeneous group of soils. Although the argument could be made that for practical interpretations, it is irrelevant whether the clay is pedogenic or lithological, we believe the present definitions do not conform to the original intent of the pale concept. We propose, therefore, that they be changed to specify clay distribution in argillic or kandic horizons that extend to a depth of 150 cm below the soil surface. Part of the problem also stems from the fact that the lower limit of the argillic horizon is not defined in Soil Taxonomy, in contrast to the upper boundary. More studies are needed to arrive at an appropriate resolution to this issue.

The Dilemma of Paddy Soils

Paddy soils, or soil used for wetland rice cultivation, have yet to be appropriately classified. The problem is shared by all classification systems. The puddling action of management that is a prerequisite for rice cultivation destroys all surface attributes in the upper 25 to 40 cm of the soil surface. In addition, in these seasonally flooded soils, the anthropic change induces changes in hydrologic properties of the soil, altering the innate pedological processes. Seasonally flooded soils are water saturated for 4 to 6 months of the year, and are subject to the climate-induced moisture stress for the remaining part of the year. The use of soil moisture regimes, as defined in Soil Taxonomy, is inappropriate. During some parts of the year, the soil has an aquic soil water condition, while during the remaining parts, the soil moisture regime is undefined (as it would not meet the definition of ustic, xeric, aridic, or udic). This periodicity in soil water conditions, either natural or human-induced, affects land use and directs the mode of genesis of the soil.

Soils converted to paddy soils from former well-drained upland soils have the previously stated difficulties in classification, hence application of the system for use and management. An example is the sequence of soils studied by Hseu and Chen (1996). The unmodified upland soil developed on basic materials is a red Paleudult. After about 30 years of rice cultivation, iron is removed from the upper 30 cm of the soil and the reduction and removal of the iron proceeds in the subsoil as vertical streaks. The present-day subsoil has the morphology of plinthite, but would not qualify as plinthite as defined by Daniels et al. (1978). This is a clear example of anthropogenesis, except that the process is not implied by the classification.

The solution rests in the concept and definition of anthropic processes and specifically as it applies to paddy soils. It also requires the identification and definition of contrasting soil moisture conditions, as in the case of paddy soils.

The Predicament of the Kandic Horizon in Oxisols

The kandic horizon (Beinroth et al., 1986; Moorman, 1985), which was introduced on the recommendation of the International Committee on the Classification of Soils with Low Activity Clays (ICOMLAC), is a hybrid between an argillic horizon and an oxic horizon. Its introduction constitutes the revival of an old concept, and it may be appropriate to cite what Guy Smith (1965) had to say about this issue: "In the 7th Approximation we tested the possibility that a single horizon might be both an oxic and an argillic horizon. This was not well received, so we have abandoned the idea." Apparently, this notion has now, in the form of the kandic horizon, become acceptable again.

As a consequence, soils that have 40% or more clay in the surface horizon and a kandic horizon that has the weatherable-mineral properties of an oxic horizon, with an upper boundary within 100 cm of the mineral soil surface, key out as Oxisols. Unlike the argillic horizon and most other horizons, which have a morphogenetic connotation, the origin of the kandic horizon is attributed to several processes.

As far as we know, no systematic testing of the implications of the kandic horizon has been conducted, and we therefore have no factual basis for assessing the faults and virtues of the new horizon. One consequence was that it resulted in the reclassification of many Ultisols and some Alfisols as Oxisols, and thus in a significant increase in the area occupied by Oxisols. In Puerto Rico, for example, the Oxisol area increased from 2.5 to 6.2%. (The reclassification may not stand close scrutiny, however, as the amount of weatherable minerals was not determined).

There is also an inconsistency in the definition of the kandic horizon, which states that "the percentage of clay is either measured by the pipette method or estimated to be 2.5 times [percentage of water retained at 1500 kPa tension minus percent organic carbon], whichever is higher, but not more than 100" (Soil Survey Staff, 1999). However, the default coefficient for the oxic horizon is 3. (We have not been able to establish the reason for this discrepancy.) Further, to qualify for Oxisols, a kandic horizon must have more than 40% clay in the surface horizon. Yet there are soils, for example the Rengam series in Malaysia (Paramanathan, 1977), which have less than 40% clay in the surface horizon and a kandic horizon with distinctly oxic properties, but nevertheless classified as Ultisols. The Rengam Series of Malaysia lacks cutans (as confirmed by micromorphological analysis by Eswaran and Wong, 1978) to a depth of 18 m. Many such studies have shown that many of the Kandiudults have properties more akin to Oxisols than to Ultisols (Beinroth et al., 1974; Eswaran and Tavernier, 1980).

Another concern is that management-induced soil erosion may selectively remove part of the fine fraction from the surface horizon. The resulting textural differentiation may cause a change in the classification from Oxisols to Ultisols or Alfisols that is anthropogenic rather than pedogenic in nature (Eswaran et al., 1986). From a management point of view, the low nutrient holding capacity and absence of a potential to supply some of the nutrients (characteristics of Oxisols) are more restricting than the small clay increase required by a kandic horizon (Herbillon, 1980). Van Wambeke (1989) also objected to the use of the kandic horizon in two opposite directions at the same level in the hierarchical structure, as the rationale is difficult to understand: The horizon is used, on the one hand, to include soils in the Oxisols, and on the other hand, to exclude soils from the Oxisols.

Although we believe that there were no compelling reasons for introducing the kandic horizon, it is now a *fait accompli*. Nevertheless, if there was a consensus that the kandic horizon properties should be diagnostic for Oxisols, they could have been considered more elegantly. It could have been accomplished by deleting item 5 (clay increase with depth) from the definition of the oxic horizon, and changing item E in the Key to Soil Orders (Soil Survey Staff, 1999) to read: "Other soils that have an oxic horizon that has its upper boundary within 150 cm of the mineral soil surface." The Oxisols that have an argillic horizon (as currently defined) would become Argi or Kandi great groups. The Alfisols and Ulltisols would then have a CEC greater than 16 cmol per kg clay, and those with CEC between 16 and 24 cmol per kg clay would become Kandi great groups. Such a change would maintain class purity and the logic of the system. Sys (1968) already proposed this option.

CONCLUSION

Soil Taxonomy is a dynamic system that must be updated periodically to accommodate new knowledge and thus to ensure that it can accomplish its stated purpose of "making and interpreting soil surveys." However, necessary revisions should only be considered after a thorough review of adequately documented proposals has confirmed their validity. Moreover, as Soil Taxonomy aspires to be an international system, such reviews should involve the global pedologic community. This protocol is not always followed, as the recent change in the definition of the iso soil temperature regimes exemplifies. Yet to abide by the proper procedures is often easier said than done. This is particularly true for users of Soil Taxonomy in less-developed tropical countries. They may have solid qualitative and intuitive evidence for needed changes, but lack the human and capital resources to document their concern with factual field and laboratory data.

Regrettably, the USDA Natural Resources Conservation Service, the custodian of Soil Taxonomy, cannot provide assistance as it is constrained by its Congressional mandate that requires it to use allocated funds only for domestic programs. In the past, the Soil Management Support Services, a project funded by the U.S. Agency for International Development (USAID), provided such technical assistance and also a forum for discussion of issues of concern. The project was set up to advance the adoption of Soil Taxonomy in the interest of global communication and agrotechnology transfer. In spite of its success, it was terminated by USAID. We consider this unfortunate, as the new users of Soil Taxonomy are now left without technical support. This may result in unilateral and haphazard changes to the system. As a consequence, Soil Taxonomy may gradually lose global relevance and degenerate into a parochial scheme of national applicability. We regret this development and believe that it would be in the interest of USAID to again contribute to a process, which will foster the global use and application of the system.

The second edition of Soil Taxonomy, which was published in 1999, is the result of enormous efforts to improve the system. Nonetheless, the process appears to have been *ad hoc* to some extent, because of a lack of a clear enunciation of the role and limits of the categories of the system. These and some other principles incorporated in the system must be periodically revisited to ensure that changes are made according to reasonable rationales. Expediency and strong views of peers are insufficient reasons to make changes.

Today's enviro- and infocentric world demands good databases and for soil resources, Soil Taxonomy is probably the best system to guide their use and management. Its hierarchical nature enables data clustering at all scales, from field to farm to watersheds and continents. The quantitative approach reduces bias in the clustering of individuals, and enhances the ability to aggregate units into higher categories. These attributes have attained a greater importance than when the system was conceived in 1960. These and issues of the kind highlighted in the paper are aspects of the system that we must bear in mind as we continue the task of making this an international soil classification system.

REFERENCES

Afanasev, I.N. 1928. Soil classification problems in Russia. *Proc. 1st Int. Congr. of Soil Sci.* 4:498. Washington, DC.

Alexander, L.T. and Cady, J.G. 1962. Genesis and hardening of laterites in soils. USDA Tech. Bull., 1282.

Aubert, G. 1958. Classification des Sols. Compte Rendu, Reunion Sous-comite, Congo, Brazzaville.

Baldwin, M., Kellogg, C.E., and Thorp, J. 1938. Soil Classification. Soils and Men. U.S. Dept. Agric. Yearbook. U.S. Government Printing Office, Washington, DC, 979–1001.

Baldwin, M. 1928. The gray-brown podzolic soils of the eastern United States. First Int. Congr. Soil Sci., Washington, DC. 4:276–282.

Beinroth, F.H. and Panichapong, S., Eds. 1978a. Proceedings of Second International Soil Classification Workshop. Part II, Thailand. Publ. Dept. of Land Development, Bangkok, Thailand.

Beinroth, F.H. and Paramanathan, S. Eds. 1978b. Proceedings of Second International Soil Classification Workshop. Part I, Malaysia. Publ. Dept. of Land Development, Bangkok, Thailand.

Beinroth, F.H., Uehara, G., and Ikawa, H. 1974. Geomorphic relations of Oxisols and Ultisols on Kauai, Hawaii. *Soil Sci. Soc. Am. Proc.* 38:128–131.

Beinroth, F.H., Neel, H., and Eswaran, H., Eds. 1983. Proceedings of the Fourth International Soil Classification Workshop, Rwanda. Part 1: Papers; Part 2: Field Trip Background and Soil Data. Agric. Editions 4, ABOS-AGDC, Brussels, Belgium.

Beinroth, F.H., Camargo, M.N., and Eswaran, H., Eds. 1986. Proceedings of the Eighth International Soil Classification Workshop: Characterization, Classification, and Utilization of Oxisols. Brazil Publ. EMBRAPA-SNLCS, Rio de Janeiro, Brazil.

Bennema, J., Lemos, R.C., and Vetturs, L. 1959. Latosols in Brazil. IIIrd Inter-African Soils Conference, Dalaba. I:273–281.

Botelho da Costa, J.V. 1954. Sure quelques questions de nomenclature des sols des regions tropicales. Compte Rendu Conf. Int. Sols Africains, Leopoldville, Congo. 2:1099–1103.

Botelho da Costa, J.V. 1959. Ferralitic, tropical fersiallitic and tropical semi-arid soils: definitions adopted in the classification of the soils of Angola. IIIrd Inter-African Soils Conference, Dalaba. I:317–319.

Buchanan, F. 1807. A journey from Madras through the countries of Mysore, Kanara and Malabar. East India Co., London, 2:436–461.

Buol, S.W. and Eswaran, H. 2000. Oxisols. *Adv. Agron.* 68:152–195.

Charter, C.F. 1958. Report on the environmental conditions prevailing in Block A, Southern Province, Taganyika Territory, with special reference to the large-scale mechanized production of ground-nuts. Ghana Department of Soil and Land Use Survey, Occasional Paper 1.

Cline, M.G. 1949. Basic principles of soil classification. *Soil Sci.* 67:81–91.

Cline, M.G. 1963. Logic of the new system of soil classification. *Soil Sci.* 96:17–22.

Cline, M.G. 1975. Origin of the term Latosol. *Soil Sci. Soc. Am. Proc.* 39:162.

Cline, M.G. et al. 1955. Soil survey of the territory of Hawaii. USDA Soil Survey Series 1939. No. 25.

Daniels, R.R., Perkins, H.F., Hajek, B.F., and Gamble, E.E. 1978. Morphology of discontinuous phase plinthite and criteria for its field identification in the Southeastern United States. *Soil Sci. Soc. Am. J.* 42:944–949.

Dokuchaev, V.V. 1896. "O novoi klassifikatsii Pochv" (On new soil classification). Trudy Imperatorskogo Volnogo Ekonomicheskogo Obshch., 6: 87 pp.

Eswaran, H. and Bin, W.C. 1978. A study of a deep weathering profile on granite in peninsular Malaysia. Part I. Physico-chemical and micromorphological Properties. Part II. Mineralogy of the clay, silt and sand. Part III. Alteration of feldspars. *Soil Sci. Soc. Am. J.* 42:144–158.

Eswaran, H. and Tavernier, R. 1980. Classification and genesis of Oxisols, in B.K.G. Theng, Ed. *Soils with Variable Charge*. Publ. Soils Bureau, DSIR, Lower Hutt, New Zealand. 427–442.

Eswaran, H., Kimble, J., Cook, T., and Mausbach, M. 1987. World Benchmark Soils Project of SMSS. *Soil Surv. Land Evaluation*. 7:111–222.

FAO-UNESCO, 1971–1976. Soil Map of the World. FAO, Rome, Italy.

Forbes, T.R. 1986. The Guy D. Smith Interviews: Rationale for Concepts in Soil Taxonomy. Soil Management Support Services, Tech. Monogr. 11. Publ. Dept. of Agronomy, Cornell University, Ithaca, New York.

Hallberg, G.R. 1984. The U.S. system of Soil Taxonomy: From the outside looking in, in R.B. Grossman, R.H. Rust, and H. Eswaran, Eds. Soil Taxonomy: Achievements and Challenges. *Soil Sci. Soc. Am. Special Publ.* 14:45–60.

Herbillon, A. 1980. Mineralogy of Oxisols and oxic materials, in B.K.G. Theng, Ed. *Soils with Variable Charge*, New Zealand Society of Soil Science, Lower Hutt, New Zealand. 109–126.

Hseu, Z.Y. and Chen, Z.S. 1996. Saturation, reduction, and redox morphology of seasonally flooded Alfisols in Taiwan. *Soil Sci. Soc. Am. J.* 60:941–949.

Jenny, H. 1941. *Factors of Soil Formation*. McGraw-Hill, New York.

Kellogg, C.E. 1949. Preliminary suggestions for the classification of great soil groups in tropical and equatorial regions. Commonwealth Bur. of Soil Sci. Tech. Commun.n No. 46:76–85.

Kubiëna, W.L. 1953. Bestimmungsbuch und Systematik der Böden Europas. Enkke, Stuttgart.

Lang, R. 1922. Über die Nomenklatur der Böden. Comptes Rendus de la Conference extraordinaire Agropedologique. Prague. 154–172.

Moormann, F.R. 1985. Excerpts from the Circular Letters of the International Committee on Low Activity Clay Soils (ICOMLAC). Soil Management Support Services, Tech. Monogr. No. 8. Washington, DC.

Paramanathan, S. 1977. Soil Genesis on Igneous and Metamorphic Rocks in Malaysia. Ph.D. thesis, Univ. Gent, Belgium.

Smith, G.D. 1963. Objectives and basic assumptions of the new classification system. *Soil Sci.* 96:6–16.

Smith, G.D. 1965. Lectures on Soil Classification. Special Bull. No. 4. Pedological Society, Ghent, Belgium.

Soil Survey Staff. 1960. Soil Classification, A Comprehensive System, 7th Approximation. USDA, Washington, DC.

Soil Survey Staff. 1975. Soil Taxonomy: A Basic System of Soil Classification for Making and Interpreting Soil Surveys. U.S. Department of Agriculture, Agric. Handbook No. 436. Wshington. DC.

Soil Survey Staff. 1999. Soil Taxonomy: A Basic System of Soil Classification for Making and Interpreting Soil Surveys. 2nd ed. U.S. Dept. Agric. Handbook 436. Government Printing Office, Washington, DC.

Sys, C. 1968. Suggestions for the classification of tropical soils with lateritic materials in the American Classification. *Pedologie*. 18:189–198.

Sys, C. et al. 1961. La Cartographie des Sols au Congo. Ses principes et ses methods. Publ. INEAC, Serie Technique No. 66, Brussels, Belgium.

Tavernier, R. and Sys, C. 1965. Classification of the soils of the Republic of Congo. *Pedologie*. 91–136.

Thorp, J. and Smith, G.D. 1949. Higher categories of soil classification: Order, suborder, and great soil groups. *Soil Sci.* 67:117–126.

Van Wambeke, A. 1989. Tropical soils and soil classification updates. *Adv. Soil Sci.* 10:171–193.

Van Wambeke, A. 1992. *Soils of the Tropics: Properties and Appraisal*. McGraw-Hill, New York.

Vysotskii, G.N. 1906. On orogenic and climatological foundations of soil classification. *Pochvovedenie*. 1–4:3–18.

CHAPTER 19

Anticipated Developments of the World Reference Base for Soil Resources

Jozef Deckers, Paul Driessen, Freddy O.F. Nachtergaele, Otto Spaargaren, and Frank Berding

CONTENTS

ABSTRACT

Since its endorsement by the IUSS in 1998, the World Reference Base for Soil Resources has established itself as a comprehensive soil correlation system. The system has so far been translated into nine languages and is used and tested all over the world. This paper summarizes the main attributes of the WRB, outlines preliminary results of its field-testing, and reflects on the place of WRB in the future. Although originally designed for general-purpose soil correlation at world scale, WRB is increasingly used as a classification system. This paper discusses how WRB correlates with the USDA Soil Taxonomy and draws attention to some major differences between the WRB and Soil Taxonomy. The issue of strict ranking of taxa and the rationale behind ranking have been major points of debate; a ranking scenario is presented here, mainly to stimulate the discussion. In addition, WRB definitions of diagnostic horizons, properties, and materials are discussed with reference to modern soil information systems. It is hypothesized that the IUSS Working Group RB might explore the zone between its present area of application and a detailed horizon classification system as elaborated by Fitzpatrick and Aitkenhead (2000).

0-8493-1339-2/03/$0.00+$1.50

INTRODUCTION

Soil is a three-dimensional body with properties that reflect the impact of (1) climate; (2) vegetation, fauna, and Man; and (3) topography on the soil's surface; (4) parent material over a variable (5) time span. The nature and relative importance of each of these five soil-forming factors vary in time and in space. With few exceptions, soils are still in a process of change; they show in their *soil profile* signs of differentiation or alteration of the soil material incurred in a process of soil formation, or pedogenesis (Driessen et al., 2001).

Unlike plants and animals, which can be identified as separate entities, the world's soil cover is a continuum. Its components occur in temporal and/or spatial successions. In the early days of soil science, soil classification was based on the (surmized) genesis of the soils. Many traditional soil names refer to the soil-forming factor considered to be dominant in a particular pedogenetic history, for instance, desert soils (climate being the dominant factor), plaggen soils (human interference), prairie soils (vegetation), mountain soils (topography), or volcanic ash soils (parent material). Alternatively, soil names referred to a prominent single factor, for instance, Brown Soils (color), alkali soils (chemical characteristic), hydromorphic soils (physical characteristic), sandy soils (texture), or lithosols (depth).

The many soil classification schemes developed over the years reflect different views held on concepts of soil formation, and mirror differences of opinion about the criteria to be used for classification. In the 1950s, international communications intensified while the number of soil surveys increased sharply both in temperate regions and in the tropics. The experience gained in those years and the exchange of data between scientists rekindled interest in (the dynamics of) the world's soil cover. Classification systems were developed, which aimed at embracing the full spectrum of the soil continuum. In addition, emphasis shifted away from the genetic approach, which often contained an element of conjecture, to the use of soil properties as differentiating criteria. By and large, consensus evolved as to the major soil bodies, which needed to be distinguished in broad level soil classification, although differences in definitions and terminology remained.

In 1998, the International Union of Soil Sciences (IUSS) adopted the World Reference Base for Soil Resources (WRB) as the Union's system for soil correlation. The structure, concepts, and definitions of the WRB are strongly influenced by the legend of the FAO-UNESCO 1/5,000,000 Soil Map of the World (FAO, 1974; FAO-UNESCO-ISRIC, 1988; 1990), which in turn borrowed the diagnostic horizons and properties approach from USDA Soil Taxonomy. At the time of its inception, the WRB proposed 30 Soil Reference Groups, accommodating more than 200 (second level) *Soil Units*.

The World Reference Base for Soil Resources provides an opportunity to create and refine a common and global language for soil classification. The WRB aims to serve as a framework through which ongoing soil classification throughout the world can be harmonized. The ultimate objective is to reach international agreement on the major soil groups to be recognized at a global scale, as well as on the criteria and methodology to be applied for defining and separating them. Such an agreement is needed to facilitate the exchange of information and experience, to provide a common scientific language, to strengthen the applications of soil science, and to enhance the communication with other disciplines and make the major soil names into household names (Dudal, 1996).

WRB CONCEPTS: DIAGNOSTIC HORIZONS, PROPERTIES, AND MATERIALS

The taxonomic units of the WRB are defined in terms of measurable and observable diagnostic horizons, the basic identifiers in soil classification. Diagnostic horizons are defined by (combinations of) characteristic soil properties and/or soil materials. Diagnostic horizons, properties, and materials and a selection of qualifiers that are used to differentiate between Reference Soil Groups are

Table 19.1 A Sample of Diagnostic Horizons in WRB

Surface horizons	
Anthropogenic horizons	Surface and subsurface horizons resulting from long-continued anthropedogenic processes, notably deep working, intensive fertilization, addition of earthy materials, irrigation, or wet cultivation
Histic horizon	(Peaty) surface horizon, or subsurface horizon occurring at shallow depth, consisting of organic soil material
Umbric horizon	Well-structured, dark surface horizon with low base saturation and moderate to high organic matter content
Yermic horizon	Surface horizon of rock fragments (desert pavement) usually, but not always, embedded in a vesicular crust and covered by a thin aeolian sand or loess layer
Subsurface horizons	
Cryic horizon	Perennially frozen horizon in mineral or organic soil materials
Duric horizon	Subsurface horizon with weakly cemented to indurated nodules cemented by silica (SiO_2), known as "durinodes"
Fragic horizon	Dense, noncemented subsurface horizon that can only be penetrated by roots and water along natural cracks and streaks
Vertic horizon	Subsurface horizon rich in expanding clays and having polished and grooved ped surfaces (slickensides), or wedge-shaped or parallelepiped structural aggregates formed upon repeated swelling and shrinking

Table 19.2 Descriptive Summary of Sample Diagnostic Properties

	Examples of Unique Qualifier Definitions
Albeluvic tonguing	Iron-depleted material penetrating into an argic horizon along ped surfaces
Alic properties	Very acid soil material with a high level of exchangeable aluminium
Permafrost	Indicates that the soil temperature is perennially at or below 0°C for at least two consecutive years
Stagnic properties	Visible evidence of prolonged waterlogging by a perched water table

summarized in Tables 19.1, 19.2, 19.3, and 19.4. For a comprehensive overview of all WRB elements, reference is made to FAO World Soil Resources Reports Number 84 (FAO et al., 1998).

Note that the generalized descriptions of diagnostic horizons, properties, and soil materials given in Tables 19.2, 19.3, and 19.4 are solely meant as an introduction to WRB terminology. For exact concepts and full definitions, reference is made to FAO Soil Resources Reports Number 84 (FAO/ISRIC/ISSS, 1998).

TWO-TIER APPROACH IN THE WRB

The WRB comprises two tiers of detail (Nachtergaele et al., 2000a):

1. The "Reference Base," which is limited to the first (highest) level, having 30 Reference Soil Groups
2. The "WRB Classification System," suggesting combinations of adjectives to the Reference Soil Groups, which allows precise characterization and classification of individual soil profiles

The Reference Base: Accommodating the World Soil Cover in 30 Reference Soil Groups

In the current text, the WRB's 30 Reference Soil Groups are summarized in a simplified key to the Reference Soil Groups in Table 19.5 (after ISSS/ISRIC/FAO, 1998b).

The following successive steps have to be taken to classify a soil:

Table 19.3 Descriptive Summary of Sample Diagnostic Materials

	Examples of Unique Qualifier Definitions
Anthropogenic soil material	Unconsolidated mineral or organic material produced largely by human activities and not significantly altered by pedogenetic processes
Organic soil material	Organic debris, which accumulates at the surface and in which the mineral component does not significantly influence soil properties
Sulfidic soil material	Waterlogged deposit containing sulphur, mostly sulphides, and not more than moderate amounts of calcium carbonate
Tephric soil material	Unconsolidated, non- or only slightly weathered products of volcanic eruptions, with or without admixtures of material from other sources

Table 19.4 Sample of WRB Qualifiers and Their Definitions

	Examples of Unique Qualifier Definitions
Carbi-	Having a cemented *spodic* horizon which does not contain enough amorphous iron to turn redder on ignition *(in Podzols only)*
Carbonati-	Having a soil solution with pH > 8.5 (1:1 in water) and $HCO_3 > SO_4 >> Cl$ *(in Solonchaks only)*
Chloridi-	Having a soil solution (1:1 in water) with $Cl >> SO_4 > HCO_3$ *(in Solonchaks only)*
Cryi-	Having a *cryic* horizon within 100 cm of the soil surface

- Identification of soil properties/materials in the field, further evidenced by laboratory analyses if needed
- Identification of the presence and kind of soil horizons
- Identification of the vertical arrangement of horizons
- Identification of the applicable WRB Reference Soil Group(s) in the WRB Key
- Final classification of the soil; considers the list of (ranked) qualifiers as described in the following paragraph for the example of a Ferralsol

The WRB as a Soil Correlation System

The World Reference Base has a modular (building-block) structure. The building blocks are the uniquely defined qualifiers discussed above; a sample set of qualifiers and their definitions are given in Table 19.4. Once the Reference Soil Group is identified, applicable qualifiers define individual Soil Units. The following example classifies a Soil Unit in the Reference Soil Group of the Ferralsols:

EXAMPLE

In Ferralsols the following qualifiers have been recognized so far (in ranking order):

Strong expression qualifiers:

- Gibbsic >25% gibbsite in fine earth fraction, within 100 cm depth
- Geric Having geric properties within 100 cm
- Posic Having a zero or positive charge within 100 cm

Intergrade qualifiers (in the order of the key):

- Histic intergrade with Histosols reference soil group
- Gleyic intergrade with Gleysols reference group
- Andic intergrade with Andosols reference group
- Plinthic intergrade with Plinthosols
- Mollic intergrade with Phaeozems
- Acric intergrade with Acrisols

Table 19.5 Simplified Key to the WRB Reference Soil Groups

No.	Criterion (↓ no)	yes →	Reference Soil Group	No.	Criterion (↓ no)	yes →	Reference Soil Group	No.	Criterion (↓ no)	yes →	Reference Soil Group
1	Organic materials > 40 cm deep	yes →	HISTOSOLS	11	Plinthite or petroplinthite within 50 cm	yes →	PLINTHOSOLS	21	Argic horizon and albeluvic tonguing	yes →	ALBELUVISOLS
2	Cryic horizon	yes →	CRYOSOLS	12	Ferralic horizon	yes →	FERRALSOLS	22	Argic horizon with $CEC_c > 24$, $Al_{sat} > 60\%$	yes →	ALISOLS
3	Human modifications	yes →	ANTHROSOLS	13	Natric horizon	yes →	SOLONETZ	23	Argic and nitic horizons	yes →	NITISOLS
4	Depth < 25 cm	yes →	LEPTOSOLS	14	Abrupt textural change	yes →	PLANOSOLS	24	Argic horizon with $CEC_c < 24$, BS < 50%	yes →	ACRISOLS
5	> 35% swell-clay vertic horizon	yes →	VERTISOLS	15	Chernic or dark mollic horizon	yes →	CHERNOZEMS	25	Argic horizon with $CEC_c > 24$	yes →	LUVISOLS
6	Fluvic materials	yes →	FLUVISOLS	16	Brownish mollic horizon and secondary $CaCO_3$	yes →	KASTANOZEMS	26	Argic horizon with $CEC_c < 24$, BS > 50%	yes →	LIXISOLS
7	Salic horizon	yes →	SOLONCHAKS	17	Mollic horizon	yes →	PHAEOZEMS	27	Umbric horizon	yes →	UMBRISOLS
8	Gleyic properties	yes →	GLEYSOLS	18	Gypsic or petrogypsic horizon	yes →	GYPSISOLS	28	Cambic horizon	yes →	CAMBISOLS
9	Andic or vitric horizon	yes →	ANDOSOLS	19	Duric or petroduric horizon	yes →	DURISOLS	29	Coarse texture > 100 cm	yes →	ARENOSOLS
10	Spodic horizon	yes →	PODZOLS	20	Calcic or petrocalcic horizon	yes →	CALCISOLS	30	Other soils	→	REGOSOLS

Adapted from ISSS/ISRIC/FAO, 1998b.

- Lixic intergrade with Lixisols
- Umbric intergrade with Umbrisols
- Arenic intergrade with Arenosols

Secondary characteristics qualifiers directly related to diagnostic horizons, properties or soil materials:

- Endostagnic stagnic properties between 50 and 100 cm
- Humic strongly humic properties
- Ferric presence of a ferric horizon within 100 cm

Secondary characteristics qualifiers not directly related to defined diagnostic horizons, properties, or soil materials:

- Vetic $ECEC_{clay}$ of less than 6 cmolc/kg
- Alumic Al saturation (ECEC) of 50% or more
- Hypereutric Having a base saturation of 80% or more
- Hyperdystric Having a base saturation of less than 50% in all parts between 20 and 100 cm and less than 20% in some part

Qualifiers related to soil colors:

- Rhodic ferralic horizon with hue redder than 5YR, etc
- Xanthic ferralic horizon with hue of 7.5YR or yellower, etc.

"Remaining characteristics" qualifier:

- Haplicother ferralic horizons

To classify a Ferralsol, one would go down the list of qualifiers and note that qualifier #2 applies. Therefore, the soil is classified as a Geric Ferralsol. On the basis of available information on clay distribution (clay increase meets minimum specifications of an argic horizon) and base saturation (less than 50% in at least part of the ferralic B horizon) that is available (qualifier #9), one would further classify this soil as an Acri-Geric Ferralsol. If more than two qualifiers apply, these can be added between brackets behind the standard name. If, for instance, the Ferralsol discussed also features strongly humic properties (qualifier #14) and a dark red color (qualifier #20), the soil would be named an Acri-Geric Ferralsol (Humic, Rhodic).

In addition, the Soil Unit name might express the depth (from shallow to deep: Epi, Endo, Bathi) and intensity (from weak to strong: Proto, Para, Hypo, Ortho, and Hyper) of features, important for management interpretations. In poly-sequential soil profiles, the qualifiers Cumuli or Thapto can be used to indicate accumulation or burial.

For each Reference Soil Group there is a defined list of qualifiers available, with suggested qualifier ranking. For a comprehensive discussion of qualifiers, reference is made to the FAO World Soil Resources Reports Number 84 (FAO et al., 1998). Note that the issue of qualifier ranking is still a point of debate (Nachtergaele et al., 2001).

CORRELATION: THE WRB (1998) WITH USDA'S SOIL TAXONOMY (1999)

Table 19.6 correlates the WRB (1998) Reference Groups and Soil Orders/Suborders of Soil Taxonomy (1999). The table demonstrates that international consensus on soil naming is growing; near-perfect correlation exists for many groups such as Histosols, Vertisols, Andosols, and Podzols. Evidently the match can only be partial for several reasons:

Table 19.6 Correlation between WRB (1998) and Soil Taxonomy (1999)

WRB	Soil Taxonomy
Histosols	Histosols pp.*.
Cryosols	Gelisols pp.
Anthrosols	Inceptisols pp., Plaggepts
Leptosols	Entisols, Lithic subgroups pp.
Vertisols	Vertisols
Fluvisols	Entisols – Fluvents
Solonchaks	Aridisols – Salorthids pp.
Gleysols	Inceptisols – Aquepts pp., Entisols – Aquents pp.
Andosols	Andisols
Podzols	Spodosols
Plinthosols	Oxisols – Plinthaquox pp.
Ferralsols	Oxisols pp.
Solonetz	Aridisols – Natrargids pp.
Planosols	Alfisols – Abruptic Albaqualf pp., Ultisols – Abruptic Albaquults pp.
Chernozems	Mollisols – Borolls pp.
Kastanozems	Mollisols – Ustolls, Xerolls pp.
Phaeozems	Mollisols – Udolls pp.
Gypsisols	Aridisols – Gypsids pp.
Durisols	Aridisols – Durids pp.
Calcisols	Aridisols – Calcids pp.
Albeluvisols	Alfisols – Fraglossudalfs
Alisols	Ultisols – Udults pp.
Nitisols	Oxisols – Kandiudox pp., Ultisols – Kandiudults pp.
Acrisols	Ultisols – Udults, Ustults, Haplustalfs pp.
Luvisols	Alfisols pp.
Lixisols	Alfisols – Paleustalfs, Ustalfs, Udalfs, Ustalfs pp.
Umbrisols	Inceptisols pp.
Cambisols	Inceptisols pp.
Arenosols	Entisols – Psamments pp.
Regosols	Entisols pp.

* pp. = *pro parte*

- In Soil Taxonomy, climatic factors are used to differentiate at Soil Order level (e.g., Gelisols, Aridisols) as indicated by 'pp.' (pro parte) in Table 19.6. WRB avoids climatic criteria and argues that an overlay with precise climatic information (as result of modelling) may result in a more accurate analysis for the purpose of land evaluation.
- Another reason for the less than complete match between Soil Taxonomy and the WRB is that the WRB distinguishes 30 Reference Soil Groups at the highest categorical level, whereas Soil Taxonomy differentiates between 12 Soil Orders. For instance, the Mollisols of Soil Taxonomy cover the WRB Reference Soil Groups of the Chernozems, Kastanozems, and Phaeozems.
- WRB uses only two categorical levels, whereas Soil Taxonomy comprises six levels.
- WRB uses more diagnostic horizons, properties, and materials than does Soil Taxonomy.
- Though there is great similarity between names of diagnostic horizons in Soil Taxonomy and WRB, the differences in definition are sometimes large (e.g., argic and salic horizons). WRB has made a special effort to simplify analytical requirements to classify soils. It should also be noted that WRB recognizes at the highest level processes such as prolonged hydromorphy or intense man-made changes, whereas these are recognized at the second level only in Soil Taxonomy.

THE RANKING OF WRB QUALIFIERS

A rather comprehensive list of lower level names for the World Reference Base has been arrived at after reviewing the uses that have been made of a second level by FAO (1988), Soil Survey Staff (1999), and WRB (ISSS-ISRIC-FAO, 1994), and at the third level in soil classifications of Botswana (Remmelzwaal and Verbeek, 1990), northeastern Africa (FAO, 1998), Bangladesh (Brammer et al, 1988), and the European Union (CEC, 1985), and by reclassifying a large number of typifying pedons of all reference soil groups. The use of these qualifiers has been tested in WRB field correlation tours in Vietnam, China, Sicily, Spain, Georgia, Benin, Hungary, and France. An issue in all meetings was the rationale behind the WRB system. Although WSR Number 84 (FAO et al., 1998) provides a rationale for WRB as a whole, the report is rather scanty on the rationale underlying the qualifier listings of the 30 Reference Soil Groups.

The question of qualifier priority ranking is not easy *inter alia* because there is often a personal bias in ranking. A road constructor and a farmer will attach different weights to certain soil qualifiers. Prominent WRB authors have suggested to do away with priority ranking altogether, and use the alphabetical order instead. Others maintain that predetermined ranking makes the system easier, e.g., for students. The following presents the provisional outcome of three years of discussions.

General Principles for Distinguishing (Second Level) Soil Units in the WRB

The general rules for differentiating Soil Units are the following:

- The second level of the WRB classification system is defined by one or more diagnostic criteria named "qualifier." Each qualifier has a unique meaning, which, as far as possible, has been applied to all Reference Soil Groups in which it occurs.
- Diagnostic criteria are derived from established definitions of diagnostic horizons, properties, and soil materials. They may include additional (new) elements, as well as criteria used for phase definitions in other taxonomies.
- Criteria related to climate, parent material, vegetation, or physiographic features such as slope, geomorphology, or erosion are not considered. Neither are criteria referring to soil-water relationships, such as depth of water table or drainage.
- Characteristics/properties of the substratum (below the control section) are not used for the differentiation of lower level units.
- Normally two qualifiers are sufficient to characterize most soils. If additional qualifiers are needed, these should be listed after the Reference Soil Group name between brackets, e.g., Geri-Acric Ferralsol (Humic and Xanthic).
- Definitions of qualifiers used may not overlap or conflict either with each other or with the Reference Soil Group definitions to which they are attached. For instance, a Dystri-Petric Calcisol is a contradiction (Dystri clashes with Calcisol), whereas a Eutri-Petric Calcisol is an overlap because the prefix "Eutri" adds no information to the Calcisol Reference Group.
- New units may only be established after being documented by soil profile descriptions and supporting laboratory analyses.
- Priority rules for the use of second level soil units. It is recommended that priority rules for the use of qualifiers follow the qualifier categories and ranking as given below:
    - first (if any) one or more "strong expression" qualifiers
    - secondly (if any) an intergrade qualifier
    - thirdly (if any) one or more secondary characteristics qualifiers, directly derived from defined diagnostic horizons, properties, or soil materials
    - fourthly (if any) one or more secondary characteristics qualifiers, not directly related to define diagnostic horizons, properties, or soil materials
    - fifthly (if any) a qualifier related to soil color
    - the "Haplic" qualifier is used for other situations not covered in the foregoing list

THE CHALLENGE OF CAPTURING THE COMPLEXITY OF SOIL SYSTEMS

As Nachtergaele et al. (2000b) rightly point out, much confusion exists about the requirements that soil classifications and map legends must meet. Soil surveys, and particularly large-scale surveys, require a measure of detail that cannot be provided by a comprehensive soil classification system. It is argued here that the morphogenetic approach to soil classification, which underlies

Qualifier Categories and Ranking per WRB Reference Soil Group

HISTOSOLS	CRYOSOLS	ANTHROSOLS	LEPTOSOLS	VERTISOLS
Glacic	Turbic	Hydragric	Lithic	Thionic
Thionic	Glacic	Irragric	Hyperskeletic	Salic
Cryic	Histic	Terric	Rendzic	Natric
Gelic	Lithic	Plaggic	Gelic	Gypsic
Salic	Leptic	Hortic	Vertic	Duric
Folic	Salic	Gleyic	Gleyic	Calcic
Fibric	Gleyic	Spodic	Mollic	Alic
Sapric	Andic	Ferralic	Umbric	Gypsiric
Ombric	Natric	Luvic	Humic	Grumic
Rheic	Mollic	Arenic	Gypsiric	Mazic
Alcalic	Gypsic	Regic	Calcaric	Mesotrophic
Toxic	Calcic	Stagnic	Yermic	Hyposodic
Dystric	Umbric	Haplic	Aridic	Eutric
Eutric	Thionic		Dystric	Pellic
Haplic	Stagnic		Eutric	Chromic
	Yermic		Haplic	Haplic
	Aridic			
	Oxyaquic			
	Haplic			

FLUVISOLS	SOLONCHAKS	GLEYSOLS	ANDOSOLS	PODZOLS
Thionic	Histic	Thionic	Vitric	Densic
Histic	Gelic	Histic	Silandic	Carbic
Gelic	Vertic	Gelic	Aluandic	Rustic
Salic	Gleyic	Anthraquic	Eutrisilic	Histic
Gleyic	Mollic	Vertic	Melanic	Gelic
Mollic	Gypsic	Endosalic	Fulvic	Anthric
Umbric	Duric	Andic	Hydric	Gleyic
Arenic	Calcic	Vitric	Histic	Umbric
Tephric	Petrosalic	Plinthic	Leptic	Placic
Stagnic	Hypersalic	Mollic	Gleyic	Skeletic
Humic	Stagnic	Gypsic	Mollic	Stagnic
Gypsiric	Takyric	Calcic	Duric	Lamellic
Calcaric	Yermic	Umbric	Luvic	Fragic
Takyric	Aridic	Arenic	Umbric	Entic
Yermic	Hyperochric	Tephric	Arenic	Haplic
Aridic	Aceric	Stagnic	Placic	
Skeletic	Chloridic	Abruptic	Pachic	
Sodic	Sulphatic	Humic	Calcaric	
Dystric	Carbonatic	Calcaric	Skeletic	
Eutric	Sodic	Takyric	Acroxic	
Haplic	Haplic	Alcalic	Vetic	
		Toxic	Sodic	
		Sodic	Dystric	
		Alumic	Eutric	
		Dystric	Haplic	
		Eutric		
		Haplic		

PLINTHOSOLS	FERRALSOLS	SOLONETZ	PLANOSOLS	CHERNOZEMS
Petric	Gibbsic	Vertic	Thionic	Chernic
Endoduric	Geric	Salic	Histic	Vertic
Alic	Posic	Gleyic	Gelic	Gleyic
Acric	Histic	Mollic	Vertic	Calcic
Umbric	Gleyic	Gypsic	Endosalic	Glossic
Geric	Andic	Duric	Gleyic	Siltic
Stagnic	Plinthic	Calcic	Plinthic	Vermic
Abruptic	Mollic	Stagnic	Mollic	Haplic
Pachic	Acric	Humic	Gypsic	
Glossic	Lixic	Albic	Calcic	
Humic	Umbric	Takyric	Alic	
Albic	Arenic	Yermic	Luvic	
Ferric	Endostagnic	Aridic	Umbric	
Skeletic	Humic	Magnesic	Arenic	
Vetic	Ferric	Haplic	Geric	

Qualifier Categories and Ranking per WRB Reference Soil Group (continued)

PLINTHOSOLS	FERRALSOLS	SOLONETZ	PLANOSOLS	CHERNOZEMS
Alumic	Vetic		Calcaric	
Endoeutric	Alumic		Albic	
Haplic	Hyperdystric		Ferric	
	Hypereutric		Alcalic	
	Rhodic		Sodic	
	Xanthic		Alumic	
	Haplic		Dystric	
			Eutric	
			Rhodic	
			Chromic	
			Haplic	

KASTANOZEMS	PHAEOZEMS	GYPSISOLS	DURISOLS	CALCISOLS
Histic	Vertic	Andic	Leptic	Petric
Gleyic	Gleyic	Ferralic	Gleyic	Hypercalcic
Alic	Andic	Mollic	Vitric	Leptic
Umbric	Plinthic	Alic	Andic	Vertic
Arenic	Nitic	Umbric	Plinthic	Endosalic
Gelic	Umbric	Humic	Umbric	Gleyic
Stagnic	Arenic	Vetic	Arenic	Luvic
Abruptic	Stagnic	Alumic	Stagnic	Takyric
Ferric	Abruptic	Dystric	Abruptic	Yermic
Fragic	Humic	Eutric	Geric	Aridic
Siltic	Albic	Rhodic	Humic	Hyperochric
Alumic	Profondic	Haplic	Albic	Skeletic
Endoeutric	Lamellic		Profondic	Sodic
Haplic	Ferric		Lamellic	Haplic
	Skeletic		Ferric	
	Hyperdystric		Hyperochric	
	Rhodic		Skeletic	
	Chromic		Vetic	
	Haplic		Alumic	
			Hyperdystric	
			Rhodic	
			Chromic	
			Haplic	

LIXISOLS	UMBRISOLS	CAMBISOLS	ARENOSOLS	REGOSOLS
Leptic	Thionic	Thionic	Gelic	Gelic
Gleyic	Gelic	Gelic	Hyposalic	Leptic
Vitric	Anthric	Leptic	Gleyic	Hyposalic
Andic	Leptic	Vertic	Hyperalbic	Gleyic
Plinthic	Gleyic	Fluvic	Plinthic	Thaptovitric
Calcic	Ferralic	Endosalic	Hypoferralic	Thaptoandic
Arenic	Arenic	Gleyic	Hypoduric	Arenic
Geric	Stagnic	Vitric	Hypoluvic	Aric
Stagnic	Humic	Andic	Tephric	Garbic
Abruptic	Albic	Plinthic	Gypsiric	Reductic
Humic	Skeletic	Ferralic	Calcaric	Spolic
Albic	Haplic	Gelistagnic	Albic	Urbic
Profondic		Stagnic	Lamellic	Tephric
Lamellic		Humic	Fragic	Gelistagnic
Ferric		Calcaric	Yermic	Stagnic
Hyperochric		Gypsiric	Aridic	Humic
Skeletic		Takyric	Protic	Gypsiric
Vetic		Yermic	Dystric	Calcaric
Rhodic		Aridic	Eutric	Takyric
Chromic		Hyperochric	Rubic	Yermic
Haplic		Skeletic	Haplic	Aridic
		Sodic		Hyperochric
		Dystric		Anthropic
		Eutric		Skeletic
		Rhodic		Hyposodic
		Chromic		Vermic
		Haplic		Dystric
				Eutric
				Haplic

the WRB, and the flexibility in qualifier use lend themselves admirably to describe, characterize, and classify the fuzzy complexity of natural soilscapes (Duchaufour, 2001). The WRB has opted for precise definitions of a large number of diagnostic horizons, most of which express one or more prominent pedogenetic processes. However, when working in the field, one inevitably comes across border cases, which triggers the question of whether it is wise to (attempt to) accommodate the world's highly variable soil resources in a simple categorical system such as the WRB. It appears that for detailed soil classification and for the purpose of computerized data management, the WRB ought to be linked to a more detailed horizon-based system, e.g., the one suggested by Fitzpatrick and Aitkenhead (2000). The latter, also called the "Fitz system," identifies soil horizons by properties of the horizons themselves, in a multidimensional space. The properties are conceived as coordinates, which intersect in space to create conceptual segments with centroids, or reference points. Each horizon is uniquely defined by its central concept and has a mathematical centroid, which can be used for quantitative assessment by models or in evaluation schemes. As the Fitz system pursues greater detail than the broad definitions used in the WRB, there are many horizons (sub-types) for each diagnostic horizon in the WRB. The Fitz system considers soils to be unique individuals whose identity (border case or not) can be established by considering the vertical arrangement and thickness of soil horizons.

The Fitz system has adopted the WRB as a reference platform for its own diagnostic horizons. This presents the WRB with the challenge of harmonizing its diagnostic horizons, properties, and materials to make the Fitz system and the WRB compatible. Both systems have strengths and weaknesses, but when used in conjunction they should greatly enhance the usefulness of large soil databases for quantitative land evaluation.

The FAO stated in its objectives (FAO et al., 1998) that it views the WRB as an easy means of communication between scientists, to identify, characterize, and name major types of soils. It is not meant to replace national soil classification systems, but to be a tool for better correlation between national systems (Deckers, 2000). National systems are invited to forge a bridge to the WRB. A logical next step might be to use the WRB as a basic framework and overlay it with the FAO Topsoil Classification. Adding information on soil texture and clay mineralogy may enhance the system further.

Nachtergaele et al. (2000a) suggest that in the European context, the WRB presents an opportunity for regional soil services to overcome their differences and adopt the WRB as a common approach to soil classification in line with IUSS recommendations. The future success of the WRB will depend on how compatible the WRB will be with modern information tools and technology, such as computer databases, Geographic Information Systems, and quantitative land evaluation procedures. Considering that the WRB is a single-level system (with a second level if one prefers), regardless of the number of qualifiers used, the WRB is not hindered by an imposed, strictly hierarchical structure, which is in line with the requirements of digital information transfer.

REFERENCES

Deckers, J. 2000. Letter to the Editor on World Reference Base for Soil Resources (WRB), IUSS Endorsement, World-Wide Testing, and Validation. *SSSA J.* 64: 2187.

Driessen, P., Deckers, J., Spaargaren, O., and Nachtergaele, F., Eds. 2001. Lecture Notes on the Major Soils of the World. World Soil Resources Reports 94, FAO, Rome, Italy.

Duchaufour, P. 2001. Introduction à la science du sol. Sol, végétation, environment. 6e édition de l4 Abrégé de pédologie. Dunot, Paris, ISBN 2 10 005440 6.

Dudal, R. 1996. A World Reference Base for Soil Resources: Background, Principles and Challenges. Proceedings International WRB Workshop 1996, Pretoria.

FAO. 1974. UNESCO, Soil Map of the World Volume 1, Legend. FAO, Rome, Italy.

FAO/ISRIC/ISSS. 1998. World Reference Base for Soil Resources. World Soil Resources Report #84. FAO, Rome.

FAO-UNESCO-ISRIC.1988. Revised Legend of the Soil Map of the World. World Soil Resources Report no 60 1990, FAO, Rome.

Fitzpatrick, E.A. and Aitkenhead, M.J., 2000. Horizon indentification, Test Version 1.1. February 2000, Interactive CD-ROM.

ISSS Working Group RB, World Reference Base for Soil Resources. Deckers, J.A., Nachtergaele, F.O., and Spaargaren, O.C., Eds. First edition. ISSS/ISRIC/FAO 1998a, Acco Leuven.

ISSS Working Group RB, World Reference Base for Soil Resources. Atlas. Bridges, E.M,. Batjes, N.H., and Nachtergaele, F.O., Eds. First edition. ISSS/ISRIC/FAO 1998b, Acco Leuven.

Nachtergaele, F.O., Spaargaren, O., Deckers, J.A., and Ahrens, B. 2000a. New developments in soil classification. World Reference Base for Soil Resources. *Geoderma*. 96:345–357.

Nachtergaele, F.O., Spaargaren, O., and Deckers, J. 2000b. An overview of the WRB Reference Soil Groups with special attention to West-Africa. Paper presented at the International Symposium on Integrated Plant Nutrient Management in Sub-Saharan Africa, Cotonou, October 2000.

Nachtergaele F., Berding, F., and Deckers, J. 2001c. Pondering hierarchical soil classification systems. Proc. Int. Symposium, Soil Classification 2001, October 8–12, 2001, Hungary.

Soil Survey Staff. 1999. Keys to Soil Taxonomy. USDA Natural Resources Conservation Service. Washington DC.

Index

A

B

C

D

E

M

N

O

P

R

S

T

U

V

W

Z

Milton Keynes UK
Ingram Content Group UK Ltd.
UKHW050452071024
449327UK00015B/345